THE EMERGENCE
OF BACTERIAL
GENETICS

THE EMERGENCE OF BACTERIAL GENETICS

THOMAS D. BROCK
University of Wisconsin, Madison

COLD SPRING HARBOR LABORATORY PRESS
1990

THE EMERGENCE
OF BACTERIAL
GENETICS

© 1990 by Cold Spring Harbor Laboratory Press
All rights reserved
Printed in the United States of America
Cover and book design by Emily Harste

Library of Congress Cataloging-in-Publication Data

Brock, Thomas D.
     The emergence of bacterial genetics / by Thomas D. Brock.
         p.   cm.
     Includes bibliographical references.
     ISBN 0-87969-350-9
     1. Bacterial genetics.   I. Title.
QH434.B76   1990   90-1828
589.9'015--dc20   CIP

All Cold Spring Harbor Laboratory Press publications may be ordered
directly from Cold Spring Harbor Laboratory Press, Box 100, Cold
Spring Harbor, New York 11724. (Phone: 1-800-843-4388). In New
York (516) 367-8325. FAX: 516-367-8432.

For many young scientists the future is more important than the past and the history of science begins tomorrow . . . many facts and theoretical views are gloriously discovered which were known a long time ago. It has seemed, therefore, desirable to credit early workers for their achievements and also to spare unnecessary efforts directed to later rediscoveries. Moreover, it is interesting to know how phenomena were discovered, how the problems were born, attacked and solved, and how and why our ideas have evolved. The danger of parachuting young enthusiastic scientists into a flower bed of selected data and fully bloomed conceptions should not be underestimated.

André Lwoff
1953

O. Avery

J. Monod, 1947

N. Zinder, 1953

S. Luria (*top*) and M. Delbrück, 1940s

A.D. Hershey and S. Benzer, 1950s

J. Lederberg, E. Zimmer Lederberg, and F. Ryan, 1947

F. Jacob, 1953

E. Wollman, 1950s

A. Lwoff, 1953

B.D. Davis and M. Demerec, 1951

Preface

There have been numerous books and articles dealing with the history of molecular biology and genetics, and many of these books have considered, at least peripherally, the field of bacterial genetics. Although all such treatments note the central importance that bacterial genetics has played in the development of modern biology, no book has dealt with the history of bacterial genetics itself or with the broader questions of the connections between bacteriology and genetics. As emphasized in this book, bacteriology and genetics developed virtually independently for many years, only converging around World War II in a series of stunning research achievements that were to influence the development of all of modern biology. By the end of the 1950s, bacterial genetics not only had emerged as an independent discipline, but also was providing the tools and concepts that became the foundation of modern biology.

In the present book, the key research findings of bacterial genetics that led to the present era of molecular genetics have been brought together. I am interested here primarily in the *experimental* work that drove bacterial genetics forward. Although scientific wisdom comes from the empiricism of experimental research, raw data are of little interest for their own sake. It is only when these data are interpreted in the light of an intelligent model that the advancement of scientific knowledge and understanding occurs.

This book is not a social history like so many contemporary books in the history of science. It is a history of science rather than of scientists. The individuals who have played such an important role in the development of the story are not ignored, however, and the social structure should be plain to all. Science is performed by individuals, and although the acceptance of research depends ultimately on the validity of the research results, the speed of acceptance of these results depends on numerous nonscientific factors. The cult of personality exists in science as it does in other fields; this will be quite obvious to one who reads the present book. One can argue indefinitely about whether a certain discovery could only have been made by a certain individual, and whether if this individual had not been born, the field would be different. Most people who have thought about this question in detail have concluded that only extremely rarely would a single person make an absolute difference, although without this person the development of the field may have been greatly delayed. But the human side of science is important and, as much as possible, I have tried in this book to present science and scientific experiments within the cultural and psychological

ix

context of the times. I have also been careful to point out important advances in related fields that made a difference (for example, the development of radioisotope technology or ultracentrifugation).

This book is concerned primarily with 20th century science. The book would have been too long if strict limits had not been placed on the coverage and emphasis. Most critically, a decision had to be made about how close to the present the book would come. Thus, certain areas whose history is extremely fascinating but which are almost today's contemporary research, such as plasmid biology, have been considered only in a secondary manner. It was also essential to decide how much that was not strictly bacterial genetics would be covered. Although genetics began as mostly a formal discipline, it soon became entwined with physiology and biochemistry; contemporary genetics is an amalgam of all three. When essential, I have brought physiological and biochemical research into the picture, but I have not been able to cover in detail many fascinating aspects of these latter fields. There are also some other limitations: At one time I planned a separate chapter on cytoplasmic inheritance, but in the interests of a reasonable length, I decided to weave this material into the other chapters where appropriate.

This book is not a chronological history. I decided early that reading (and perhaps writing!) such a history would be dull. In addition, many of the fields covered in this book developed virtually independently and combining them would be misleading. One hazard of the way the book is organized is that the connections between separate areas (for instance, transduction and transformation) might be missed. In an attempt to show these connections, extensive cross-referencing between chapters has been included. A fairly extensive chronology has also been provided. Many of those reading this book will not read it from cover to cover, but will dip in where their interests lie. For those individuals, the organization used may be more appealing than a straight chronological history.

Some effort has been made to include the key data from the published sources. In some cases, original figures or tables have been used, but in many cases it was necessary, in the interest of clarity, to restructure tables or figures. However, unless stated otherwise, all of the data presented are exactly as in the original papers.

The bibliography for this book is quite large. After careful consideration, I decided to place the references at the ends of each chapter, rather than in a large reference section at the end of the book. Although this leads to a small amount of repetition, most of the papers apply so specifically to certain chapters that the overlap is not excessive. The extensive author index can be consulted for guidance to particular papers or individuals.

It is hoped that this book will be of interest to both scientists and historians. Scientists should find it interesting because it deals with the history of an extremely important area of modern biology, one that has not been covered in other books. I hope historians will find that the technical detail and analysis, presented in a historical context, provides insights not obtained in a conventional historical volume.

Some of the key scientists discussed in this book are alive, even still active in their fields. As much as possible, I have tried to obtain their current insights, and I have questioned them about key experiments. However, to a great extent, this book is based not on personal recollections, but on a close reading of the individual research papers and review articles published at the time the research was carried out.

I have also attempted to examine nonpublished sources that would give insight into the field. I examined the available records and unpublished reports of the Cold Spring Harbor Laboratory. I have examined in detail the very extensive Max Delbrück archives at the California Institute of Technology, although the restriction on copying imposed by this archives has presented me with numerous problems. I have also examined several major archives that have material of S.E. Luria, including those of the Rockefeller Foundation, Guggenheim Foundation, American Philosophical Society, and Indiana University. I also thank Professor Luria personally for responding to requests for permission to examine all of this material.

I am greatly in debt to Joshua Lederberg for sending me large amounts of material from his personal archives. The extensive correspondence which he and I have had over the years that this book was in preparation has been absolutely invaluable. The archives of the Research Administration-Financial and the Department of Genetics of the University of Wisconsin-Madison have also been useful in providing insights on Lederberg's early career. Others who have been generous in correspondence, advice, and criticism include Julius Marmur of Albert Einstein University, Bernard Davis of Harvard University, Melvin Cohn of the Salk Institute, Norton Zinder of the Rockefeller University, and Elie Wollman and Francois Jacob of the Pasteur Institute. In addition, I also thank William Hayes, Richard Burian, Bentley Glass, Waclaw Szybalski, and William Dove for conversations or correspondence that helped in setting the final approach of the book.

The first draft of this book, considerably longer than the final version, has been read in detail by a number of individuals whose advice and editorial comments have been invaluable. They are listed here: Edward Adelberg, Seymour Cohen (Chapter 6), Melvin Cohn (Chapter 10), Bernard Davis, Philip Hartman (Chapter 8), Rollin Hotchkiss (Chapter 9), Lily Kay (Chapters 1–3), Joshua Lederberg (Chapters 1–5), S.E. Luria (Chapters 4 and 6), Julius Marmur, Aaron Novick, Robert Olby, Arthur Pardee (Chapter 10), Eli Siegel (Chapters 4, 5, and 9), Elie Wollman, and Norton Zinder.

Except for some of the archive material mentioned above, all of the library work was done in the libraries of the University of Wisconsin-Madison. I can think of no better place to work on a book such as this, because these libraries are absolutely superb in their holdings of 19th and 20th century material in bacteriology and genetics. The two libraries that I used most were the Steenbock Library of the College of Agricultural and Life Sciences, and the Middleton Health Sciences Library; I am grateful to their staffs for numerous kind assistances.

The preparation of this book was supported by funds provided to me by the Graduate School of the University of Wisconsin-Madison. Computer equipment, software, and some copy facilities were provided without charge by Science Tech Publishers of Madison, Wisconsin.

I am especially grateful to my wife, Kathie, and my children Emily and Brian, for their tolerance of my absences for so many hours while this book was in preparation.

T.D. Brock

Contents

4
MUTATION, *45*

5
MATING, *75*

10
GENE EXPRESSION AND REGULATION, *265*

11
FROM BACTERIAL GENETICS TO RECOMBINANT DNA, *325*

THE EMERGENCE
OF BACTERIAL
GENETICS

1
INTRODUCTION

It sometimes happens that two fields of science appear to their participants to be so disparate that there is no discourse between them, even though such discourse would be of great benefit. In such cases, each separate field may develop and mature virtually in isolation from the other, and they only come together by chance at some late date. This is apparently what happened in the fields of bacteriology and genetics, which coalesced into the field of bacterial genetics long after each separate field had matured as an independent discipline. As a result of this lengthy separation, certain bacteriological discoveries of genetic relevance were ahead of their time, or what has been called *premature*. On the other hand, because of their insularity, bacteriologists ignored important genetic concepts.

This slow initial development of bacterial genetics was especially unfortunate because bacterial genetics has probably had more profound impact on the development of modern biology than any other biological discipline. Almost every area of modern biology under active research study today owes a major debt to bacterial genetics. Molecular biology, immunology, cancer research, medical virology, epidemiology, genetics of higher organisms, evolution, taxonomy, cell biology, and developmental biology all depend on concepts that arose first from studies in bacterial genetics. The proof that DNA is the genetic material came out of studies in bacterial genetics. Recombinant DNA techniques, so crucial to contemporary biology, arose directly out of research in bacterial and bacteriophage genetics. Most of the important studies in molecular biology, generally occurring since the early 1950s, have had their roots in a few key experiments in bacterial genetics that were done just a few years before. Once the basic concepts became clear, a virtual flood of new ideas poured out.

1.1 THE FOUNDERS OF BACTERIAL GENETICS

The purpose of this book is to describe in some detail the history of bacterial genetics and to show the loose and frequently confusing connections that existed between this field and conventional genetics. The period covered extends through the mid-1960s and ends with the discoveries of messenger RNA and the mechanisms of gene regulation. It does not cover the history of the recombinant DNA era nor, except indirectly, research on the genetic code. Even with these restrictions, a vast field of research is under consideration, and numerous areas had to be covered briefly or not at all. Although biochemistry and genetics are now intimately woven, this was not the case during most of the period under consideration here. The focus of this book is genetics and the interrelationships between this field and bacteriology.

With a few exceptions, geneticists did not believe that phenomena occurring in bacteria had any relevance to their discipline, and bacteriologists in turn did not believe that an understanding of genetics would be of much importance for their own fields. Until the late 1950s, textbooks of genetics contained little about bacteria, and textbooks of bacteriology had almost nothing about genetics.

Although a few geneticists thought about bacteria, none of the principal contributors to bacterial genetics had a "classical" genetics background. Likewise, bacteriologists, immersed in their immensely technical and highly specialized studies, cared little about broader questions in biology. A few exceptional scientists, some of whom are discussed below, were working in the borderline area between genetics and bacteriology and made the key contributions that brought the fields together. The training of these scientists is interesting because it was so disparate.

Max Delbrück was a physicist who turned toward biology because of a fascination with the possible applications of theoretical physics to an understanding of the gene. Salvador Luria was medically trained, but he also had a physics background and brought quantitative thinking to bear on bacterial populations. Oswald T. Avery was a physician who had turned to immunochemistry and medical bacteriology. André Lwoff had a protozoology background but was pulled toward bacteria by an interest in the evolution of nutritional requirements. Jacques Monod, a protozoologist interested in growth, turned to bacteria because they offered more favorable research material. Joshua Lederberg was a young medical student with a strong interest in genetics who abandoned his medical training because he conceived of a technique for carrying out genetic analysis in bacteria. Alfred D. Hershey was trained as a chemist but became involved in bacteriophage research after he was employed in a bacteriology department. (Hershey's initial interest was in using phage for immunochemical studies.) None of these key players in the drama of bacterial genetics had a background in conventional genetics or bacteriology. However, Delbrück, Luria, Lwoff, Monod, and Lederberg all did work first on eucaryotic systems and then turned to bacteria (or bacteriophage) because these systems provided more suitable experimental material for answering basic biological questions.

Given that Mendel's work was ignored until 1900, it is instructive that bacteriology, as a discipline, had actually matured before legitimate genetics research started. The agar colony technique, so central to bacterial genetics, was developed by Robert Koch in the early 1880s primarily to obtain pure cultures of pathogenic bacteria. Although geneticists would later realize that a pure culture of bacteria is a clone and can therefore be thought of as a population of dividing and occasionally mutating cells, bacteriologists were generally incapable of thinking in terms of populations of cells. What was a clone to the geneticist was simply a culture to the bacteriologist, and any changes that took place through successive transfers were thought to be due to ill-defined concepts of "training" or "adaptation."

As late as the mid-1940s, the eminent geneticist Julian Huxley had the following to say about bacteria:

Bacteria have no genes in the sense of accurately quantized portions of hereditary substance...the entire organism appears to function both as soma and germ plasm and evolution must be a matter of alteration in the reaction system as a whole...there is no ground for supposing that mutations are similar in nature to those of higher organisms (quoted from Burnet, 1945).

The principal founder of classical genetics, Thomas Hunt Morgan, never grasped the significance of bacteria, or even of the importance of gene chemistry. In his Nobel Prize address, he stated:

At the level at which the genetic experiments lie, it does not make the slightest difference whether the gene is a hypothetical unit, or whether the gene is a material particle. In either case, the unit is associated with a specific chromosome and can be localized there by purely genetic analysis (Morgan, 1933).

Since bacteria were thought not to have chromosomes, they could hardly be thought to contribute anything significant to the central problems of genetics.

Hermann J. Muller, one of the major figures in classical genetics, came as close as anyone to recognizing the significance of bacteria. Forced by his interest in mutation to consider genes as more than beads on a string, Muller perceived that simple living systems such as viruses could provide experimental material for reaching the gene itself. In a pioneering paper that was to be widely quoted only in the molecular biology era, Muller discussed the possible relevance of d'Herelle's then recent discovery of bacteriophage:

...there is a phenomenon...of...striking nature, which must not be neglected by geneticists. This is the d'Herelle phenomenon...if these d'Herelle bodies were really genes, fundamentally like our chromosome genes, they would give us an utterly new angle from which to attack the gene problem...we cannot categorically deny that perhaps we may be able to grind genes in a mortar and cook them in a beaker after all. Must we geneticists become bacteriologists, physiological chemists, physicists, simultaneously with being zoologists and botanists? Let us hope so (Muller, 1922).

Unfortunately, Muller's advocacy was unappreciated; it would be many years before research on bacteria and their viruses would shed light on the nature of the gene.

1.2. KEY EXPERIMENTS OF BACTERIAL GENETICS

The key experiments (listed below) that can be said to constitute a turning point in the development of bacterial genetics all came within a few years of each other in the 1940s.

1. The isolation by George Beadle and Edward L. Tatum of nutritional (biochemical) mutants in the fungus *Neurospora crassa*, making possible the study of rare genetic events because of the power of nutritional selection. The Beadle/Tatum experiments also led to the one-gene/one-enzyme concept, the first real insight into the connection between genotype and phenotype (Beadle, 1945).
2. The Luria/Delbrück experiment, subsequently known as the fluctuation test, which became the basis for quantitative studies on bacterial mutation (Luria and Delbrück, 1943).
3. The demonstration by Avery, MacLeod, and McCarty (1944) that the transforming principle of pneumococcus was DNA, and the experiment of

Hershey and Chase (1952) showing that it was the DNA of the bacteriophage particle that entered the cell during infection.

4. The discovery of mating (sex) in *Escherichia coli* K-12 by Lederberg and Tatum (1946).

5. The demonstration that bacteriophage is capable of undergoing genetic recombination (Delbrück and Bailey, 1946; Hershey, 1946).

6. The demonstration by Lwoff and Gutmann (1950) that phage production by lysogenic bacteria is a cellular event rather than a population event.

Although the paper by Luria and Delbrück is generally considered to be the starting point of bacterial genetics, this work did not arise de novo, since numerous mutation studies in bacteria had been reported earlier, and some were actually rather convincing. What made the Luria/Delbrück paper so important is that it was published by two workers who retained a strong connection to the *genetics* community. Although it was the starting point of bacterial genetics, this study was a detour for Luria and Delbrück from their studies of bacteriophage as model viruses. Gradually, Luria and Delbrück came to realize that bacteriophage provided not only models for animal and plant viruses, but also a means to get a handle on the physical gene. Bacteria were to them just the hosts that were used to grow bacterial viruses. Delbrück, in particular, became a vigorous champion of bacteriophage as a model for genetics research. Through lectures, summer courses, and the sheer force of his personality, he lit a fire under numerous workers. Through the study of lysogeny by Lwoff and others, primarily at the Pasteur Institute in Paris, the connection between phage and bacterial genetics was made, and the phage and bacterial fields began to coalesce (Cairns, Stent, and Watson, 1966). Fortuitously, Delbrück's promotion of phage came at a time when research could flourish in the financial vigor of post-World-War-II United States.

The paper by Avery, MacLeod, and McCarty on pneumococcus transformation focused attention on DNA, and is demonstrably the grandfather of the Watson and Crick structure. The Avery work approached the problem in a strictly biochemical manner, with the fractionation of the active transforming principle, using its biological activity as an assay. Avery and co-workers handled the transforming principle as a biochemist handles an enzyme. However, this work had little influence on *genetics* itself, either classical or bacterial. The pneumococcus transformation system was a "difficult" experimental system and only became susceptible to sophisticated genetic analysis in the late 1950s.

The Avery work had a profound effect on Lederberg's thinking about experimental approaches to genetic phenomena. For Lederberg, the significance of the Avery experiment was not so much that the active substance was DNA, but that it provided an experimental approach for getting at the genetic substance (see Section 5.3). It was only when Lederberg and his emulators were able to carry out genetic analysis with *Escherichia coli* K-12 that gene structure in bacteria could be studied. Thus, Lederberg's work became the cornerstone of bacterial genetics. Once bacterial mating could be carried out, it was possible to do genetic analysis in the classical sense. The first genetic map in bacteria was that of Lederberg, and his "classical" genetic approach provided a critical foundation for subsequent work, even though some aspects of it were erroneous or improperly interpreted.

Lederberg's work on "sex" in bacteria was controversial, not the least because it was based on indirect techniques. However, the technique used was not only simple and reproducible, but also brilliant in concept. It was "among the most fundamental advances in the whole history of bacteriological science" (Luria, 1947). Lederberg's technique made possible the study of extremely rare genetic events. It was soon realized that this approach had wide applications. It led to the discovery of virus-mediated genetic transfer (transduction) by Zinder and Lederberg and also led to fine-structure genetic analysis (using bacteriophage), most closely identified with Seymour Benzer. Through such analysis, the detailed connection between the gene and the DNA molecule, i.e., between genetics and biochemistry, could be made.

As noted, the most important outcome of the Beadle/Tatum work was the one-gene/one-enzyme hypothesis, which became the theoretical underpinning for the whole field of genetics. Gene physiology, or what is now called *gene expression*, derived from the Beadle/Tatum work, but only after research had turned from *Neurospora* to bacteria. Jacques Monod, the principal champion of gene expression, is in close debt to Beadle and Tatum. There is also an interesting connection between Beadle and Lwoff via the French geneticist Boris Ephrussi. Lwoff's early research in Paris was on the nutrition of protozoa and pathogenic bacteria. Beadle spent a year in the mid-1930s working in Paris with Ephrussi, who was a close friend of Lwoff. This was the time when Lwoff was showing the growth requirements of bacteria. Beadle's idea of inducing biochemical mutants in *Neurospora* was in part motivated by his knowledge of Lwoff's work.

Biochemistry is, in a sense, a completely independent field, but one that played a major role in the development of modern molecular biology. At the beginning of the bacterial genetics era, biochemists were interested primarily in *pathways* and *enzymes*, an area that was often called *intermediary metabolism*. However, the techniques that biochemists developed for the preparation of cell-free systems soon became applied to the more central and much more complicated studies of gene expression and protein synthesis. Although much of the initial work on the biochemistry of protein synthesis was done using mammalian systems, the concept of messenger RNA arose strictly from bacterial and phage studies, most closely identified with Jacques Monod, Francois Jacob, Sydney Brenner, Francis Crick, and Sol Spiegelman. The stunning syntheses of Monod and Jacob using genetic and biochemical concepts are presented in the finale of this book. Almost immediately, the concept of messenger RNA led to the development of successful cell-free systems for deciphering the genetic code. The code itself was first worked out in bacteria, and it was tested every step of the way by genetic studies on bacteria and phage.

The differences between genetic and biochemical thinking need to be emphasized. The gene as such is only inferred from the results of mutation or recombination experiments. In Mendel's original work, the gene was a mere formalism to describe the results of hybridization experiments. (The term "gene" itself was not coined until the 20th century.) Bacterial genetics, specifically, deals most commonly with rare events, whether these are mutations or recombinational events. Biochemistry, on the other hand, is much more concrete. It deals primarily with common events, the kinds that can be studied in the average molecules isolated from average cells. Specific cell constituents (e.g., the amino acids) can be isolated and purified, and their chemical structures can be determined. Enzymes can be purified, even crystallized, and both their structures and functions can be studied independently of the organism from which they came. Genetics, on the other hand, because it is based on hybridization experiments, is always linked to the whole living organism.

The linkage between genetics and biochemistry is physiology, the discipline that focuses on the connection between genotype and phenotype. Bacteria are the ideal experimental organisms for physiological study and hence for bringing biochemistry and genetics together. Bacteria can be easily grown in defined culture media to high cell densities, and their rapid growth rates make it possible to accomplish many experiments in a short period of time. Because bacterial mutants can be readily isolated and rare recombinational events can be studied, bacteria present the preeminent material for analyzing the genotype. Because they can be cultured so readily, bacteria are the best organisms for studying physiology. Finally, the large batches of bacterial cells that can be so readily obtained provide the necessary material for the biochemist.

The "cellular" thinking that has arisen out of studies in bacterial genetics has been applied in a striking manner to central problems of integration and differentiation in higher organisms. Modern research in immunology, cancer, development, hormonal action, and many other areas has depended on concepts that first arose from bacterial studies. It is not accidental that many of the workers who made the principal contributions to the development of bacterial genetics turned later to research on immunology and cancer.

REFERENCES

Avery, O.T., MacLeod, C.M., and McCarty, M. 1944. Studies on the chemical nature of the substance inducing transformation of pneumococcal types. Induction of transformation by a desoxyribonucleic acid fraction isolated from Pneumococcus Type III. *Journal of Experimental Medicine* 79: 137–158.

Beadle, G.W. 1945. Biochemical genetics. *Chemical Reviews* 37: 15–96.

Burnet, F.M. 1945. *Virus as Organism*. Harvard University Press, Cambridge, Massachusetts. 134 pp.

Cairns, J., Stent, G.S., and Watson, J.D. 1966. *Phage and the Origins of Molecular Biology*. Cold Spring Harbor Laboratory of Quantitative Biology, Cold Spring Harbor, New York.

Delbrück, M. and Bailey, W.T. 1946. Induced mutations in bacteriophage. *Cold Spring Harbor Symposia on Quantitative Biology* 11: 33–50.

Hershey, A.D. 1946. Spontaneous mutations in bacterial viruses. *Cold Spring Harbor Symposia on Quantitative Biology* 11: 67–77.

Hershey, A.D. and Chase, M. 1952. Independent functions of viral proteins and nucleic acid in growth of bacteriophage. *Journal of General Physiology* 36: 39–56.

Lederberg, J. and Tatum, E.L. 1946. Novel genotypes in mixed cultures of biochemical mutants of bacteria. *Cold Spring Harbor Symposia on Quantitative Biology* 11: 113–114.

Luria, S.E. 1947. Recent advances in bacterial genetics. *Bacteriological Reviews* 11: 1–40.

Luria, S.E. and Delbrück, M. 1943. Mutations of bacteria from virus sensitivity to virus resistance. *Genetics* 28: 491–511.

Lwoff, A. and Guttman, A. 1950. Recherches sur un *Bacillus megatherium* lysogène. *Annales de l'Institut Pasteur* 78: 711–739.

Morgan, T.H. 1933. The relation of genetics to physiology and medicine, pp. 3–18. Reprinted in *Nobel Lectures in Molecular Biology*. Elsevier North-Holland Inc., Amsterdam.

Muller, H.J. 1922. Variation due to change in the individual gene. *American Naturalist* 56: 32–50.

2
ROOTS IN CLASSICAL GENETICS

Although genetics has had antecedents going back to ancient times, its real origin can be said to begin with the work of Charles Darwin and Gregor Mendel in the mid-19th century. Darwin's theory of the origin of species, based on natural selection, naturally focused on the origin of variation and the means by which variations were transmitted from one generation to the next. Mendel's work arose out of the more mundane activity of plant breeding and the observations on variation during domestication.

2.1 MENDEL'S WORK

The plant breeding work of Mendel, carried out in the 1850s and 1860s, was published in 1865 (Mendel, 1865). The object of Mendel's work was to construct a generally applicable law for the formation and development of hybrids and to determine their numerical (statistical) relationships. In brilliant fashion, Mendel formulated his questions in such a way that they could be answered by direct experimentation. He selected for his experimental material a species of plant, peas, that was very suitable for the work. He recognized the importance of using a plant that could be crossed without having to worry about the influence of foreign pollen and one whose seeds could themselves produce plants that could be hybridized. He also recognized the importance of selecting starting material that possessed constant differing traits. Perhaps the most brilliant aspect of Mendel's work is that he studied clearly identifiable traits (seed color and shape, stem length, position of the flowers) and ignored traits that were less suitable for tracing from one generation to the next. An English translation of Mendel's paper is given by Stern and Sherwood (1966).

Among other things, Mendel recognized that the traits he studied behaved as discontinuous characteristics, without any intermediate forms. In addition, certain traits appeared to disappear in the hybrids (Mendel called them "latent") and to reappear in the next generation, whereas other traits appeared in the hybrids virtually unchanged. He termed the traits that were unchanged in the hybrids *dominating* and those that became latent *recessive*. He established the terminology (still used) of expressing the dominant form as a capital letter and the recessive form as a lowercase letter. An important aspect of Mendel's work was statistical; he counted the number of offspring of different types and calculated their ratios. In the first generation (the one we would call the F_1), all of the offspring possessed the dominant trait, and the recessive trait appeared to vanish. When the hybrids were self-fertilized to produce an F_2 generation, two phenotypes were obtained, with three quarters of the offspring possessing the dominant trait and one quarter the recessive character. The latter always bred true in the F_3 and subsequent generations, but among the dominants, one quarter bred true and one half again produced offspring with three quarters dominant and one quarter recessive traits. He thus established the 1:2:1 ratio between pure dominants, heterozygotes, and pure recessives (although he did not use the term heterozygote).

Another important aspect of Mendel's work was the hybridization of plants differing by more than one trait. He showed that in such crosses, each trait behaved independently of the others. (None of the traits in Mendel's published work were linked.) Mendel also interpreted his results in terms of the formation of gametes, although Mendel himself did not use the word *gamete*, referring to two types of plants as *seed plants* and *pollen plants*.

Mendel's experimental tests also showed that during the formation of egg and pollen, there was a partitioning or splitting of the traits, a process we would now call *segregation*. Thus, except for the phenomenon of mutation, Mendel not only discovered the basic facts of genetics, but also developed the experimental procedures and terminology with which genetics could be pursued. A critical aspect of Mendel's work was the use of "paired" characters, one dominant and the other recessive, characters that we would now call *alleles*. Without the availability of two alleles for a gene, proper genetic analysis cannot be carried out. In most genetics research, such alleles are obtained through a mutation program. Although he lacked mutants, Mendel had available paired allelic forms in his collection of cross-fertile pea stocks.

Mendel's work, as is well known, lay buried for many years after it was published in an obscure journal, only to be rediscovered in the year 1900 by no less than three separate workers (see below).

2.2 DARWIN'S WORK

The other major antecedent of modern genetics arose from Darwin's theory of evolution. Although Mendel's work was ignored by others, Darwin's theories and writings became widely known and were highly influential. Darwin (1875) first considered hereditary phenomena in detail in his book entitled *The Variation of Animals and Plants under Domestication*. This book, first issued in 1868, was revised several times and was widely translated. In attempting to explain the manner by which the various forms of inheritance and variation take place, Darwin advanced a provisional hypothesis that he called *pangenesis*. He postulated the existence in tissues of minute granules capable of multiplication by self-division; these granules were present in all parts of the organism and were collected together to constitute the sexual elements involved in the development of the new generation. He called these granules *gemmules*.

Darwin's pangenesis hypothesis strongly influenced thinking in the late 19th century and was the forerunner of Hugo de Vries's hypothesis of *intracellular pangenesis* (see below), which ultimately led to the concept of the gene. It is interesting that Darwin, when attempting to justify the self-reproducing properties of gemmules that he had hypothesized, used microbial analogies:

> Nor does the extreme minuteness of the gemmules, which can hardly differ much in nature from the lowest and simplest organisms, render it improbable that they should grow and multiply. A great authority...says 'that minute yeast cells are capable of throwing off buds or gemmules, much less than the 1/100000 of an inch in diameter,' and these he thinks are 'capable of subdivision practically ad infinitum.' A particle of small-pox matter, so minute as to be borne by the wind, must multiply itself many thousandfold in a person thus inoculated; and so with the contagious matter of scarlet fever (Darwin, 1875).

Darwin was not the only thinker to hypothesize elements analogous to what we would now call genes. Independently, Herbert Spencer had published a hypothesis of "physiological units" in 1864, postulating structures that "possess the property of arranging themselves into the special structures of the organisms to which they belong...The germ cells are essentially nothing more than vehicles in which are contained small groups of the physiological units in a fit state for obeying their proclivity towards the structural arrangement of the species they belong to" (Dunn, 1965).

Other biologists who developed similar hypotheses were August Weismann and Carl Nägeli. In the meantime, the "nuclear" theory of heredity became established through the cytological work of Eduard Strasburger, Walther Flemming (who coined the words *mitosis* and *chromatin*), and W. Waldeyer (who coined the word *chromosome*). Much of this work related to the general problems of embryology and differentiation and was admirably summarized in the influential book by Edmund B. Wilson (1899).

2.3 DE VRIES'S WORK

One of the most important figures in the development of modern genetics in the late 19th century was Hugo de Vries, a professor of botany at the University of Amsterdam. de Vries's work was motivated in the first instance by an interest in developing a mechanism for evolution. Carrying out extensive plant hybridization studies, he independently discovered Mendel's laws. He conceived a mechanism of inheritance that seems, today, surprisingly modern. Following Darwin, de Vries developed what he called the hypothesis of *intracellular pangenesis*, postulating living, self-replicating units that he called *pangenes*. de Vries's pangene subsequently provided the model for the modern gene. de Vries postulated that a species was constituted from a large number of separate and more or less independent factors. "Just as physics and chemistry go back to molecules and atoms, the biological sciences have to penetrate to these units in order to explain, by means of their combinations, the phenomena of the living world" (translated from de Vries, 1889).

de Vries's great contribution, however, was to be his study of variation of plants by mutation. Carrying out extensive hybridization studies, primarily with plants of the genus *Oenothera* (evening primrose), de Vries observed that occasionally a new plant would arise that was dramatically different from the parent and from the other offspring. de Vries called these "sports" *saltations* or *mutations*.[1] Two key characteristics of mutations were that they occurred without any transitional gradations and that they were rare, whereas ordinary variations were seen to be continuous and common.

Completely independently of Mendel, de Vries created two broad general principles of genetics: (1) independent segregation of characters during reproduction and (2) the concept of mutability as the source of variation. (It should be noted that the mutations observed by de Vries were actually changes in ploidy rather than direct changes in single genes.) Although many of de Vries's examples of mutation turned out not to be mutations in the modern sense, his work was so carefully done and convincing that it is generally considered the starting point for modern genetics.

de Vries was also one of three scientists who, in the year 1900, rediscovered Mendel's work (the other two were Carl Correns and Erich von Tschermak-Seysenegg). In a paper published in 1900 and entitled "The Law of Splitting of Hybrids," de Vries mentioned that after his researches were completed and his conclusions drawn, he discovered Mendel's paper of 1865.[2] de Vries's main point in this paper was that when carrying out crosses, one should not look at species or their hybrids in the aggregate, but in relation to species-specific traits that are being followed through the cross. In a cross, only a single character should be followed, the others being disregarded temporarily, and crossing experiments should be limited to these opposing (what de Vries called *antagonistic*) characters. The lack of transitional forms between two antagonistic characters provided for de Vries the best evidence that such characters are well-delimited units. He noted that this idea had long ago been advanced by Mendel, when he developed his ideas of dominant and recessive characters. "In the hybrid the two antagonistic characters lie next to each other as anlagen. In vegetative life only the dominating one is usually visible...*The pollen grains and ovules...are not hybrids* but belong exclusively to one or the other of the two parental types" (translated from de Vries, 1900). The German word "anlage" (plural, anlagen), used first by de Vries, was often used later in genetics writings in English without translation. It can be translated in this context as "plan" or "design."

2.4 WILLIAM BATESON

Another central figure in the creation of the science of genetics after 1900 was the English zoologist William Bateson. One of the major controversies in the late 19th century concerned the relative importance of continuous and discontinuous variation. Bateson's hybridization work on plants had convinced him that variation was *discontinuous*. When he first read Mendel's paper, in the spring of 1900, he at once incorporated it into his thinking. Bateson quickly became one of the most ardent proponents of what he called "Mendelism," and he was more responsible than anyone else for the popularization of Mendel's work. He arranged for the publication of an English translation of Mendel's paper and wrote a spirited book, *Mendel's Principles of*

---

[1] Although de Vries coined the term "mutant," the terms "mutation" and "mutability" were not original with him, as they can be found in literary as well as medieval writings.
[2] See Section 4.1 for a note on the role of the bacteriologist Martinus Beijerinck in the rediscovery of Mendel's paper.

Heredity (Bateson, 1909). According to Bateson: "[Mendel's] experiments are worthy to rank with those which laid the foundation of the Atomic Laws of Chemistry."

It was also Bateson who introduced important new terminology for the developing field. He coined the term "genetics" itself, as well as the terms "zygote," "homozygote," "heterozygote," and "allelomorph" (later shortened to *allele*). Bateson also showed that the inheritance of a biochemical trait, alcaptonuria, behaved as a simple Mendelian recessive (Bateson, 1909).

2.5 WILHELM JOHANNSEN AND THE CONCEPT OF THE GENE

Mendel had used the German word *Merkmal*, which is usually translated as *character*, often used as *unit character*. Bateson, who developed so much of genetics terminology, usually used the word *factor*. None of these terms were sufficiently specific, and the term "gene" was thus coined by Wilhelm L. Johannsen, a Danish botanist, in an important book published in 1909. *Gene* was actually a shortened version of de Vries's *pangene*, which itself had a vague antecedent in Darwin's *pangenesis*. Johannsen also clearly saw the important distinction between the gene and the trait as expressed in the organism and devised the terms *phenotype* and *genotype*. According to Johannsen: "Phenotypes are measurable realities, just what can be observed as characteristic, in variation distributions of the 'typical' measurement, the center around which the variants group themselves. Through the term phenotype the necessary reservation is made, that the appearance itself permits no further conclusion to be drawn. A given phenotype may be the expression of a biological unit, but it does not need to be...[In addition,] the phenotype of an individual is thus the sum total of all of his expressed characters" (Johannsen, 1909; the English translation is taken from Dunn, 1965).

It is instructive that of the two terms, genotype and phenotype, Johannsen defined phenotype first, as an expression of what is actually observed. The phenotype was that which could be measured and studied and hence was susceptible to an operational definition. The underlying structure in the organism, that which was transmitted during hybridization, was then called a "gene." "The gene is thus to be used as a kind of accounting or calculating unit. By no means have we the right to define the gene as a morphological structure in the sense of Darwin's gemmules...Nor have we any right to conceive that each special gene...corresponds to a particular phenotypic unit-character or...a 'trait' of the developed organism." Johannsen then coined the word "genotype to express the underlying constitution of the organism from which development of the organism begins. This constitution we designate by the word genotype. The word is entirely independent of any hypothesis; it is *fact*, not hypothesis that different zygotes arising by fertilization can thereby have different qualities, that, even under quite similar conditions of life, phenotypically diverse individuals can develop" (Johannsen, 1909; translation from Dunn, 1965).

According to Dunn (1965), the speed and vigor with which genetics developed in the first decade of the 20th century owed much to Johannsen's critical sense and clear thinking. Relating genetics to evolution, Johannsen made a strong case for the noninheritability of acquired characteristics. From

Johannsen on, approaches to experimental genetics were well established so that intelligent and productive research could be done.

2.6 A.E. GARROD: INBORN ERRORS OF METABOLISM

An antecedent of biochemical genetics that indirectly influenced the development of bacterial genetics was the perception by the British physician Archibald E. Garrod that certain inherited defects in humans had a metabolic basis (Carlson, 1966; Olby, 1974). Although conventional genetic analysis is difficult in humans, medicine provides a vast array of interesting "defects," some of which are attributable to enzymatic deficiencies. In 1902, Garrod and Bateson suggested that the disease *alkaptonuria* was due to a single recessive gene, and subsequent analysis of family medical histories by Garrod provided strong confirmation. Alkaptonuria was revealed as a pronounced darkening of the urine when exposed to the air. Garrod showed that alkaptonuria resulted from an inability to utilize the amino acids tyrosine and phenylalanine. It is now known that patients with alkaptonuria excrete large amounts of homogentisic acid, a normal intermediate in the oxidation of aromatic amino acids. The enzyme homogentisic acid oxidase is lacking in these patients. Another example of a metabolic deficiency, recognized by Garrod, was albinism, due to a deficiency of the enzyme tyrosinase.

In 1908, Garrod published a paper describing his concept of an "inborn error of metabolism" (Garrod, 1908); this was followed in 1909 by a small book with the same title (Garrod, 1909). However, despite Garrod's attempts to popularize his genetic theory of metabolism, he failed to stimulate the interest of geneticists. It was not until George W. Beadle and Edward L. Tatum initiated research on the biochemical genetics of *Neurospora* in the early 1940s (see Sections 2.8 and 5.1) that the significance of Garrod's work was appreciated. As noted by Alfred Sturtevant, "few geneticists were well enough grounded in biochemistry to be willing to make the moderate effort required to understand what he was talking about" (Sturtevant, 1965).

2.7 T.H. MORGAN: CHROMOSOME THEORY OF HEREDITY

During the period from 1910 to 1920, genetics matured rapidly, mostly through the efforts of Thomas Hunt Morgan and his "school" in the Department of Zoology of Columbia University. Morgan, who introduced the fruit fly *Drosophila* into genetics, was the first geneticist to win the Nobel Prize (in 1933). The "Fly Room" on the sixth floor of Schermerhorn Hall became the world center of genetics research.[3] Among many things, Morgan and his students demonstrated unequivocally the physical reality of the gene by showing that it was part of the chromosome. Confirming Mendel's laws from studies of a vast number of mutants, the Morgan group discovered genetic linkage and crossing-over. Using crossing-over as a tool for genetic mapping, they demonstrated the linear order of genes. They also demonstrated that sex was chromosomally determined, and through extensive cytogenetic studies, they identified duplications, deletions, translocations, inversions, and trip-

---

[3] In a sense, bacterial genetics also began at Columbia University because Joshua Lederberg did his first hybridization experiments on *Escherichia coli* K-12 at Schermerhorn Hall in the laboratory of Francis Ryan.

loids. They also discovered the position effect, multiple alleles, and pleiotropy (more than one phenotypic effect caused by a single gene). When Morgan began his work on *Drosophila*, the reality of the gene was uncertain at best, but Morgan's work put the gene and genetics on a firm footing. The acclaim with which Morgan's work was received can be appreciated from the following quotes: "...the chromosome theory begins to appear as one of the great miracles in the history of human achievement" *C.D. Darlington*. "Morgan's evidence for crossing over and his suggestion that genes further apart cross over more frequently was a thunderclap: hardly second to the discovery of Mendelism, which ushered in that storm that has given nourishment to all our modern genetics" *H.J. Muller* (quoted in Shine and Wrobel, 1976). The chromosome theory of the gene would eventually lead to the chemical theory of the gene, although Morgan himself never participated in this development.

A passionate materialist, Morgan entered genetics through his interest in embryology. He fancied his work in the broader context of "experimental zoology." His first genetics research arose from an interest in the role of heredity in evolution. Initially, Morgan was reluctant to accept Mendelism (as it was then called) because he viewed it as a modern version of a theory he abhorred: preformationism. As an embryologist, Morgan put great emphasis on the fact that most of the characteristics of an organism arose during its development from a fertilized egg. However, although Morgan resisted Mendelism and its proneness to postulate unknown "factors" (the term then used for gene), he accepted enthusiastically de Vries's mutation theory of evolution. Morgan had actually visited de Vries in The Netherlands in 1900 and became convinced of the validity of de Vries's work. Mutation satisfied Morgan's need for an explanation of the origin of species. But Morgan, the experimentalist, needed an experimental tool to study this process. At this time, zoology at Columbia was under the dominant intellectual influence of E.B. Wilson. Research was emphasized over teaching, and numerous able students were attracted to graduate work. Among them was Fernandus Payne,[4] who had studied blind fish in Indiana caves. The mechanism by which these blind races of fish arose was completely unknown, but the genetic materialists needed a mechanism other than the inheritance of acquired characteristics. Morgan gave Payne the project of raising an animal for many generations in the dark to see if the eyes were lost, and Payne chose *Drosophila* because of its ease of culture and short generation time. Although this project yielded negative results, it did introduce *Drosophila* as an experimental organism into Schermerhorn Hall.

In a 1904 lecture at Cold Spring Harbor, de Vries suggested that radiation, newly discovered, might induce mutation. *Drosophila* was therefore used in an attempt to induce mutation by means of radiation. Morgan's attempts to induce mutation in *Drosophila* with radiation failed, but he did discover the first *Drosophila* mutant, a white-eyed specimen (wild type has red eyes). Although the exact origin of this specimen was never determined, it clearly

---

[4] A tenuous, albeit interesting, connection exists between Fernandus Payne and bacterial genetics! As a faculty member at Indiana University, I was privileged to know Payne as a retired Dean of the College of Arts and Sciences. It was Payne who brought Muller, Sonneborn, and Luria to Bloomington, and (so the story goes) he was also responsible for converting James Watson from ornithology, his original interest when he came to Indiana as a student, to genetics.

was a mutant. Morgan propagated the white-eyed character by mating the mutant fly with wild type. Red eye was dominant, but segregation occurred in the F_2 in Mendel's classic ratio of 3:1 red:white.

There was, however, an anomaly in the crosses, since red-eyed flies were mostly females and white-eyed flies were mostly male. Further crosses quickly showed that the factor for white eyes was unlike other recessive Mendelian characters in that the outcome of the cross was influenced by the sex of the parents. Morgan hypothesized that the eye-color factor and the sex-determining factor were combined or, as would be said later, linked. The cytological work on *Drosophila* that was taking place at this time showed that females had two X chromosomes, whereas males had only one. Although this made attractive the hypothesis that the eye-color factor was on the X chromosome, Morgan was cautious about drawing this conclusion.

An important characteristic of *Drosophila* is that it has only four chromosomes, so that the probability of genetic linkage is high. Effort was made to find more mutants, and the genetic analysis of these mutants soon showed that the genes were inherited in groups that corresponded to the number of chromosomes. "Since the number of chromosomes is relatively small and the characters of the individual are very numerous, it follows on the theory that many characters must be contained in the same chromosome. Consequently many characters must Mendelize together" (Morgan, 1910).

An important aspect of Morgan's work was the accumulation of mutants, and *Drosophila* was especially favorable for such work because, with a generation time of about 10 days, it could be raised quickly in large populations in relatively small space. In 1910, when genetics was being done with higher plants, mice, or pigeons, *Drosophila* must have appeared to biologists to be an extremely favorable organism for research, although once microorganisms were introduced into genetics, much of the appeal of *Drosophila* for mutant selection and genetic analysis vanished (Pontecorvo, 1958).

Although Morgan had always worked in the laboratory himself, he soon attracted an outstanding group of co-workers, among whom the most important were A.H. Sturtevant, Calvin B. Bridges, and Hermann J. Muller. Sturtevant and Bridges joined Morgan in 1910 and continued their association with him for the rest of their lives. Muller, who always had a strong independent streak (Carlson, 1981), joined in 1912 but left when he completed his doctoral research in 1915 to become a professor at the newly established Rice University in Houston. All three made major contributions to the developing chromosome theory of the gene.

Among the most important discoveries of the Morgan group was the phenomenon of crossing-over. It turned out that two genes that occasionally failed to segregate together were located farther apart than those that always appeared linked. Morgan postulated that during meiosis, the two chromosomes of a pair exchanged genes. If this were so, then it seemed evident that the farther apart two genes were, the more likely exchange could take place. A cytological basis for crossing-over was soon found in the occurrence of cross figures at meiotic synapsis, termed *chiasmata*. These studies led, within a short time, to the theory of the linear order of gene loci and to the construction, by Sturtevant, of genetic maps. By 1915, Morgan and his co-workers were able to present the locations on a genetic map for 30 distinct genes of the four *Drosophila* chromosomes (Morgan, Sturtevant, Muller, and Bridges, 1915).

It is interesting that in Morgan's 1915 book, the term "Mendelian unit" or "factor" was used instead of the word "gene." Morgan first used the term gene in his 1919 book on the physical basis of heredity (Morgan, 1919).

On the basis of genetic analysis, Morgan could present a number of characteristics of genes.

1. A gene could have more than one effect. For instance, insects that had the white-eye gene not only had white eyes, but also grew slower and had a lower viability.
2. The effects of the gene could be modified by external conditions, but these modifications were not transmitted to future generations. The gene itself was stable; only the character that the gene controlled varied.
3. Characters that were indistinguishable phenotypically could be the product of different genes.
4. At the same time, each character was the product of many genes. For instance, 50 different genes were known to affect eye color, 15 affected body color, and 10 affected length of wing.
5. Heredity was therefore not some property of the "organism as a whole," but rather of the genes.
6. Genes of the pair did not jump out of one chromosome into another, but changed when the chromosome thread broke as a piece in front of or else behind them. Thus, crossing-over affected linked genes as groups and was a product of the behavior of the chromosome as an entity.

Morgan's studies were based, to a great extent, on the availability of a large number of mutants, but the nature of the mutation process itself remained a mystery. Morgan (1919) summarized what was known: (1) Mutants appear rather infrequently; (2) the mutation does not arise gradually, but rather all at once; (3) some mutations are recurrent; and (4) the difference between the mutant and the wild type could be either small or large. Multiple alleles were known to be due to modifications at the same locus on the chromosome. Lethal genes, hiding within heterozygotes, were known to be fairly frequent. Some genes caused lethality of gametes, destroying eggs or sperm before they developed. Zygotic lethals were known that affected the embryo, larva, or adult. The ability to analyze lethals genetically led to a vastly increased estimate of the number of genes in *Drosophila*, but the nature of the mutation process itself remained outside Morgan's research interest. It remained for one of Morgan's students, H.J. Muller, to study this process.

2.8 H.J. MULLER AND THE CONCEPT OF MUTATION

Throughout most of his career, Hermann J. Muller was not on good terms with his former colleagues Morgan, Sturtevant, and Bridges, but his experimental and theoretical work in *Drosophila* genetics was superb and was to have far-reaching influence on the development of the field. Muller's life and work have been admirably described by Carlson (1981). Realizing that the problem of genetic transmission had been solved, Muller turned to a study of the nature of the gene itself. "Muller saw in the gene the enigma of life itself. By determining its properties he hoped to learn how genes worked. How often did mutations occur? Could the frequency be measured? How many varieties of mutation can occur for any individual gene? Did the mutant states reveal anything about the function or structure of the gene?" (Carlson, 1981).

It should be emphasized at this point that de Vries's idea of mutation and the idea that developed out of the genetic work of Morgan and his associates were two different things. de Vries saw mutation not as a change in the nature of a gene, but as something that was ill-defined. But in the 20 years since de Vries's book was published, genetic studies provided a completely different interpretation of mutation. Thus, the use of the word mutation by Morgan and Muller was quite different from the use of the word by de Vries.

By the 1920s, so many *Drosophila* mutants had been mapped that it was possible to perform rather sophisticated genetic experiments. Muller in particular became adept at constructing new *Drosophila* stocks that could be used to test particular genetic hypotheses. Much of this work was carried out with his good friend Edgar Altenburg, who had also gone to Rice University from Columbia. In 1922, Muller published an important paper that should have had far-reaching influence but turned out to be ahead of its time. Entitled "Variation Due to Change in the Individual Gene," the paper foresaw the developments of bacterial and bacteriophage genetics, and molecular biology, in an amazingly prescient way (Carlson, 1971).

> It is commonly said that evolution rests upon two foundations—inheritance and variation; but there is a subtle and important error here. Inheritance by itself leads to no change, and variation leads to no permanent change, unless the variations themselves are heritable. Thus it is not inheritance *and* variation which bring about evolution, but the inheritance *of* variation, and this in turn is due to the general principle of gene construction which causes the persistence of autocatalysis despite the alteration in structure of the gene itself (Muller, 1922).

Muller postulated that the gene must be of a unique chemical nature, able to change (mutate), but able to reproduce its changes. Thinking about how genes mutate led to a focus on the chemical nature of the gene in a way that Morgan's work on transmission genetics never could. "To Muller the thought was exhilarating, and years later he listed it as his chief discovery in theoretical genetics. It represented a break with the past; it freed the gene from the undefined status of a merely transmissible unit to a functional unit. The thought spurred him to seek in the gene's unique structure the basis of its evolution in a never-ending sequence of mutations..." (Carlson, 1981).

It was in this 1922 paper that the passage on bacteriophage quoted in Chapter 1 was published. According to Carlson (1981), when Muller completed his lecture that included the mention of phage as a possible gene analog, his colleagues thought they had been treated to a fanciful hoax. Henry Fairfield Osborn, who had given the welcoming address at the Congress, congratulated Muller on his sense of humor.

Unfortunately, the times were not ripe for geneticists to study bacteriophage, although they remained of interest to bacteriologists. However, it was not until Luria and Delbrück began their seminal work (see Section 6.2) that phage studies really began to be applied to genetics.

One of Muller's most important studies was the proof that mutation rate could be vastly increased by the use of X-rays. Although such an experiment would be relatively trivial if done now with haploid bacteria, it required heroic efforts when carried out with diploid *Drosophila*. The key to the problem was the construction of *Drosophila* stocks with sex-linked lethal heterozygotes that, for complex genetic reasons, greatly increased the probability of detecting mutants. Muller showed that radium and X-rays vastly increased the mutation rate. In his first paper, he reported a rise of about 15,000% percent in the mutation rate (Muller, 1927)! The experiments were so cleverly designed and carefully executed that there was no room for question. According to Carlson (1981), Muller's discovery of artificial mutation created a sensation. "Man's most precious substance, the hereditary material which he could pass on to his offspring, was now potentially in his control. X rays could 'speed up evolution,' if not in practice, at least in the [news] headlines. Like the discoveries of Einstein and Rutherford, Muller's tampering with a fundamental aspect of nature provoked the public awe."

The fact of mutation, and its susceptibility to artificial alteration, focused as never before on the chemical nature of the gene. Radiation was obviously a useful probe for unraveling the secrets of the gene, and in the 1930s, radiation genetics became a highly developed field. Indeed, two of the major researchers in bacterial genetics, Max Delbrück and Salvador Luria, began their careers studying radiation genetics (see Chapter 4). However, as it developed, radiation genetics by itself could not provide any critical information about the nature of the gene, although it developed into a sophisticated and highly quantitative discipline.

2.9 SOME CONNECTIONS BETWEEN HIGHER ORGANISMS AND BACTERIA

Haeckel and the Phylogeny of Bacteria

Ever since Ernst Haeckel's writings in the 19th century, the primitive nature of bacteria had been emphasized by biologists. Haeckel was a passionate Darwinian who derived vast evolutionary and taxonomic schemes. Before Haeckel, single-celled organisms such as protozoa had been considered to be "perfect" organisms, with fully developed nerves, muscles, and other organs, characteristic of higher forms. From his microscopic studies, Haeckel showed that this idea was false and that these organisms really consisted of free-living single cells. This idea had implications not only for the general theory of the cell, but also for evolution. To Haeckel, these simple creatures were the lowest forms of life and, because of their "imperfect" nature, could be considered representatives of "original life forms." To reflect this idea, Haeckel created the Protista, a third kingdom of life separate from the plant and animal kingdoms. Haeckel's Protista contained the protozoa and other unicellular microorganisms. Descending further on the evolutionary scale, Haeckel reached the bacteria, an even simpler group of organisms. With the imperfect microscopy available to him, Haeckel could distinguish no features in a bacterium and considered it to be simply a "glob of protoplasm, which in opposition to true cells even lacks a nucleus" (translated from Haeckel, 1878). He considered bacteria to represent an early stage in the development of life on earth, the oldest representatives of all other organisms. Because a nucleus could not be seen, he concluded that bacteria were not true cells, but nucleus-free structures which he called *cytodes*. He designated the group that contained the bacteria *Monera*. He believed that the first organism to arise on earth (by spontaneous generation) was a type of moneran, a formless blob of protoplasm, devoid of nucleus and membrane.

Because of Haeckel's powerful position in biology and his polemical writing style, his ideas had enormous influence on thinking about the place of bacteria in the living world. Haeckel's classification continues to have influence; even today, the term Protista finds wide use among biologists, although generally to refer to unicellular eucaryotic organisms.

Troland and the Enzyme Theory of Life

By the time World War I ended, biochemistry had started to flourish, and the chemical steps of intermediary metabolism were being pinpointed. At the same time, the discovery of radioactivity and the nature of the atom had led to startling developments in chemistry and physics. A materialist view of life based on chemistry and physics thus seemed possible, although *vitalism* still held sway among biologists who were unable to grasp the newly developing theories of the physical world. In a paper that was to be widely read and extensively quoted, the Harvard University biologist Leonard T. Troland brought together the new physics, biochemistry, and genetics in an attempt to dispel ideas of vitalism (Troland, 1917; for a detailed discussion of Troland's work, see Olby, 1974).

At this time, physical chemists had defined the *colloidal state*, and it appeared that colloid theory had important things to say about cell theory. Troland saw that a strong link between biology and chemistry could be forged through the concept of *enzyme action*. According to him, "*specific catalysis* provides a definite, general solution for all of the fundamental biological enigmas: the mysteries of the origin of living matter, of the source of variations, of the mechanism of heredity and ontogeny, and of general organic regulation...Catalysis is essentially a determinative relationship, and the *enzyme theory of life*, as a general biological hypothesis, would claim that all intra-vital or 'hereditary' determination is, in the last analysis, catalytic" (Troland, 1917).

At the time of Troland's writings, the importance of enzymes was established, but the nature of enzymes was still uncertain. The first enzyme would not be crystallized until the 1920s. The relationship between genes and enzymes was also uncertain, although to most it appeared that these two kinds of entities were likely related, possibly identical. "Several Mendelians have even hinted that the 'unit characters' themselves are enzymes..." Thus, to Troland the term genetic enzyme seemed not inappropriate.

Among the most important concepts Troland was to advance was the distinction between two kinds of catalysis, *autocatalysis* and *heterocatalysis*. Some catalysts were able to generate more quantities of themselves, a process that could be called *autocatalytic*. The formation of a crystal in a supersaturated solution was viewed as an autocatalytic process. By this time, William H. Bragg had already begun to study the structure of crystals by X-ray diffraction, making the autocatalytic formation of crystals seem intelligible in physical terms. According to current theory, crystal formation was akin to polymerization, and autocatalysis could readily explain the "synthesis of polymeric molecules from individual units which are all alike."

Heterocatalysis, on the other hand, is the promotion of an unrelated chemical reaction. Troland saw heterocatalysis as an extension of autocatalysis, but with the important distinction that heterocatalysis may be the *stronger* process. "Indeed, the catalytic effect which is based upon direct similarity of structure between two systems should be much weaker than that which accompanies certain types of structural *correspondence*, such as that existing between a body and its mirror-image, or between a lock and a key." The lock and key analogy, essentially an idea of complementarity, had already been developed by immunologists, notably Paul Ehrlich, to explain antibody-antigen reactions. Similar ideas would frequently turn up in theories of the autocatalytic and heterocatalytic roles of the gene (Emerson, 1945; see Section 10.3).

Troland then considered the application of the above ideas to the life processes. The fundamental biological process of *growth* was obviously an expression of autocatalytic chemical reactions, but the internal control of cellular life rested primarily with the nucleus, "or with the chromatin substance of the cell, when no well-defined nucleus is present. Even in the highly organized cell, this substance can be seen to possess a mosaic structure, and it can be shown that for a given species this structure is sensibly constant, so that it is necessary to suppose that a reduplication of chromatin units occurs with each cell-division. This process of reduplication is apparently made visible to us in mitosis. The simplest hypothesis to account for such reproduction lies in the supposition that each unit can give rise to another unit substantially identical with itself."

Morgan and co-workers had shown the connection between genes and chromosomes, and it was thus reasonable to conclude that the genes themselves exhibited autocatalysis. But how are the genes to bring about the chemical life processes? Troland concluded: "Although the fundamental life-property of the chromatin units is autocatalysis, it is necessary and legitimate to suppose that the majority of them sustain specific heterocatalytic relationships to reactions occurring in living matter. This is because nuclear material makes up a relatively small percentage of protoplasm, and because the reactions governed by enzymes are ordinarily heterocatalytic."

Troland even considered the chemistry of the gene, in a passage of exceptional clarity:

> It is a remarkable fact that the chemistry of the cell-nucleus has reached a stage of advancement superior to that attained by the chemistry of the cytoplasm. It appears that the essential constituent of chromatin is a substance called nuclein, which is composed of a basic, protein factor and nucleic acid. The facts indicate that the [nucleic] acid factor is the permanent and essential component of the nucleus, and organic chemical analysis seems to prove that only one kind of nucleic acid exists in animal tissues, although a different variety is to be found in the cells of plants. If, as now seems probable, the genetic enzymes must be identified with the nucleic acids, we shall be forced to suppose that these substances, although homogenous—in animal or plant—from the point of view of ordinary chemical analysis, are actually built up in the living chromatin, into highly differentiated colloidal, and colloidal-molar, structures. The apparent homogeneity results from the fact that ordinary chemical analysis provides us only with the *statistics of the fundamental radicles* which are involved (Troland, 1917).

Erwin Schrödinger and his little book *What Is Life* (Schrödinger, 1944) have been very influential and widely quoted (Stent, 1966), but Troland's remarkable statement places him almost 30 years ahead of Schrödinger (see also Portugal and Cohen, 1977).

Among other important things, Troland made a definite connection between genes and viruses:

> There is considerable evidence that free autocatalytic enzymes exist in our biological universe even at the present day. Such an hypothesis would serve to account for the specific contagious diseases, such as measles, rabies, and smallpox, which have been demonstrated to possess "filterable viruses"...The single cell, and so-called simple protoplasm, must be regarded as the products of a detailed process of evolution, and hence can not form the ultimate explanatory units in biology. Next to the free autocatalytic particle, the simplest typical life-structure would consist of a single particle of this sort surrounded by an envelope of semi-liquid and chemically homogeneous substance with which it sustains a heterocatalytic relationship. The most primitive substance of this kind might be called *eoplasm*, to distinguish it from complex protoplasm, and the physical system made up of protase and eoplasm would represent a living cell in its most reduced form (Troland, 1917).

Troland's paper was to be widely read and quoted in the 1920s and 1930s, as the development of classical genetics proceeded. Hermann Muller, in particular, was heavily influenced by it. Unfortunately, Troland's ideas did not lead directly to biochemical *experiments*, but the possible relevance of viruses and simple cells to an understanding of genetic phenomena was retained and returned to from time to time. Although Delbrück never cites Troland, the latter is certainly his intellectual grandfather (Delbrück, 1941).

Beadle and Tatum

Another important area, gene action, developed slowly throughout the 1920s and 1930s. Because of the vast amount of genetic information on *Drosophila*, it was natural that the first work should be carried out with this organism. In 1928, Morgan moved from Columbia to the California Institute of Technology to set up a new Division of Biology, with an emphasis on physicochemical genetics. Among the numerous outstanding scientists who joined Morgan's department was George W. Beadle, trained at Cornell as a corn geneticist, who went to CalTech as a postdoctoral student. After several years, Beadle moved to Stanford University and turned from *Drosophila* to *Neurospora*, a more favorable organism for biochemical studies; he was joined at Stanford by the bacteriologist Edward L. Tatum. The *Neurospora* work of Beadle and Tatum, discussed in Section 5.1, became a critical link between "classical" and bacterial genetics.

Milislav Demerec

Milislav Demerec, a distinguished geneticist and the long-term director of the Cold Spring Harbor Laboratory of Quantitative Biology, was among the most important links between classical and bacterial genetics (Dobzhansky, 1971; Hartman, 1988). Demerec, a Yugoslav with an agricultural background, came to the United States around World War I. After a brief period in the Department of Plant Breeding at Cornell University, he joined in 1923 the Department of Genetics of the Carnegie Institution of Washington, which was housed at Cold Spring Harbor. Demerec remained at Cold Spring

Harbor throughout his long and productive career. During his early years at Cold Spring Harbor, he studied maize genetics, but through an interest in mutation he turned to *Drosophila*. After Muller's discovery of the mutagenic action of X-rays, Demerec began a lengthy series of studies on X-ray-induced mutations. He collaborated with Calvin Bridges and others of the Morgan school of genetics. Among the important contributions of Demerec to *Drosophila* genetics was his establishment, over Morgan's objections,[5] of a stock center where mutants were maintained and distributed to interested scientists. This idea of stock centers of genetic mutants was to take firm root and to play a great role in the development of bacterial genetics. In 1941, Demerec became director of the Long Island Biological Association Laboratory, which was the founding institution of the Cold Spring Harbor complex. In 1943, he also became Director of the Department of Genetics of the Carnegie Institution. Under his leadership, the Department of Genetics and the Biological Association Laboratory were in effect combined, and from 1943 until he retired in 1960, Demerec was the head of both institutions.

In the early 1940s, both Luria and Delbrück began work at Cold Spring Harbor Laboratory (see Section 6.6), and it is likely that Demerec was introduced to microbial genetics through their influence. Demerec's first microbial genetics work came out of wartime necessity. His expertise in mutation brought to Cold Spring Harbor during World War II an important project to use mutation to increase the yields of penicillin in the fungus *Penicillium chrysogenum*. The Demerec group found a mutant that would produce penicillin in submerged culture, an important step in the large-scale production of the antibiotic. Simultaneously, Demerec began to study the development of penicillin resistance in *Staphylococcus*, and this work led him into bacterial genetics. In the late 1940s, he abandoned *Drosophila* work and carried on extensive studies on the effect of radiation and chemical mutagens on bacteria. This work eventually led to Demerec's final project, the study of genetic fine structure in *Salmonella* by the use of transduction (see Section 8.5).

Among other things, Demerec developed what was to become the standard system of genetic nomenclature for bacterial genetics (Demerec, Adelberg, Clark, and Hartman, 1966).

Demerec recognized early the qualities of Luria and Delbrück and gave them support for their bacteriophage work. Although neither Luria nor Delbrück was ever on the permanent staff at Cold Spring Harbor, they had both been offered permanent positions there and spent many summers there in collaboration. Among Demerec's contributions to phage work was the paper with U. Fano that first defined the T series of phages (discussed in Section 6.6).

Between 1941 and his retirement in 1960, Demerec organized the summer courses and symposia at Cold Spring Harbor. Delbrück's phage course was organized with Demerec's support, and later Demerec himself organized a

---

[5] Morgan's viewpoint on the establishment of a *Drosophila* stock center is discussed by Margaret Miller in the introductory notes to the Milislav Demerec papers at the American Philosophical Society Library in Philadelphia. "In a letter to Walter M. Gilbert dated November, 1934, Demerec recounts a conversation with Morgan in which Morgan felt that Demerec had no business applying for money [from the Rockefeller Foundation] for a stock center, inasmuch as such a project might affect the status of his [Morgan's] own grant applications."

similar course in the genetics of bacteria. Throughout these years, Demerec also organized all of the Cold Spring Harbor Symposia on Quantitative Biology, including the ones that were to be so important for the development of modern genetics: (1941) Genes and Chromosomes: Structure and Organization; (1946) Heredity and Variation in Microorganisms; (1947) Nucleic Acids and Nucleoproteins; (1951) Genes and Mutations; (1953) Viruses; (1956) Genetic Mechanisms: Structure and Function; (1958) Exchange of Genetic Material: Mechanisms and Consequences.

Throughout his lengthy and productive career, Demerec maintained strong ties to the genetic community. He was continually active in the International Congresses of Genetics, the Genetics Society of America (President, 1936), and the American Society of Naturalists. He was a member of the National Academy of Sciences and sat on many of their special committees dealing with genetics. He was also a member of the American Academy of Arts and Sciences and the American Philosophical Society. He founded *Advances in Genetics* and edited it for many years. Demerec's support for bacterial and phage genetics was critical for the development of these fields, especially in the vital period right after World War II.

2.10 DELAYED ESTABLISHMENT OF MENDELIAN GENETICS IN FRANCE

It is ironic that Mendelian genetics found little official acceptance in France, since it was in France that modern molecular genetics really came of age (see Chapter 10). The extreme resistance of French biologists to Mendelian genetics is discussed in detail by Burian, Gayon, and Zallen (1988), who attribute it to four factors: (1) the strong influence of ideas on the inheritance of acquired characteristics from Lamarck and his followers; (2) the power of Claude Bernard, who championed physiological thinking; (3) the strong tradition of embryological research at a time when embryologists were reluctant to believe that genetics could explain the fact of development (see Section 2.7); and (4) the well-developed microbiological tradition from Louis Pasteur. It is interesting that these last two influences led in a positive way to the development of molecular genetics in France in the 1940s and 1950s (see Chapter 10).

Genetics did not appear in the official university curriculum in France until the late 1940s. After the rediscovery of Mendel's laws in 1900, some genetics research was initiated in France, but early French geneticists seemed not to develop a lasting influence on French biology. Between the two world wars, research of a genetic nature was carried out primarily at two French institutions: at the Pasteur Institute, by Eugéne Wollman and André Lwoff, and at the Institut de Biologie Physico-chimique, by Boris Ephrussi. However, if Mendelian genetics is defined as any study employing classical Mendelian methodology (involving as part of the experimental protocol the making of crosses and the analysis of the phenotypes of progeny), most of this research does not count as Mendelian. Wollman and Lwoff worked on bacterial lysogeny, which at that time was thought to have something to do with cytoplasmic inheritance (see Chapter 7), and Ephrussi worked on physiological genetics. Although Ephrussi's work did initially involve a classical genetic system (*Drosophila*), he turned later to research on cytoplasmic inheritance in yeast. It is also noteworthy that George Beadle and Boris Ephrussi worked closely together in the 1930s.

2.11 CONCLUSION

Genetics research requires two important factors: (1) the availability of stocks with alternate states (alleles) of a gene and (2) the availability of a system of mating or hybridization, so that genes from distinct organisms can be brought together in the same organism. Only through hybridization can a gene be traced from one generation to the next. In both plants and animals, these requirements were not difficult to obtain. As we have discussed, Mendel was able to select pea plants with alternate states from various seed stocks, and Morgan was able to obtain alternate states in *Drosophila* by searching for mutants. Both peas and insects reproduce sexually, so that a hybridization system was readily available.

In bacteria, on the other hand, the situation was quite different. At the time classical genetics was developing, sexual reproduction in bacteria was unknown. Mutation, of course, occurs readily, but because of the minute size and large population densities, it was initially difficult to easily select strains that were demonstrably mutant. Indeed, since the single bacterial cell is equivalent to the whole pea plant or *Drosophila* adult, it is almost impossible to study bacterial genetics effectively in single cells (single organisms). In bacteria, one is generally dealing with population genetics and must draw conclusions about cellular genetics from population events. For many bacteriologists, the distinction between a cell and a population was not even clear. For these reasons, and some others that will become evident in the course of this book, the study of bacterial genetics lagged greatly behind the study of the genetics of higher organisms.

REFERENCES

Bateson, W. 1909. *Mendel's Principles of Heredity*. The University Press, Cambridge. 396 pp.

Burian, R.M., Gayon, J., and Zallen, D. 1988. The singular fate of genetics in the history of French biology, 1900–1940. *Journal of the History of Biology* 21: 357–402.

Carlson, E.A. 1966. *The Gene. A Critical History*. Saunders, Philadelphia. 301 pp.

Carlson, E.A. 1971. An unacknowledged founding of molecular biology: H.J. Muller's contribution to gene theory, 1910–1936. *Journal of the History of Biology* 4: 149–170.

Carlson, E.A. 1981. *Genes, Radiation, and Society. The Life and Work of H.J. Muller*. Cornell University Press, Ithaca. 457 pp.

Darwin, C. 1875. *The Variation of Animals and Plants under Domestication*, volume II, pp. 371–372. John Murray, London.

Delbrück, M. 1941. A theory of autocatalytic synthesis of polypeptides and its application to the problem of chromosome reproduction. *Cold Spring Harbor Symposia on Quantitative Biology* 9: 122–126.

Demerec, M., Adelberg, E.A., Clark, A.J., and Hartman, P.E. 1966. A proposal for a uniform nomenclature in bacterial genetics. *Genetics* 54: 61–76.

de Vries, H. 1889. *Intracellulare Pangenesis*. Fischer, Jena.

de Vries, H. 1900. Das Spaltungsgesetz der Bastarde. *Berichte der deutschen botanischen Gesellschaft* 18: 83–90.

Dobzhansky, T. 1971. Milislav Demerec. *Advances in Genetics* 16: xv–xx.

Dunn, L.C. 1965. *A Short History of Genetics*. McGraw-Hill, New York. 261 pp.

Emerson, S. 1945. Genetics as a tool for studying gene structure. *Annals of the Missouri Botanical Garden* 32: 243–249.

Garrod, A.E. 1908. Inborn errors of metabolism. *Lancet* July 4, pp. 1–7.

Garrod, A.E. 1909. *Inborn errors of metabolism*. H. Frowde, London.

Haeckel, E. 1878. *Das Protistenreich*. Ernst Günther's Verlag, Leipzig.

Hartman, P.E. 1988. Between Novembers: Demerec, Cold Spring Harbor, and the Gene. *Genetics* 120: 615–619.

Johannsen, W. 1909. *Elemente der Exacten Erblichkeitslehre.* G. Fischer, Jena.

Mendel, G. 1865. Versuch über Pflanzen-Hybriden. *Verhandlungen des naturforschenden Vereines in Brünn* 4: 3–47.

Morgan, T.H. 1910. Chromosomes and heredity. *American Naturalist* 44: 449–496.

Morgan, T.H. 1919. *The Physical Basis of Heredity.* Lippincott, Philadelphia. 305 pp.

Morgan, T.H., Sturtevant, A.H., Muller, H.J., and Bridges, C.B. 1915. *The Mechanism of Mendelian Heredity.* Henry Holt, New York. 262 pp.

Muller, H.J. 1922. Variation due to change in the individual gene. *American Naturalist* 56: 32–50.

Muller, H.J. 1927. Artificial transmutation of the gene. *Science* 66: 84–87.

Olby, R. 1974. *The Path to the Double Helix.* University of Washington Press, Seattle. 510 pp.

Pontecorvo, G. 1958. *Trends in Genetic Analysis.* Columbia University Press, New York. 145 pp.

Portugal, F.H. and Cohen, J.S. 1977. *A Century of DNA.* MIT Press, Cambridge, Massachusetts. 384 pp.

Schrödinger, E. 1944. *What Is Life.* Cambridge University Press, Cambridge.

Shine, I. and Wrobel, S. 1976. *Thomas Hunt Morgan. Pioneer of Genetics.* The University Press of Kentucky, Lexington.

Stent, G.S. 1966. Introduction: waiting for the paradox. pp. 3–8 in Cairns, J., Stent, G.S., and Watson, J.D. (ed.), *Phage and the Origins of Molecular Biology.* Cold Spring Harbor Laboratory of Quantitative Biology, Cold Spring Harbor, New York.

Stern, C. and Sherwood, E.R. 1966. *The Origin of Genetics. A Mendel Source Book.* W.H. Freeman, San Francisco.

Sturtevant, A.H. 1965. *A History of Genetics.* Harper and Row, New York. 165 pp.

Troland, L.T. 1917. Biological enigmas and the theory of enzyme action. *American Naturalist* 51: 321–350.

Wilson, E.B. 1899. *The Cell in Development and Inheritance.* Macmillan, New York.

3
ROOTS IN BACTERIOLOGY

Because the study of bacterial genetics is so linked to the techniques of bacteriology, bacterial genetics could not develop until the discipline of bacteriology had been firmly established. Although some of Louis Pasteur's work served as a precursor, bacteriology as a discipline can be said to have started with Robert Koch's work in the late 1870s and early 1880s (Brock, 1988a). Koch developed the techniques for obtaining pure cultures of bacteria, especially the plate technique that permitted the development of pure strains of bacteria by single-colony isolation on gelatin or agar plates.

3.1 IDEAS OF CONSTANCY OF BACTERIAL SPECIES

A critical idea in Koch's early work was that a single bacterial species was responsible for a single infectious disease. This idea is implicit in the use of "Koch's postulates" for identifying the causal agents of infectious diseases. In fact, many workers of Koch's time insisted that bacteria underwent extreme variability, but Koch, rightly in most cases, attributed this variability to poor technique. For instance, one of the vigorous champions of extreme variability at this time was the Munich botanist Carl Nägeli, who made the following statement: "For over 10 years I have examined thousands of different fission organisms and (with the exception of the Sarcinae) I have been completely unable to distinguish even two distinct species" (translated from Nägeli, 1877).

It is clear that if Nägeli had been correct, the whole experimental method that Koch was developing would have been impossible. If it could be shown that one bacterium could turn, in an unpredictable manner, into another, then it would be difficult to accept the fact that a *specific* disease was caused by a *specific* bacterium. Koch realized that in determining pathogenicity, animal experimentation must be done with pure cultures, always under careful microscopical control to ensure that the cultures had not become contaminated. Nägeli and others of his school were casual in their culture work and were also inexpert at using the microscope to look at organisms as small as bacteria.

One important basis of Koch's viewpoint was the taxonomic system for bacteria developed by the eminent botanist Ferdinand Cohn (1875). Cohn's system, based primarily on morphological characteristics, assumed that bacteria retained constant (or at least consistent) shapes. This idea and Cohn's systematics were strongly supported by Koch's work and eventually came to be known as the Koch-Cohn theory of monomorphism. These ideas culminated, in the mid-1880s, in the enunciation of Koch's postulates, and the stunning successes in the characterization of the causal agents of major infectious diseases caused by bacteria.

It should be emphasized that although this rigid viewpoint of the constancy of species was wrong, it was in the early days of bacteriology an idea that was essential, because only in that way could sense be made out of the research. Even today, numerous errors are made by researchers working with impure cultures, but in the late 19th century, when the techniques of bacteriology were still developing and had not yet become widely taught, errors were much more likely to occur. The book by Flügge (1890), which became the principal textbook of bacteriology, analyzes some of these erroneous experiments in considerable detail. One oft-cited case concerned the work of Hans Buchner, a follower of Nägeli, who claimed to have converted the anthrax bacillus (*Bacillus anthracis*) into the hay bacillus (*Bacillus subtilis*) by cultivation of the former on rich culture media.

Although the extreme rigidity of Koch's viewpoint was essential at this time, it was later seen as a hindrance to certain kinds of bacteriological research:

> Looking back on the road over which we have traveled, it is impossible to estimate the loss sustained by bacteriology, especially in latter years, through the repressive and misguiding influence of the strict monomorphic conceptions, or to appreciate the often serious biological blunders that are to be laid at its door...The doctrine of monomorphism has descended to us from the early conceptions of the nature of bacteria maintained by Cohn, Koch, and others of the early school. Under its influence...there were set up strict notions of 'normal' bacterial cell types, 'normal' colony forms, and 'normal' cultures. Whatever departed from the expected normality was at once relegated to the field of contaminations; or to the weird category of 'involution forms', 'degeneration forms', or pathological elements possessing neither viability, interest, nor significance...The dictum was then laid down that 'the mode of reproduction of bacteria is by simple fission'—a view which has descended through two generations of bacteriologists and through numerous generations of textbooks, even to the year 1927 (Hadley, 1928).

Despite this impassioned statement, written at a time when the genetics of higher organisms had become almost classical, another 20 years would pass before bacterial genetics would come of age!

By the 1890s, a few undisputed examples of bacterial variation had been reported, but it was not until the mutation concept of Hugo de Vries had been described in higher organisms (see Section 2.3) that a theoretical basis for understanding variation in bacterial cultures became available. And it was a Dutch bacteriologist familiar with de Vries's work, Martinus Beijerinck, who first related mutational phenomena in higher organisms to those in bacteria (see Section 4.1).

3.2 THE PURE CULTURE TECHNIQUE BEFORE ROBERT KOCH

Although bacteriology can be said to have begun with the work of Louis Pasteur (Dubos, 1988), pure cultures in the sense that we know them today were not obtained by Pasteur or any members of his school. Pasteur grew bacteria in transparent liquid media. When growth occurred, as evidenced by the development of turbidity in the culture vessel, a minute quantity of the culture was inoculated into a fresh medium, and so on in series. By means of serial transfer, and occasional checking in the microscope, Pasteur assumed that a "pure" culture of one type of microorganism would ultimately result. It was possible in many cases to select a medium that was the most appropriate for a single type of organism and hence obtain some degree of "purity." However, if a pure culture was obtained with such procedures, it was usually fortuitous. Pasteur's cultures were equivalent to what would today be called "enrichment cultures."

Joseph Lister was the first to obtain a pure culture in liquid medium (Lister, 1878). In Lister's work, serial dilutions were made into culture media. After the cultures were incubated, the vessel from the highest dilution that showed evidence of bacterial growth was selected as the source of inoculum for his subsequent cultures. This procedure, ultimately refined in bacteriology to yield quantitative estimates of viable counts in the *most-probable-number technique*, was quickly superseded by Koch's plate technique (see below). Lister's method is certainly the great-grandfather of the Luria-Delbrück fluctuation test that represents the founding of modern bacterial genetics (see Section 4.8), but it apparently had little influence on Pasteur or his school.

3.3 KOCH'S PLATE TECHNIQUE

The basis of Koch's plate technique is the development of isolated colonies on solid or semisolid surfaces. In a colony, all the progeny of a single cell remain associated and are kept separate from the progeny of other cells. The colony, if derived from a single cell, is thus a pure culture and can be used as inoculum for more cultures.

The development of pigmented colonies on the cut surfaces of incubated potatoes was first reported by Joseph Schroeter, a student of Ferdinand Cohn (Schroeter, 1875). (For many years, potatoes were used even in medical bacteriology for the isolation of pure cultures.) Schroeter pointed out that the colors of the pigmented colonies were a constant characteristic of the bacterium and that each organism exhibited its own characteristic color. Schroeter not only used potatoes, but he also used solid media made from starch paste, egg albumin, bread, and meat. Koch, a frequent visitor to Cohn's laboratory at this time, was certainly familiar with Schroeter's work, although he does not cite it.

Another important predecessor of Koch was Oscar Brefeld, a mycologist who made many important contributions to the understanding of fungi. In 1875, Brefeld laid down precisely the principles that must be followed for obtaining pure fungal cultures (Brefeld, 1875): (1) The inoculation of the medium should be made from a single spore; (2) the medium should be clear and transparent and should yield optimal growth of the organism; (3) the culture should be kept completely protected from external contamination throughout its existence.

Brefeld's paper dealt primarily with the study of life cycles of the fungi, where microscopic control from spore through vegetative culture to spore was essential in order to prove that the structures seen were legitimate parts of the fungal life cycle. It was for this reason that inoculation with a single spore was necessary to ensure that a pure culture was a population derived from a single cell. However, although Brefeld's procedure worked satisfactorily with large-celled microorganisms such as fungi, it was unsuitable for the much smaller bacteria. Brefeld's writings had wide influence on the bacteriologists of the day, and Koch certainly built his technique around Brefeld's principles.

Koch's paper describing pure culture methods for bacteria was published in 1881 and became, for late 19th century workers, the "Bible of Bacteriology" (Koch, 1881). Discussing the development of colonies on sterile potato surfaces, Koch emphasized that each colony is a pure culture, but he recognized that a potato surface is not suitable for the growth of all organisms. Instead of trying to find a solid surface that would support the growth of pathogens, Koch took a medium, nutrient broth, that supported the growth of pathogens and made it solid by adding gelatin. This provided a completely general procedure for culturing any microorganism, since one simply solidified an appropriate culture medium by adding gelatin. Since gelatin did not remain solid at body temperature, agar was introduced to culture certain pathogens, such as *Mycobacterium tuberculosis*.

Koch quickly realized that the plate technique had many uses besides its value for the isolation of pure cultures. Most important, the technique could be used to assess the effect of various chemical and physical treatments on the viability of bacteria. Systematic studies of the sterilization process using the plate technique permitted the development of reproducible methods for preparing culture media and sterile equipment. The plate technique was also used to assess environmental contamination, which helped to improve the technique of workers doing bacteriological studies. Among other things, Koch was able to make a clear distinction between sterilization, which involved the complete killing of both spores and vegetative cells, and disinfection, in which vegetative cells are killed but spores are not necessarily affected. Koch and co-workers carried out extensive studies with heat sterilization, under both dry and moist conditions, and with both spore-forming and non-spore-forming organisms. Using the quantitative procedures made possible by the use of the plate technique, Koch and co-workers were able to show that under dry heat conditions, spore-forming bacteria were only completely killed after 3 hours treatment at 140°C. With moist heat, lower temperatures and shorter heating times were necessary, although under these conditions, it was necessary to heat under pressure. The standard heat sterilization procedures used in all microbiological laboratories came out of Koch's work in the early 1880s.

3.4 CHARACTERIZATION OF BACTERIAL PATHOGENS

Using the plate technique, his postulates, and the culture and sterilization procedures that he developed, Koch was quickly able to isolate and characterize several of the important causal agents of infectious disease in humans. The most important discoveries, and those which catapulted Koch into world fame, were isolation of the causal agents of tuberculosis, *Mycobacterium tuberculosis*, and cholera, *Vibrio cholerae*. These two clinically relevant bacteria had been responsible for the deaths of vast numbers of humans. Tuberculosis at that time was responsible for one seventh of all human deaths, and cholera, endemic in India, had entered Europe several times in the 19th century and had caused huge population losses. The successes of Koch's methods in the study of these two major diseases quickly established bacteriology as an important discipline. It was clear that bacteriological work had vast implications for the newly arising fields of hygiene and public health. Between 1882, when Koch announced the isolation of *Mycobacterium tuberculosis*, and the beginning of the 20th century, virtually all of the bacteria that caused serious infectious disease in humans and domestic animals were isolated. Bacteriology became one of the most important branches of medicine as well as an important science in its own right. Koch's ideas of monomorphism and the fixity of species were not only completely accepted, but became accepted dogma (Smith, 1932).

3.5 PASTEUR'S WORK ON ATTENUATION

While the relatively young Robert Koch was developing his methods and career, the venerable Louis Pasteur was ending his scientific life with great successes in the control of infectious disease. Pasteur and Koch became, in the mid-1880s, bitter enemies and violent opponents. Pasteur did not actually begin research on infectious diseases of higher animals and humans until late in his life, about the time that Koch's work was starting. Soon, these two giants of 19th century bacteriology became violent antagonists. This was due, in part, to a residue of resentment, especially on Pasteur's part, over the Franco-Prussian War of 1870, which France had lost.

Over the years from 1877, when Pasteur published his first paper on anthrax, until 1881, when he demonstrated great success in the control of infectious disease by vaccination, Pasteur pursued his studies on bacterial diseases. After 1881, Pasteur concentrated on rabies, and his work had no further importance for the development of bacterial genetics. During those brief four years that he studied bacterial diseases, Pasteur not only isolated new pathogens (notably, the causal agent of fowl cholera), but also, most importantly, developed methods for reducing the virulence of bacterial pathogens (Pasteur, 1880). He coined the term *attenuation* to describe the phenomenon. In this work, which is certainly a foundation of bacterial genetics, Pasteur developed cultures whose virulence had lessened but which in other respects remained the same species. Such attenuated cultures could be used to treat animals so that they would survive a challenge with a virulent culture. Attenuation thus became the basis for a rational development of immunization procedures, and Pasteur soon dramatized his work by developing a vaccine for anthrax and for a feared disease of humans, rabies.

Pasteur's procedure for developing an attenuated strain involved merely leaving cultures exposed to the air for some period of time, after which the virulence had diminished. Pasteur believed that exposure to oxygen alone was responsible for the reduction in virulence, although this idea subsequently proved to be erroneous. However, since Pasteur only cultured his organisms in liquid medium, it is not clear exactly how his attenuated cultures arose. Presumably, he selected avirulent mutants from the parent culture, but in the absence of colony isolation to ensure purity, it is not possible to be certain. It is also clear from a close reading of Pasteur's papers that he did not distinguish between changes in properties of the bacterial cell and the bacterial culture or population—but, of course, neither have many bacteriologists over the many years since Pasteur's work.

Some years later, Émile Roux, one of Pasteur's closest associates and later successor as director of the Pasteur Institute, wrote a semipopular review of Pasteur's medical work and gave Pasteur's discovery of attenuation a genetic interpretation:

> The attenuated viruses...may reproduce themselves for successive generations, transmitting their qualities to their descendants. Attenuation is *hereditary*. The viruses are microscopic plants; they may be modified by culture just like the higher plants. Pasteur has obtained races of virus, as the gardeners obtain races of plants. The methods which gave the vaccine for cholera of chickens have furnished those for anthrax, for rouget of hogs and for still other diseases (Roux, 1925).

The Germans were quick to discount Pasteur's work on attenuation. Flügge stated flatly that Pasteur's results could not be confirmed by others and implied that Pasteur's work was erroneous because he did not grow his cultures on solid media. "It is possible...that Pasteur's cultivations, which were always made in fluids, were gradually overgrown by other bacteria, and that thus the diminution in activity [that is, attenuation] was produced" (Flügge, 1890). Although Flügge was correct to criticize Pasteur's methods, Pasteur's work on rabies quickly dramatized the phenomenon of attenuation and silenced any serious objections.

Pasteur's work stands as one of the milestones of immunology, but its impact on bacterial genetics was minimal. Although the idea of variation in virulence became well established, it was just one of many vague facts about bacterial variability that remained part of the bacteriologist's lore through the years (Dubos, 1945).

3.6 COLONY CHARACTERISTICS AND LIFE CYCLES OF BACTERIA

The frequent observation of variable cell morphology in culture prompted searches for bacterial life cycles. At this time the bacteria were considered to be "fission fungi," related to the true fungi, and since sexual life cycles were well known in the true fungi, it was natural to search for them in bacteria. A number of workers, among whom the most prolific was W. Zopf, claimed to have observed life cycles in bacteria, even sexual cycles. Zopf studied the large filamentous iron and sulfur bacteria and deduced that they exhibited life cycles in which small spherical cells, rod-shaped cells, and spirilla all formed, as well as the normal filamentous forms (Zopf, 1879, 1881).

Of greatest concern was the significance of life cycles in the classification of the bacteria. Cohn considered his classification to be provisional and only intended it as a general guide until "a suitable classification is obtained analogous to that of the higher plants, in which special regard is paid to the peculiarities of fructification, and to the natural processes of development" (Flügge, 1890). Flügge then stated:

> *A priori* the possibility of the mutability of form of the bacteria asserted by Zopf...must at any rate be admitted; but before we accept it as a matter of fact, and as so general, we must demand complete proof...[since] by the admission of extensive changeability of form, [we would be denied] an important diagnostic aid which is of great value to us in medical and hygienic practice (Flügge, 1890).

In expanding on this idea, Flügge noted how important it was to be able to diagnose tuberculosis from sputum samples by the morphological examination of the tubercle bacilli. Although many bacteria do exhibit cyclic changes in morphology (for instance, the fruiting myxobacteria), these are never part of a sexual cycle, and such sexual cycles are absent from the pathogenic bacteria that were of interest to late 19th century bacteriologists. However, the crux of the matter, so far as Flügge was concerned, was that most reports of morphological life cycles in bacteria were erroneous because the workers were dealing with contaminated cultures. The final conclusion was "...one may now with justice assert that variation in form does not occur to any large extent in the majority of the bacteria; that on the contrary most bacteria pass through only the limited cycle of forms which are easily observed and permit a diagnosis of the individual varieties" (Flügge, 1890). This forthright statement did not, of course, lay the matter to rest, and over the next 50 years, numerous reports of bacterial life cycles, frequently involving sexual interactions, were reported. These papers, reviewed in detail by Löhnis (1921), contributed actually very little to the development of bacterial genetics and seem not to have had much influence on Joshua Lederberg in his fundamental studies on bacterial recombination.

3.7 DARWIN OR LAMARCK?

More important for the present discussion are considerations on hereditary variation in bacteria that Flügge presented in an extensive Appendix to his book. Darwinian evolution was by this time virtually a dogma in Germany, under the strong intellectual leadership of Ernst Haeckel (see Section 2.9). Following Darwin's lead, Flügge noted that among many similar plants, some individuals may show at times slight differences, subsequent generations of plants retain these differences, and after a considerable time, one sees the gradual formation of new varieties and species. "The varieties are not produced by external conditions. As a matter of fact it seems to be impossible, by selecting the external conditions, to produce other species at will..." Flügge then applied this knowledge to the situation in the bacteria. "...in the lower fungi [bacteria] we must assume that the formation of modifications, varieties, and species must as a whole occur in a similar manner...To what extent the formation of varieties occurs will probably depend on the tendency of the lower fungi to undergo variation...in the bacteria, sexual processes are absent, and thus one of the most important factors in the formation of varieties is wanting. Thus, we must *a priori* expect a greater constancy in the species, and a slower variation. On the other hand, the rapid growth and the quick succession of new generations can lead to more rapid occurrence of variations than in the higher plants, and to their formation within measurable periods of time, and, as it were, before our eyes. Nevertheless, it is a question how we should define the individual generations of the fungi. Is every bacteric cell to be regarded as an individual, and does each individual colony represent an innumerable number of generations, or are the cells of such a colony analogous to those of higher plants, and is it only when fructification and spore formation occurs that a new generation is formed?" (Flügge, 1890).

It is hard to imagine, with such clear thinking, that bacterial genetics did not develop for another 60 years Yet, in 1947 Salvador E. Luria was to write:

> Bacterial genetics is today at a singular point of development. Scant knowledge and lack of agreement have until recently prevailed even on the most elementary facts of reproduction and character transmission in bacteria. The occurrence of sexual reproduction, although denied by most workers, was accepted by several others, mainly on the basis of suggestive but inconclusive cytological evidence. Variation in bacteria was interpreted by some as developmental, by others as genetic; and further complications resulted from attempts to explain by strictly physicochemical theories the supposed specific induction of bacterial variation, making bacteriology one of the last strongholds of Lamarckism (Luria, 1947).

3.8 NUTRITION AS A BACKGROUND FOR BACTERIAL GENETICS

Because mutants possessing growth factor requirements were to play a central role in the development of bacterial genetics, research on bacterial nutrition provided important background. Among the technical achievements of bacteriology was the development of appropriate culture media for the successful growth of a diversity of organisms. In the post-Koch era, and with medical microbiology in general, complex culture media were used to cultivate bacteria, since the goal was primarily to obtain successful growth, rather than to understand the metabolism or physiology of the organism. However, efforts in agricultural and ecological bacteriology in the late 19th and early 20th centuries, deriving primarily out of the work of Sergei Winogradsky and Martinus Beijerinck, focused on the development of synthetic culture media, especially for the study of phototrophic and lithotrophic bacteria (Kluyver and van Niel, 1956; Brock and Schlegel, 1989). Eventually, research turned to the development of synthetic media for heterotrophic bacteria. For those bacteria that had relatively simple growth factor requirements, research focused initially on inorganic nutrient requirements (Stephenson, 1930; Knight, 1936). Although there was uncertainty about the exact minimum inorganic requirements of these bacteria, the requirement for the major inorganic ions could be easily defined, and the trace elements needed generally came along as impurities of the major ions and did not need to be specifically determined. Although *Escherichia coli* had no organic requirements except a simple source of carbon and energy, bacterial pathogens often had complex requirements, and during the 1920s and 1930s, research began to focus on these.

During the early part of the 20th century, several lines of nutrition research led to the discovery of the essential vitamins and amino acids. The existence of an unsuspected nutritional factor in yeast was first uncovered by Wildiers (1901) from studies on the growth of yeast in synthetic media. Wildiers showed that at low inocula, certain strains of yeast would not grow unless the medium was supplemented with a water-soluble extract of living yeast. Wildiers named this unknown growth factor "bios." It was subsequently shown that the bios of Wildiers was related to the water-soluble or B-vitamin complex that had been discovered from studies on animal nutrition. A brief history of these discoveries is given by Gortner and Gortner (1949). The following lists the vitamins and the dates they were first chemically synthesized: thiamine (vitamin B_1), 1936; riboflavin, 1935; niacin, 1935; pyridoxine, 1939; biotin, 1943; pantothenic acid, 1940; folic acid, 1946; vitamin B_{12}, 1949.

Early work by the school of Marjory Stephenson in England focused on the nutrient requirements of certain pathogens (Stephenson, 1930). For instance, Fildes (1934) put forward the hypothesis that parasitism is caused by loss of enzymes necessary for the synthesis of cellular material, "forcing" the organism to live on nutritionally complex materials such as the animal host. Fildes's theory could be characterized as retrograde evolution. It assumed that the first organism was an autotroph, which in the initial stage became a heterotroph by losing the ability to synthesize its own organic carbon. The second stage was thought to occur when the cell was no longer able to live on simple synthetic media containing only an organic carbon and energy source, but required also at least one other compound supplied by the activity of another organism. *E. coli* was thought to exemplify stage one and *Proteus vulgaris* (requiring nicotinic acid) and *Salmonella typhosa* (requiring tryptophan) were thought to be examples of stage two. From stage two, a whole series of stages could be envisioned of organisms with increasingly complex nutritional requirements, the end member of the series being one of the lactic acid bacteria, which required a large number of vitamins and other growth factors.

> The view that nutritional exactingness is correlated with enzymic deficiency now admits of no doubt; this seems to occur by the continued growth of an organism in surroundings where nutritional units occur in great quantity; if in such surroundings a mutant arises which has lost an enzyme necessary for the production of an essential compound, it is then able to supply its need from its surroundings and survive (Stephenson, 1949).

This hypothesis of retrograde evolution stimulated extensive research during the 1930s and early 1940s, of which the most significant was that of André Lwoff (1944). Lwoff, who was later to provide a major connection to the mainstream of bacterial genetics (see Chapter 7), began his nutritional research as a protozoologist, but then studied the nutrition of blood-requiring bacteria to parallel earlier research on hemoflagellates. According to Knight (1971), the work of Lwoff arose independently of that of Fildes and Stephenson, but collaboration developed through exchange of the synthetic vitamins and their building blocks. Lwoff showed that blood-requiring *Hemophilus* species required two factors, a water-soluble substance, the so-called V factor (V for vitamin), which was actually coenzyme I (diphosphopyridine nucleotide, DPN, later called nicotinamide adenine dinucleotide, NAD), and the so-called X factor, which was hemin (the early work is reviewed extensively in Porter, 1946). A brief analysis of the connection between Lwoff's nutrition research and the development of microbial genetics is given by Burian, Gayon, and Zallen (1988).

The other major aspect of bacterial nutrition stems from studies on amino acids. During the 19th century and the early part of the 20th century, research on the chemistry of proteins led to the discovery and characterization of most of the amino acids. By 1930, all of the amino acids found in protein had been characterized (Gortner and Gortner, 1949) and were either available commercially or could be readily prepared from natural sources. As noted, studies on bacterial nutrition by Fildes and Knight had led to the discovery of amino acid requirements in pathogenic bacteria, and the extensive work of Snell and colleagues on the nutrition of the lactic acid bacteria had shown that at least one or another species of this group required each of the known amino acids (Snell, 1951). In the days before amino acid analyzers, the lactic acid bacteria were extensively used for the bioassay of amino acids in natural products. Another important basis for an understanding of amino acid nutrition was the work of Beadle and Tatum on *Neurospora* mutants, which is discussed in Section 5.1.

Intermediary Metabolism

Research on nutrition led naturally to studies on the biosynthesis of various small molecules of the cell, an area called *intermediary metabolism*. This area is in the domain of biochemistry rather than genetics, but the two areas worked symbiotically to a great extent. Intermediary metabolism flourished in the post-World-War-II period when radioactive tracers became available. Although bacteria provided important study organisms, extensive research on intermediary metabolism also was carried out using the fungus *Neurospora crassa* (see Section 5.1), animal organ and tissue slices, and yeast. Textbooks of biochemistry of the period (for example, Fruton and Simmonds, 1958) provide insights into the experimental approaches to intermediary metabolism.

A classical example of how research in "pure" intermediary metabolism operated in the bacterial world is the work of Roberts, Abelson, Cowie, Bolton, and Britten (1955). Using radioactive carbon and sulfur, these workers studied the biosynthesis of the common amino acids of *Escherichia coli*. They were able to show by analysis of incorporated radioactivity that amino acids could be grouped in "families" that were related by biosynthesis, such as the glutamic acid family, which contained the amino acids proline and arginine, and the aspartic acid family, which contained the amino acids lysine, methionine, threonine, and isoleucine. Ultimately, this information would lead to ideas of feedback inhibition and coordinate regulation (see Section 10.15), but the interest of the Roberts group was primarily biochemical.

Once nutritional mutants became available (see Chapter 4), workers following the Beadle/Tatum lead from *Neurospora* began to study intermediates that accumulated as a result of genetic blocks (Adelberg, 1953). Bacterial mutants provided important experimental material for purification of particular intermediates of biosynthetic pathways; the pure intermediates could then be used in the assay and purification of the relevant enzymes. During most of the 1950s, genetics and biochemistry worked together: The intermediates characterized by the biochemist could be used by the bacterial geneticist to characterize nutritional mutants, and the mutants isolated by the geneticist could be used by the biochemist to characterize intermediates and enzymes. By the end of the 1950s, most of the major biosynthetic pathways of *Escherichia coli* had been worked out. Although the study of genetics and metabolism is a fascinating field, it is mostly beyond the scope of the present book. A useful summary can be found in Wagner and Mitchell (1964).

3.9 THE BACTERIAL NUCLEUS

The concept of the nucleus and its role in the hereditary continuity of the plant and animal cell was firmly established in the latter part of the 19th century, but it was not until the 1960s that the essential nature of the bacterial equivalent was discerned (Brock, 1988b). In actuality, there is no bacterial analog of the nucleus of higher organisms, but most workers who studied the question worked hard to find a bacterial parallel. Because of the central role of the chromosome in the life history of the higher organism, and its clearly demonstrable role in heredity, most workers sought a chromosome or chromosomes in bacteria. Even today the term chromosome is often used in bacterial genetics to refer to what is essentially a naked DNA molecule. Current usage seems to assume an equivalence in bacteria between the terms chromosome and nucleus. Other terms used are nuclear region, nuclear body, nucleoplasm, and nucleoid, which seems to be most widely used now (Pettijohn, 1976; Kleppe, Övrebö, and Lossius, 1979).

The Eucaryotic Nucleus

The history of the eucaryotic nucleus has been covered in detail in many sources (Nordenskiöld, 1936; Singer, 1950; Portugal and Cohen, 1977) and will be presented here only in outline form. The central body in cells was readily seen by microscopists of the early 19th century. The term "nucleus" itself was coined by the Scottish botanist Robert Brown in 1833, and used extensively by Mathias Schleiden and Theodor Schwann in their presentation of the cell theory. An important advance in cytological research came with the introduction in the 1870s by the Carl Zeiss Company of the Abbe condensor and the oil immersion lens, which made possible observation (and photography) of chromosomes. The continuity of the nucleus and its behavior during cell division were first clearly described by the plant cytologist Eduard Strasburger in 1875–1880. Strasburger showed that a new nucleus arises from a preexisting nucleus by division, rather than de novo. In animals, the process of nuclear division was described by Walther Flemming in the same decade, and it was Flemming who coined the term "mitosis." Strasburger coined the term "cytoplasm," as well as terms for some of the key stages in the mitotic process: "prophase," "metaphase," and "anaphase." DNA was first characterized in 1869 by Friederich Miescher, who purified it from cell-free nuclei and termed it "nuclein." The use of coal-tar dyes for staining cells and cell components was introduced by Paul Ehrlich in the 1870s and soon led to the visualization of chromosomes. Flemming coined the term "chromatin" for the characteristic staining material seen during the mitosis process, and the term "chromosome" was coined by W. Waldeyer in 1888 for the structure containing chromatin. By the turn of the century, the constancy of chromosome number and the behavior of chromosomes during the life cycle of the organism had been firmly established. E.B. Wilson, in his highly influential book (1899), described the alternation of generations and the concept of reduction division. By the second decade of the 20th century, genes had been localized to chromosomes by Morgan and his school, and the role of chromosome pairing during genetic crossing-over had been described (see Section 2.7).

Early Studies on the Bacterial Nucleus

During this exciting period for biological research, bacteriology was developing primarily as an applied science. Although certain structural features of bacteria, such as toxins and cellular antigens, were extensively studied, general aspects of bacterial physiology and cytology were pursued primarily in the context of methodology, since the key goal was to cultivate bacterial pathogens and characterize their behavior in the host and in the environment.

Ideas about the bacterial nucleus at the peak of this "classical" period of bacteriological research can be summarized from the influential textbook by Benecke (1912):

> The present situation on the nucleus question is as follows: A number of experienced bacteriologists believe that the bacterial cell is devoid of a nucleus, that it has neither a typical nucleus nor is there any evidence of any sort of morphological equivalent. They believe that structures that have been considered by others to be nuclei are something else, perhaps reserve materials...Some researchers, on the other hand, believe that bacteria do possess a nucleus which differs from that of the higher fungi only in being smaller and because of this is visible only as small homogeneous particles, even when the best staining procedures are used...

A bacterial cell is much smaller than the cell of a eucaryote. In fact, a bacterial cell is even smaller than the nucleus of many eucaryotes. Thus, development of methods for specifically staining bacterial nuclei presented real challenges. Insight into early ideas on the nature of the bacterial nucleus can be obtained from the textbook on bacterial cytology by Knaysi (1951) and the highly influential and widely read book by Dubos (1945).

Many attempts were made to stain bacterial nuclei using procedures that worked with eucaryotes. However, bacteria presented formidable obstacles to cytologists because of their small size, their high content of cytoplasmic RNA (which made staining with basic dyes difficult), and the frequent occurrence of polyphosphate (volutin) granules, which are of nuclear size and which react strongly with basic dyes. Because of these problems, many erroneous papers appeared in the literature. Thus, by the time of the review on bacterial cytology by Lewis (1941) and Dubos's book, over 50 years of research had been carried out on the bacterial nucleus by a wide variety of workers, using a wide variety of techniques in a wide range of organisms. It is not surprising that opinions varied markedly on the nature of the bacterial nucleus.

Dubos (1945) lists no fewer than *eight* distinct theories regarding the nature of the bacterial nucleus: (1) The bacteria do not possess a nucleus or its equivalent. (2) The cell is differentiated into a chromatin-containing central body and peripheral cytoplasm. (3) The bacterial body is a nucleus devoid of cytoplasm—a naked nucleus or nuclear cell. (4) The nucleus consists of several chromatin bodies, a chromidial system, scattered throughout the cytoplasm. (5) The nucleus may occur as a discrete spherical body, an elongated chromatin thread, or scattered chromidia, depending on the stage of development. That is, bacteria have a polymorphic nucleus. (6) The nuclear substance consists of fine particles of chromatin dispersed uniformly in the cytoplasm but is not distinguishable as morphological units—a diffuse nucleus. (7) The protoplast contains one or more true vesicular nuclei. (8) The nucleus is a naked invisible gene string, or a chromatin-encrusted gene string analogous to a single chromosome.

Dubos's theory 8, that the bacterial nucleus consists of a single gene string, sounds the most modern and perhaps closest to our current understanding.

This theory was apparently first proposed by Carl C. Lindegren (1935). However, a close reading of Lindegren's papers shows that his idea only superficially resembles current understanding. At the time Lindegren was working, the dogma of classical genetics viewed the genes as particles arranged in a linear fashion ("beads on a string"), and Lindegren's hypothetical drawings of the partitioning of the "gene string" during cell division clearly show this idea. That the chromosome and the gene string might be the same thing was not in Lindegren's thinking. Although even the idea of a naked gene string might have proved useful to those doing research on bacterial genetics, it was abandoned when C.F. Robinow's work (see below) provided such apparently strong evidence for the existence of a real bacterial equivalent of the nucleus of higher organisms.

Carl F. Robinow's work was based on the staining methods developed by Piekarski (1937, 1939), who had studied dividing chromatin bodies in bacteria using ultraviolet microscopy as well as light microscopy with the Feulgen reaction (which is specific for DNA). Piekarski had also demonstrated chromatin structures by Giemsa staining after acid treatment, and it was this latter procedure that Robinow systematically employed in his work. This procedure, which came to be called Robinow's acid-Giemsa technique, had the advantage over the Feulgen reaction that it not only stained the "chromatinic" structures more deeply, but at the same time showed the outlines of the bacteria as well as their internal cell boundaries. This appeared important because it allowed a regular demonstration of "chromatin" bodies at any stage of the cell division cycle. Robinow concluded that *Escherichia coli* had two pairs of chromosomes.

Electron Microscopy of the Bacterial Nucleus

The light microscope did not provide sufficient resolution for critical studies of the bacterial nucleus. When the electron microscope became available, it was natural that researchers would turn to this important tool to study bacterial cytology. Initially, observations were made of whole cells, but it soon became clear that although bacteria were too small to observe well in the light microscope, they were too *thick* to observe well in the electron microscope. It was only after the technique of thin sectioning was developed that serious study of bacterial cytology became possible.

The thin-section technique was developed independently by J. Hillier at the RCA Research Laboratories in Princeton, New Jersey (Chapman and Hillier, 1953) and by Keith Porter and J. Blum at Rockefeller Institute for Medical Research in New York (Porter and Blum, 1953). The Porter-Blum microtome soon became a widely available commercial device, and the thin-section technique rapidly became a standard procedure in bacterial cytology. However, despite the high resolution provided, much of the early work on bacterial thin sections was of poor quality because of inadequate attention to the fixation and embedding process.

The first electron micrographs of thin sections of bacteria were published by Chapman and Hillier (1953). Although they were a major advance in bacterial cytology, these micrographs did not provide completely satisfactory images of the bacterial nucleus. The main conclusion was that there was no obvious nuclear membrane.

A major advance in understanding of the nature of the bacterial nucleus came from the work of Edouard Kellenberger and Antoinette Ryter. Motivated by a desire to understand the nature of DNA-containing plasms in bacteriophage-infected cells, these authors also carried out extensive studies on uninfected cells in various physiological states. By the time Kellenberger and Ryter initiated their work, it was widely accepted that there were functional (and probably structural) relationships between the DNA of bacterial viruses and the DNA of bacterial cells (see Chapter 6). Kellenberger and Ryter showed that during the phage replication process there is a breakdown of the bacterial nucleus, and the bacterial DNA is hydrolyzed and recycled into phage DNA. A pool of phage DNA develops within a few minutes of infection and later becomes packaged into mature virus particles. Kellenberger and Ryter developed preparation techniques that permitted preservation of the phage DNA in an unaltered state for electron microscopy. They showed that pH, divalent cations, and other solutes present in the medium during the fixation process, as well as the conditions of the embedding process, strongly influenced the appearance of the bacterial DNA in the thin sections. By making a systematic study of the conditions of fixation and embedding, they developed standard procedures that preserved the DNA in a fine-stranded fibrillar state (Kellenberger, Ryter, and Séchaud, 1958; Ryter, Kellenberger, Birch-Anderson, and Maaløe, 1958). The term "nucleoplasm" or "nucleoid" was adopted to refer to the region under study. Kellenberger and co-workers showed that the coarse coagulation of DNA-containing plasms seen by many workers was a fixation artifact. Considering the implications of their observations, Kellenberger, Ryter and Séchaud (1958) made the following statement:

> All evidence is now against complicated, complete mitosis in bacteria, i.e. morphologically the homogeneous nucleoplasm, and chemically the continuous production of DNA [during growth]...These facts lead us to postulate that the genetic material of bacteria may multiply following the same mechanism as phage. The nucleoid would be simply a pool containing one or several bacterial genomes in a form similar to vegetative phage. To explain the genetic continuity of the bacteria...we have to assume the existence either of only one linkage group, or, when more linkage groups exist, of a high number of identical strands...Nothing then prevents us from considering the bacterial nucleoid as one single multistranded chromosome. The main difference between chromosomes would be in the organization and the moiety which is not DNA...Our mind is still a prisoner of the eloquent picture of the mitotic cycle. Between this and the DNA molecule, however, there is a very great gap which we have to fill in the coming years (Kellenberger, 1960).

Electron Microscopy of Isolated DNA

Another approach that was to have marked impact on understanding of the bacterial nucleus was that of Kleinschmidt and colleagues (Kleinschmidt and Zahn, 1959; Kleinschmidt, Lang, and Zahn, 1961; Kleinschmidt, Lang, Jachertz, and Zahn, 1962) on the electron microscopy of isolated DNA (see Kleinschmidt, 1989, for a brief recounting of the background to the 1962 paper). In the Kleinschmidt technique, cells or viruses are gently lysed in place on monolayers of protein-salt solutions, and the resulting films are picked up onto electron microscope grids. After shadowing with a heavy metal, the preparations are examined at high magnification. The images obtained by Kleinschmidt revealed extremely long molecules of uniform diameter, with no branches and very few free ends. The picture that the Kleinschmidt technique revealed, now classical, was of extremely long thread-like structures protruding from the remains of the lysed cells. Relating these images to the electron micrographs of thin sections obtained by Kellenberger and Ryter (discussed earlier), Kleinschmidt and Lang (1961) concluded: "...the whole DNA content is present in undivided filamentous forms. Depending on the bacterium, these consist of one or only a few very long structures. The DNA filament is arranged in the cell interior as a yarn-like form which can be called 'nucleoplasm'" [translated from the German]. The Kleinschmidt technique was also to have implications considerably beyond its use to study the arrangement of DNA in the bacterial nucleus. It provided an important tool for studying the detailed structure of DNA molecules and showed the presence of circular and supercoiled plasmids, loops, replicating forms, and many other forms of DNA.

Autoradiographic Studies with DNA

At the same time that Kleinschmidt's work was showing that the nucleus was a single long DNA molecule, John Cairns (independently of Kleinschmidt) was using an autoradiographic technique to measure the lengths of DNA molecules. The procedure involved labeling the DNA with tritiated thymidine and gently layering the liberated DNA onto autoradiographic emulsion. After exposure and development of the emulsion, the configuration of the DNA could be observed under the light microscope. Cairns's work not only provided an independent confirmation of Kleinschmidt's conclusions, but also permitted a study of the cellular DNA replication process itself.

In his first work, Cairns (1961) estimated the length of the DNA of bacteriophage T2 (released by gentle osmotic shock). The length obtained, 52 μm, agreed with the molecular weight of T2 DNA. Cairns then proceeded to a determination of the molecular weight of *E. coli* DNA, a much more difficult problem (Cairns, 1962). Knowing from the work of Levinthal and Davison (1961) that lengthy DNA molecules are subject to extensive hydrodynamic shear, Cairns handled his DNA carefully. The lengths obtained in the first work were 400 μm, somewhat shorter than reality, but much longer than the lengths calculated from the molecular weight determinations of earlier workers. Cairns concluded: "Until the existence in DNA of non-nucleic acid links has been demonstrated, it is probably legitimate to think of these threads as molecules." Cairns then added in a footnote: "Since this paper was written, Kleinschmidt, Lang and Zahn have produced beautiful electron micrographs showing that protoplasts of *M. lysodeikticus*, lysed at an air-water interface, release their deoxyribonucleoprotein in the form of a tangled skein which has no visible free ends. Thus two dissimilar procedures suggest that bacterial DNA may exist as a single molecule."

Using an improved version of his autoradiography procedure, Cairns then proceeded to study the replication process of the bacterial chromosome itself. This work was published shortly after the circular chromosome model of Jacob and Wollman had been developed from genetic studies (see Section

5.11) and provided a dramatic confirmation of the Jacob/Wollman model. Cairns's classic image, which was to appear in numerous textbooks, is shown in Figure 3.1 (Cairns, 1963a,b). The total length of the molecule was now found to be at least 1100 μm, equivalent to a molecular weight of 2.8×10^9.

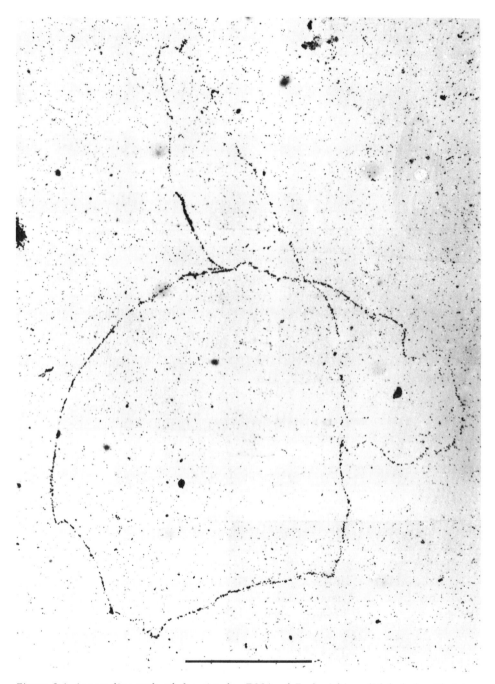

Figure 3.1 Autoradiograph of the circular DNA of *Escherichia coli* labeled with tritiated thymidine for two generations. (Reprinted, with permission, from Cairns, 1963a.)

Although slightly lower than the molecular weight calculated from the DNA content per cell nucleus, this value was close enough (given the vagaries of the spreading technique) to make it seem likely that the complete *E. coli* genome was in one circular molecule.

Cairns was also able to confirm that the DNA replication process begins at a single point and moves progressively around the circle. Although the molecular details of this process, especially how the double helix unwinds, had to await studies on the biochemistry of DNA replication (and especially the discovery of enzymes involved in winding and unwinding DNA), the overall model that Cairns presented in 1963 is still the accepted model today.

3.10 CONCLUSION

The development of the plate culture technique by Robert Koch ushered in the era of bacteriology. It also presented the opportunity for a study of bacterial genetics, but bacteriologists found little motivation to carry out such studies. Although Darwinian theory was well established for higher organisms, the main thrust in the 19th century was not on genetics but on the origin of species. Bacteriologists were not primarily concerned with this topic, since they were heavily involved in hygienic and medical matters. Indeed, Flügge even implied that it would be inconvenient for bacteriology if variation in species occurred, since this would render unacceptable the standard diagnostic procedures.

Only Pasteur's work on attenuation seemed to be of relevance to genetics, but since Pasteur only used liquid cultures, his work was suspect by bacteriologists of the Koch school. Pasteur himself soon left bacteria for his work on the rabies vaccine and never returned to the question of the origin of variability. It was only after the beginning of the 20th century, with the work of de Vries and the rediscovery of Mendel's laws, that a theoretical basis for research on bacterial genetics became apparent. This is discussed in the next chapter.

An increasingly sophisticated understanding of bacterial cytology, physiology, and nutrition developed during the first half of the 20th century. Eventually, this work would be related to the facts of bacterial genetics, but for the most part, the key aspects of bacterial genetics, mutation and genetic recombination, had to develop independently. The rest of this book will deal with the development of these concepts.

REFERENCES

Adelberg, E.A. 1953. The use of metabolically blocked organisms for the analysis of biosynthetic pathways. *Bacteriological Reviews* 17: 253–267.

Benecke, W. 1912. *Bau und Leben der Bakterien*. B.G. Teubner, Leipzig. 650 pp.

Brefeld, O. 1875. Methoden zur Untersuchungen der Pilze. *Beiträge zur Biologie der Pflanzen* 8: 43–62.

Brock, T.D. 1988a. *Robert Koch. A Life in Medicine and Bacteriology*. Science Tech Publishers, Madison, Wisconsin. 365 pp.

Brock, T.D. 1988b. The bacterial nucleus. A history. *Microbiological Reviews* 52: 397–411.

Brock, T.D. and Schlegel, H.G. 1989. Introduction. pp. 1–35 in Schlegel, H.G. and Bowien, B. (ed.), *Autotrophic Bacteria*. Science Tech Publishers, Madison, Wisconsin.

Burian, R.M., Gayon, J., and Zallen, D. 1988. The singular fate of genetics in the history of French biology, 1900–1940. *Journal of the History of Biology* 21: 357–402.

Cairns, J. 1961. An estimate of the length of the DNA molecule of T2 bacteriophage by autoradiography. *Journal of Molecular Biology* 3: 756–761.

Cairns, J. 1962. A minimum estimate for the length of the DNA of *Escherichia coli* obtained by autoradiography. *Journal of Molecular Biology* 4: 407–409.

Cairns, J. 1963a. The chromosome of *Escherichia coli*. *Cold Spring Harbor Symposia on Quantitative Biology* 28: 43–46.

Cairns, J. 1963b. The bacterial chromosome and its manner of replication as seen by autoradiography. *Journal of Molecular Biology* 6: 208–213.

Chapman, G. and Hillier, J. 1953. Electron microscopy of ultra-thin sections of bacteria. *Journal of Bacteriology* 66: 362–373.

Cohn, F. 1875. Untersuchungen über Bacterien. *Beiträge zur Biologie der Pflanzen* 1: 127–222.

Dubos, R. 1945. *The Bacterial Cell. In Its Relation to Problems of Virulence, Immunity and Chemotherapy.* Harvard University Press, Cambridge. 460 pp.

Dubos, R. 1988. *Pasteur and Modern Science.* Science Tech Publishers, Madison, Wisconsin.

Fildes, P. 1934. Some medical and other aspects of bacterial chemistry. *Proceedings of the Royal Society of Medicine* 28: 79.

Flügge, C. 1890. *Micro-organisms with special reference to the etiology of the infective diseases.* English translation of *Fermente und Mikroparasiten.* New Sydenham Society, London. 826 pp.

Fruton, J.S. and Simmonds, S. 1958. *General Biochemistry.* John Wiley, New York. 1077 pp.

Gortner, R.A. and Gortner, W.A. 1949. *Outlines of Biochemistry.* John Wiley and Sons, New York.

Hadley, P. 1928. The dissociative aspects of bacterial behavior. p. 84 in E.O. Jordan and I.S. Falk (ed.), *The Newer Knowledge of Bacteriology and Immunology.* University of Chicago Press, Chicago.

Kellenberger, E. 1960. The physical state of the bacterial nucleus. pp. 39–69 in *Tenth Symposium, Society for General Microbiology,* Cambridge University Press, United Kingdom.

Kellenberger, E., Ryter, A., and Séchaud, J. 1958. Electron microscope study of DNA-containing plasms. II. Vegetative and mature phage DNA as compared with normal bacterial nucleoids in different physiological states. *Journal of Biophysical and Biochemical Cytology* 4: 671–678.

Kleinschmidt, A.K. 1989. Preparation and length measurement of the total DNA content of T2 bacteriophages: a commentary. *Biochimica et Biophysica Acta* 1000: 35–40.

Kleinschmidt, A.K. and Lang, D. 1961. Intrazelluläre formationen von Bacterien-DNS. pp. 690–693 in *Proceedings of the European Regional Conference on Electron Microscopy Delft 1960.* Academic Press, New York.

Kleinschmidt, A.K. and Zahn, R.K. 1959. Über Desoxyribonucleinsäure-Molekeln in Protein-Mischfilmen. *Zeitschrift für Naturforschung* 14b: 770–779.

Kleinschmidt, A.K., Lang, D., and Zahn, R.K. 1961. Uber die intrazelluläre Formation von Bakterien-DNS. *Zeitschrift für Naturforschung* 16b: 730–739.

Kleinschmidt, A.K., Lang, D., Jachertz, D., and Zahn, R.K. 1962. Darstellung und Längenmessungen des gesamten Desoxyribonucleinsüre-Inhaltes von T2-Bacteriophagen. *Biochimica et Biophysica Acta* 61: 857–864.

Kleppe, K., Övrebö, S., and Lossius, I. 1979. The bacterial nucleoid. *Journal of General Microbiology* 112: 1–13.

Kluyver, A.J. and van Niel, C.B. 1956. *The Microbe's Contribution to Biology.* Harvard University Press, Cambridge.

Knaysi, G. 1951. *Elements of Bacterial Cytology,* 2nd edition. Comstock Publishing Co., Ithaca. 375 pp.

Knight, B.C.J.G. 1936. *Bacterial Nutrition.* Medical Research Council, Special Report Series, No. 210. His Majesty's Stationery Office, London.

Knight, B.C.J.G. 1971. On the origins of "growth factors." pp. 16–18 in Monod, J. and Borek, E. (ed.), *On Microbes and Life.* Columbia University Press, New York.

Koch, R. 1881. Zur Untersuchungen von pathogenen Organismen. *Mittheilungen aus dem Kaiserlichen Gesundheitsamte.* 1: 1–48.

Levinthal, C. and Davison, P.F. 1961. Degradation of deoxyribonucleic acid under hydrodynamic shearing forces. *Journal of Molecular Biology* 3: 674–683.

Lewis, I.M. 1941. The cytology of bacteria. *Bacteriological Reviews* 5: 181–230.

Lindegren, C.C. 1935. Genetical studies of bacteria. I. The problem of the bacterial nucleus. *Zentralblatt für Bakteriologie II* 92: 40–47.

Lister, J. 1878. On the lactic fermentation and its bearing on pathology. *Transactions of the Pathological Society of London* 29: 425–467.

Löhnis, F. 1921. Studies upon the life cycles of the bacteria. Part I. Review of the literature, 1838–1919. *Memoirs of the National Academy of Sciences* 16: 1–252.

Luria, S.E. 1947. Recent advances in bacterial genetics. *Bacteriological Reviews* 11: 1–40.

Lwoff, A. 1944. *L'evolution physiologique. Études des pertes de fonction chez les microorganismes.* Hermann, Paris.

Nägeli, C. 1877. *Die niederen Pilze in ihren Beziehungen zu den Infectiouskrankheiten und der Gesundheitspflege.* R. Oldenbourg, München.

Nordenskiöld, E. 1936. *The History of Biology.* Tudor Publishing Co., New York. 629 pp.

Pasteur, L. 1880. De l'attenuation du virus du choléra des poules. *Comptes rendus de l'Academie des Sciences* 91: 673–680.

Pettijohn, D.E. 1976. Prokaryotic DNA in nucleoid structure. *CRC Critical Reviews of Biochemistry* 4: 175–202.

Piekarski, G. 1937. Cytologische Untersuchungen an paratyphus und colibakterien. *Archiv für Mikrobiologie* 8: 428–439.

Piekarski, G. 1939. Lichtoptische und übermikroskopische Untersuchungen zum Problem des Bakterienzellkerns. *Zentralblatt für Bakteriologie, I*, Originale 144: 140–148.

Porter, J.R. 1946. *Bacterial Chemistry and Physiology.* John Wiley, New York. 1073 pp.

Porter, K.R. and Blum, J. 1953. A study in microtomy for electron microscopy. *Anatomical Record* 117: 685–708.

Portugal, F.H. and Cohen, J.S. 1977. *A Century of DNA.* MIT Press, Cambridge. 384 pp.

Roberts, R.B., Abelson, P.H., Cowie, D.B., Bolton, E.T., and Britten, R.J. 1955. *Studies of Biosynthesis in* Escherichia coli. Carnegie Institution of Washington Publication 607, Washington, D.C. 521 pp.

Roux, E. 1925. The medical work of Pasteur. *Scientific Monthly* 21: 364–389.

Ryter, A., Kellenberger, E., Birch-Andersen, A., and Maaløe, O. 1958. Etude au microscope électronique de plasmas contenant de l'acide désoxyribonucléique. I. Les nucléoides des bactéries en croissance active. *Zeitschrift für Naturforschung* 13b: 597–605.

Schroeter, J. 1875. Ueber einige durch Bacterien gebildete Pigmente. *Beiträge zur Biologie der Pflanzen.* 1: 109–126.

Singer, C. 1950. *A History of Biology.* Henry Schuman, New York. 579 pp.

Smith, T.E. 1932. Koch's views on the stability of species among bacteria. *Annals of Medical History*, N.S. 4: 524–530.

Snell, E.E. 1951. Bacterial nutrition—chemical factors. pp. 214–255 in Werkman, C.H. and Wilson, P.W. (ed.), *Bacterial Physiology.* Academic Press, New York.

Stephenson, M. 1930. *Bacterial Metabolism.* Longmans, Green and Co., London. 320 pp.

Stephenson, M. 1949. *Bacterial Metabolism*, 3rd edition. Longmans, Green and Co., London.

Wagner, R.P. and Mitchell, H.K. 1964. *Genetics and Metabolism*, 2nd edition. John Wiley, New York.

Wildiers, E. 1901. Nouvelle substance indispensable au développement de la levure. *La Cellulle* 18: 313–332.

Wilson, E.B. 1899. *The Cell in Development and Inheritance.* Macmillan, New York.

Zopf, W. 1879. *Entwicklungsgeschichtliche Untersuchungen über Crenothrix polyspora, etc.* Berlin. 21 pp.

Zopf, W. 1881. Über den genetischen Zusammenhang von Spaltpilzformen. *Monatsberichte der Akademie der Berlin* 1881: 277–284.

4
MUTATION

The central role that bacterial genetics has played in developing our understanding of the chemical basis of heredity is paradoxical when it is considered that for many years after the rediscovery of Mendel's laws in 1900, it was even doubted that bacteria had genetic mechanisms similar to those of higher organisms. There were lengthy, rather heated, debates in the bacteriological literature between the so-called "mutationists" and "adaptationists" as to the mechanism by which bacteria responded to environmental change. Bacteriologists were accustomed to thinking of the behavior of *cultures* rather than of the individual bacterial cells and their descendants. The idea that a bacterial colony was actually a *clone* of cells was difficult for bacteriologists to grasp, and hence the techniques necessary for decisive experiments developed slowly. Animal and plant geneticists did not have this difficulty because they dealt primarily with individuals, but the individual bacterium is a short-lived and rather unapproachable experimental object. Variation occurs often in the bacterial culture, but the reasons for this variation were difficult for bacteriologists to fathom. Ignorance led to fallacious ideas, such as that bacteria were so primitive that they had no genes, or that they varied much more extensively than higher organisms (Hayes, 1964).

It is not clear now why most bacteriologists were unable to grasp clonal thinking, since some bacteriologists had seen the correct view early. For instance, in the year of the rediscovery of Mendel's laws, the Dutch bacteriologist Martinus Beijerinck had the following to say about microbial genetics:

> Though the culture of microbes, compared to that of higher plants and animals, is subject to many difficulties, it cannot be denied that, these once mastered, microbes are extremely useful material for the investigation of the laws of heredity and variability. The starting from the single individual, which of course is required here, is commonly almost as simple as for the higher organisms...The generations succeed each other quickly; hundreds, nay thousands of individuals can be very easily surveyed in their posterity...it is easy to perform with microbes experiments of competition, which is difficult or impossible with higher plants and animals...the individual microbe can be compared to the whole individual of the higher organism, or to a single tissue-cell of it... (Beijerinck, 1900).

For the first 40 years of genetics research, however, the difficulties of working with bacteria appeared to geneticists to outweigh the advantages. The small size of bacteria made it difficult to study characteristics of individuals, so that these characteristics had to be inferred from studies on populations. This not only greatly complicated genetic study, but also made it especially difficult to interpret cellular events. The lack of sexual reproduction was perceived as the greatest difficulty, since a hybridization process of some sort is essential if unit genetic characters are to be followed.

We now know that mutation rates in bacteria are no higher than mutation rates in plants and animals: between 10^{-5} and 10^{-8} per cell per generation. However, the high population density of a bacterial culture means that in any culture there will always be a significant number of mutant cells. Most of these mutants will grow less well than wild type, but if selection pressure is applied, for example, by changing the culture medium or environmental conditions, one or another of these mutants may be favored. Because of the high growth rates of bacteria, after several transfers, a genetically different culture may easily develop. Thus, mutation and selection, the two major factors in Darwinian evolution, operate rapidly and frequently in bacterial cultures to cause genetic change. If the bacteriologist is not attentive to the conditions in which the culture is grown, an improper interpretation of such change is possible. This explains the wide variety of fallacious observations reported in the 1920s under the heading of "bacterial dissociation" (see Section 4.5). Even as late as 1953, some well-known bacteriologists still had not understood this fact and were devising adaptation nongenic hypotheses to explain bacterial variability (Yudkin, 1953).

Another difficulty in the study of bacterial variability arises from the fact that bacteria, more than higher organisms, regulate enzyme synthesis in dramatic ways. The phenomena of enzyme induction and repression (see Chapter 10) are dramatically developed in bacteria, and lead to extensive physiological change when culture conditions are changed. Although these changes are phenomena of whole populations rather than of individuals, they confused many workers. It was not until the work of Marjory Stephenson and Jacques Monod (see Chapter 10) that the distinction between change due to enzyme regulation and change due to mutation and selection was made clear (see, e.g., Monod and Audureau, 1946).

With the above ideas in mind, let us now turn to the early studies on bacterial variation.

4.1 MARTINUS BEIJERINCK AND THE APPLICATION OF DE VRIES'S MUTATION THEORY TO BACTERIA

In 1900, Hugo de Vries announced his mutation theory of biological variability in a series of books, papers, and talks (see Section 2.3). One of the talks, given at the Royal Academy of Sciences in Amsterdam, was heard by the distinguished Dutch microbiologist Martinus Beijerinck. Beijerinck had been trained originally as a botanist (van Iterson, den Dooren de Jong, and Kluyver, 1983) and had a broad interest in biological questions. He immediately understood the importance of de Vries's ideas.[1] "The interesting lecture

---

[1] One story of how de Vries discovered Mendel's work involves Beijerinck. According to Stomps (1954), Beijerinck had a copy of Mendel's paper in his files. When he heard de Vries talk, he sent him Mendel's paper with the note: "I know that you are studying hybrids, so perhaps the enclosed reprint of 1865 by a certain Mendel which I happen to possess is still of some interest to you." Beijerinck had his first position as a botanist at the Agricultural University at Wageningen. It is likely that Mendel had sent a reprint of his paper to a botanist at Wageningen, from whom Beijerinck had acquired the paper.

of Prof. Hugo de Vries...on the origin of new forms in higher plants induces me to draw attention to some observations regarding the same subject in microbes." Beijerinck then wrote the words quoted at the beginning of this chapter and proceeded to present some of his observations on microbial variability. In addition to some work on yeast that is not relevant here, Beijerinck discussed his observations on pigment formation in *Bacillus prodigiosus* (now known as *Serratia marcescens*) and luminescence (light production) in *Photobacter indicum*. He clearly showed that these two organisms could throw off variants that were permanently altered in these key phenotypic characteristics. He distinguished three classes of hereditary variability: (1) *degeneration*, in which *all* individuals by a slow process of variability lose their ability to grow; (2) *transformation*, in which *all* individuals lose a specific characteristic and acquire either another or none (he thought this was rare); (3) common hereditary variability or *variation*, in which during growth some individuals are "thrown off" that differ from the parent strongly in a single "salient characteristic." It was this latter type of variation, most analogous to what we understand today as mutation, which Beijerinck thought was the most common. He then presented a model for how such variability might arise, which he called "heterogene cell-partition." Beijerinck's model clearly implied some defect arising during cell division, but he did not speculate about hereditary units, Mendelian characters, or any of the other possible hypotheses to explain genetic change. He made it clear, however, that he thought *one* character at a time was changing. "I perfectly agree with Professor de Vries, that the origin of species should often be sought in the almost suddenly produced variants, or mutants, as he calls them."

An argument that was current at that time, derived originally from Rudolf Virchow's classic dictum that each cell arises from a preexisting cell, was that during cell division, each cell produces two cells that are *identical* to the original. Beijerinck quotes O. Hertwig to this effect: "Gleiches erzeugt nur Gleiches" (*Like only produces like*) and "Art erzeugt stets seine Art" (*A species always produces only its own species*). According to Hertwig: "Thus, in no way does the cell division process in unicellular organisms appear to be a means by which new species arise" (my translation). Beijerinck concluded that these ideas were incorrect.

Beijerinck also clearly saw the significance of mutation for the origin of bacterial species. "Especially in the microbes, where the want of crossing must strongly favour the prolonged continuing of the once formed variants, it is to be foreseen that in nature will often be found variants, which will long maintain themselves at their habitat. If they are isolated, the discoverer will at first be almost sure to see new species in them, and only after an accurate investigation recognise them as variants of another species" (Beijerinck, 1900).

In 1910, Beijerinck reported more detailed observations of variation in the ability of *Bacillus prodigiosus* to produce its characteristic red pigment. He continued to use the term "variant" rather than mutant, but emphasized that the variants were "thrown off" and existed besides the parent. He stated that the variants remained constant in characteristic but could, under certain conditions, revert to wild type, a condition he called *atavism* (Beijerinck, 1910).

In 1912, Beijerinck published a long review paper entitled "Mutation bei Mikroben," which related his and other observations on microorganisms to the growing body of knowledge about genetics of higher organisms. By now, he was using the term mutation rather than variant, citing the growing acceptance of de Vries's ideas (see Chapter 2) as justification (Beijerinck, 1912). In this paper, Beijerinck clearly related microbial genetics phenomena in bacteria, fungi, and algae to those in higher organisms. After an extensive discussion of how mutants are recognized in microbial cultures, and the kinds of mutants that can be found, Beijerinck turned to a discussion of gene theory. By now, Morgan's work on the theory of the gene (see Section 2.6) was well advanced. According to Beijerinck: "The genes or hereditary units are the carriers of the visible traits and can be thought of as being part of the protoplasm or cell nucleus. They exist in two distinct conditions in the cell, either as active genes, which bring about the production of the externally observed characters of the organism, and in dormant form or *progenes*, which lie at the basis of the phenomena of mutability. During mutation, progenes are converted into active genes. In certain cases, there is reason to believe that the genes can be considered to be 'protoplasmic germs' which reproduce and bring about the various protoplasmic functions. In this regard it must be concluded that there are as many genes and as many protoplasmic functions as there are distinct and independent inheritable characteristics" (Beijerinck, 1912). This idea of the progene harkens back to Darwin's gemmule hypothesis (see Section 2.2) and foreshadows the work of Eugene Wollman and André Lwoff on lysogeny (see Chapter 7).

Throughout his long career, Beijerinck had worked extensively with yeast, higher fungi, and algae, as well as bacteria, and many of his genetic observations were with microorganisms other than bacteria. It seems evident from reading Beijerinck's work that he did not make any fundamental distinction between the genetics of bacteria and the genetics of other microorganisms. In fact, many of his thoughts about sexual reproduction in bacteria were derived by analogy with the known processes in the fungi. Similar ideas about the relationships between bacterial and fungal genetics were advanced throughout the next 40 years, until the work of Lederberg led to the understanding that sex in bacteria was a distinct process.

4.2 THE PURE LINE CONCEPT

An interesting connection between bacterial genetics and that of higher organisms was made by the distinguished Wisconsin geneticist L.J. Cole. In an insightful paper, Cole and Wright (1916) made a strong case for beginning a study of bacterial genetics with cultures that had been carefully selected to ensure their uniformity. This so-called "pure line concept" had developed from studies on higher organisms, especially plants, where marked genetic variability can exist in a sample of seeds collected from nature. Analyzing the situation in bacteriology, Cole and Wright concluded that care had not previously been taken to ensure that the initial culture used was uniform. "...within recognized species [of bacteria] there are distinct cultural races or varieties, each with its own characteristics and range of variability, and... these may exist side by side, independent of the environmental conditions. This belief is a distinct modification of the earlier view that one type could be transformed into the other directly by a change in the environment, or by selection." Note that Cole and Wright were not considering impure cultures, but cultures of a *single* species that contained various intermingled genotypes.

> The most common fact in the experience of the bacteriologist is that variation occurs in cultures in response to changes in the numerous factors in the environment—temperature, light, composition and reaction of the medium, and the like. There is not absolute agreement as to the heritability of variations produced in this manner, but one may state that the majority of bacteriologists consider them at least partially heritable. In fact, it is among bacteriologists and paleontologists, of all biologists, that a belief in the inheritance of acquired characters has its strongest hold. In the case of the bacteria, the belief is based on the not uncommon observation that when a particular condition produces a change in a culture of organisms, and cultivation is continued a sufficient time under the condition, the variation may become so thoroughly fixed that there is no return to normal, even tho [sic] cultivation under the original conditions is resumed. In other words, it is believed that the change has been impressed on the organism by the conditions of the environment, and such changes are called impressed variations...The views [of bacteriologists] as to bacterial variation most commonly held are presented by Winslow..."With unicellular organisms...there is no bar to the persistence of acquired characters. Bacteria respond in many ways to the direct influence of environmental conditions; and each strain transmits to a degree the impress of its recent history" (Cole and Wright, 1916).

Countering this Lamarckian view, Cole and Wright pointed out that such variability only has meaning if the observations are made on pure lines. "A pure line is defined by Johannsen as 'the descendants from one single homozygotic organism, exclusively propagating by self-fertilization;' but for our purposes in the case of bacteria, in which reproduction is asexual, it may be taken as the descendants from any single cell. (The term clone is coming to be largely used by biologists to designate the aggregate line of descendants from a single individual when reproduction is vegetative or asexual...)" (Cole and Wright, 1916).

Cole and Wright then proceeded to adduce evidence that bacterial cultures, as originally isolated, are frequently genetically mixed populations of a larger or smaller number of distinct hereditary forms or lines. "Different cultural methods or other environmental changes *act as a sieve* through which the multiplex populations are sifted" (emphasis added). They then showed that if proper care was taken to begin a study with a uniform culture, evidence of mutation could be found.

> Mutations apparently do occur in pure lines, both spontaneously and in response to certain environmental changes. Such mutations give rise to new races, or biotypes, which vary about modes of their own, and within which selection is ineffective. Mutations may be of a magnitude to be readily appreciated, or they may be so small as to be determinable only by statistical methods. The biotypes which make up the ordinary bacterial culture have undoubtedly arisen in times past as distinct mutations, and have lived on side by side, retaining their individuality because of the absence of amphimixis. It is only by the most refined methods that the foregoing kinds of variations can be recognized and differentiated (Cole and Wright, 1916).

It should be noted that these ideas were published in a respected bacteriology journal. However, an influential British student of bacterial variation in the next decade, Joseph A. Arkwright, insisted that the pure line concept did not apply to bacteria:

The use of the term "pure line" in bacteriology for the descendants of a single bacterial cell does not appear to be very happy, since, although undoubtedly different forms or races may be separated from apparently pure cultures, these are not exactly comparable to Johanssen's "pure lines" in beans and wheat. In the first place, the different forms often arise again readily from the "pure lines," and any resemblance as regards natural and artificial crossing cannot be verified, since conjugation in bacteria cannot be assumed to occur. At any rate, the occurrence of syngamy and conjugation cannot be observed at will, have never been satisfactorily demonstrated, and the evidence in their support is very slight (Arkwright, 1930).

The publication in which Arkwright's review was published was one of the most influential treatments of bacteriology in the 1930s and 1940s.

4.3 ARE BACTERIA MORE MUTABLE THAN OTHER ORGANISMS?

One property of mutation that has just been understood over the past decade should be mentioned at this point, since it helps in understanding the historical events. In general, the gene is an exceedingly stable structure, and mutation rates of 10^{-6} or less per generation are the rule. However, under some conditions, and with certain kinds of characters, much higher mutation rates can be observed. Many of these changes are due to the interaction of mobile genetic elements capable of inducing gene rearrangements with resident, stable genes. In some cases, they are due to interactions of plasmids with chromosomal genes or to the loss of plasmids. Tandem duplication and amplification of regions of the *Escherichia coli* chromosome have been detected at relatively high frequency (10^{-3}/cell/generation), so that in a clonal population arising after 20–30 generations of growth under unselective conditions, individual cells may differ with respect to the number of copies of specific genes (Simon and Silverman (1983).

The point is that there are well-known molecular explanations for the high variability exhibited under some conditions. Simon and Silverman (1983) have presented a partial list of bacterial characters where rapid genetic change has been reported. This list includes most of the characters that bacteriologists interested in variability studied in the first three decades of the twentieth century: surface antigens, flagellar phase variation, colony morphology, pigment variation, bioluminescence, phage sensitivity. The best studied, flagellar phase variation, is due to a region in the genome that undergoes inversion at a high rate. The switch from one orientation to the other occurs at a frequency of 10^{-3} to 10^{-5} per cell generation. Thus, there is no question that high variability for certain characteristics exists in bacteria. Note that these kinds of variations also exist for higher organisms (immunoglobulin diversity in mammals is the best example), but since they occur in somatic cells they are not passed on to succeeding generations. In addition, in bacteria, the enormous population densities involved make them easier to detect.

4.4 *ESCHERICHIA COLI MUTABILE*

An organism that played a considerable role in ideas about bacterial variability during the period 1900–1945 was the bacterium first called *Escherichia coli-mutabile* by Neisser and Massini. This organism, which can be frequently isolated from nature, is lactose-negative but throws off lactose-positive variants. Thus, when streaked on an agar medium that contains an indicator of lactose fermentation (for instance, Endo's or eosin methylene blue), the colonies that develop are initially white or light pink, but dark red sectors or papillae arise within the colonies (Fig. 4.1). If cells from one of these dark red papillae are picked, the culture breeds true, producing only dark red colonies, whereas the original wild type continues to form colonies that exhibit the papillae.

Escherichia coli mutabile was first isolated by Massini (1907), a student of I.M. Neisser at Frankfurt. When Neisser himself presented a preliminary report of Massini's work, he clearly related it to mutation in the sense of de Vries (Neisser, 1906). An important tenet of the de Vries mutation doctrine, its suddenness and the absence of an intermediate stage, was emphasized. Either white or red colonies were obtained, nothing in between. The presence of red papillae on the white colonies showed clearly the *discontinuous* nature of the change, although later workers (see below) would not see this idea so clearly. Among other things, Massini noted that only certain strains of *Escherichia coli* produced papillae and that they only appeared when the culture was streaked on a medium containing lactose. He noted that the appearance of the papillae required *growth* and that the variation was a sudden event. Massini characterized some of the lactose-positive variants taxonomically and showed them to be identical to the parent in all characteristics except lactose fermentation. In regard to lactose fermentation, Massini noted that this ability was linked to the ability of the organism to produce a specific enzyme that he called *lactase*. The mutation to this newly

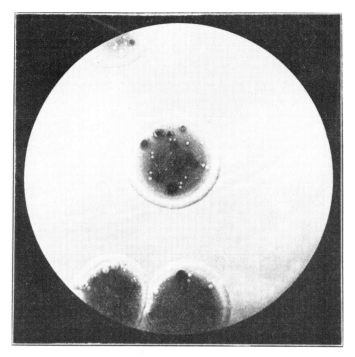

Figure 4.1 Massini's photograph of a colony of *Escherichia coli mutabile* showing lactose-positive papillae. This colony, enlarged about 6 × , had grown on Endo's agar. It was 6 days old when photographed. (From Massini, 1907.)

acquired property was a rare event, but once it occurred the mutant continued to breed true.

The mutation concept of Massini and Neisser was not to prevail for long. Several years later, Burri, in a long paper, studied *Escherichia coli mutabile* and other bacteria that exhibited similar phenomena and claimed that the *mutabile* phenomenon was not due to mutation, but to an *adaptation* to ability to use lactose (Burri, 1910). Examples of enzyme adaptation were by this time already known in other microorganisms (see Section 10.1). Burri attempted to repeat Massini's work using his own isolates, but it appears that the organism he isolated was not, actually, an *Escherichia coli mutabile* strain. Confusing the utilization of the sugar and the inheritance of the ability to use it, Burri insisted that a bacterium acquired a new enzyme when grown on the substrate of this enzyme. The ability to use the sugar was thought to be "latent" in the culture, and this latent ability could be "awakened" by cultivating the organism on the substrate. According to Burri, the conversion from lactose-negative to lactose-positive occurred in "all cells" of the culture, but once the property had been acquired, it continued to be inherited indefinitely. The change was fairly rapid but not discontinuous. It was not, therefore, a mutation in the sense of de Vries. Burri ignored the significance of Massini's papillae, which cannot be explained by his hypothesis, but his own culture did not exhibit these papillae. The confusion between acquisition of a new enzyme by adaptation (the process that Jacques Monod would later call *induction*) and mutation was to exist until the late 1940s when it was finally clarified by the demonstration that *both* enzyme adaptation and mutation occurred (see Section 10.4). Burri's paper was cited by all subsequent workers who believed in Lamarckian or neo-Lamarckian mechanisms of bacterial variability; Massini's paper was frequently ignored or discounted because it had not been confirmed by Burri.

Most bacteriologists at the time had little interest in broader questions of genetics or biology, being mainly involved in their extremely technical applied work. To such bacteriologists, the main significance of Massini's work was in diagnostic bacteriology (Henderson-Smith, 1913).

The Work of I.M. Lewis on Escherichia coli mutabile

The essential nature of *Escherichia coli mutabile* was not finally clarified until the important work of Lewis (1934). Originally trained as a botanist, Lewis began to study bacterial formation of secondary colonies, first with *Bacillus mycoides* and then with *Escherichia coli mutabile*. By the time Lewis carried out his work, techniques for culture of bacteria, and especially for culture media preparation, had advanced considerably. Synthetic culture media had been constructed and used in extensive studies on bacterial physiology and nutrition (see Section 3.8), and much information was available about enzymes and enzyme formation in bacteria.

The question Lewis set out to answer was whether the variation in ability to utilize lactose occurred spontaneously in rare members of the population, or whether it was a phenomenon of the whole population. Most workers since Massini had concluded the latter, but without really good evidence. Lewis prepared a synthetic medium containing either glucose or lactose and then determined by viable count the frequency of cells present in the *mutabile*

Table 4.1   Number of Variant and Original Type Cells in Plain Agar Colonies of *B. coli-mutabile* as Determined by Capacity for Growth in Lactose Synthetic, Glucose Synthetic, and Beef Extract Peptone Agar

	Kind of medium and number of cells per colony capable of growth		
Colony	lactose synthetic agar	glucose synthetic agar	beef extract agar
1	6,300	3,420,000,000	3,540,000,000
2	3,300	1,640,000,000	1,610,000,000
3	2,900	710,000,000	730,000,000
4	2,200	860,000,000	840,000,000
5	2,080	1,100,000,000	1,200,000,000

Colonies were suspended in 10 ml water, and dilutions of the suspensions were plated on the agar media listed. (Data from Lewis, 1934.)

culture that would grow on each of these sugars (Table 4.1). The colonies that developed on the lactose plates were seen to derive from variants that had been present in the initial inoculum. Although large numbers of colonies developed on the glucose plates, about 10^{-6} fewer colonies developed on the lactose plates (see Table 4.1). By picking colonies from lactose synthetic agar, it could be readily shown that these strains bred true as lactose-positives, whereas colonies from the glucose synthetic agar plates were lactose-negative and continued to throw off lactose-positive variants. These data showed another important thing: "The fact that variation to specific carbon compounds occurs in mutabile strains independently of the compound has not been generally recognized. Burri's...conclusions have been frequently quoted but seem to be mainly erroneous...We are obliged to conclude, therefore, that sugars or alcohols act as specific selective agents rather than as stimulators of variation...an organism which has no capacity for spontaneous variation to a carbon compound is not caused to vary to it by cultivation in contact with the compound no matter how long the contact may be prolonged...Variation, when beneficial, may be preserved by selective action of the medium. Non-beneficial variation, such as poor adaptation to nutrients, would be eliminated by overgrowth of the more vigorous original cells" (Lewis, 1934).

Although these clearly presented ideas were right on the mark, it would be almost ten more years before they would be generally accepted, after the work of Luria and Delbrück (see below), which seems to have developed independently of a knowledge of Lewis.

4.5 BACTERIAL DISSOCIATION

A peculiar concept called *bacterial dissociation* was discussed extensively in the bacteriological community during the 1920s and 1930s. It is of interest here primarily because it bears on the transformation phenomenon in bacteria (see Section 9.2). Although dissociation was sometimes discussed in relation to life cycles in bacteria, it was mainly of interest because of its connection with variation in colony morphology in bacteria. Bacterial cultures often exhibit more than one kind of colony, of which three common types described were (1) mucoid (M) colonies, which have a viscous consistency, probably related to the amount and nature of the capsular material produced by the culture; (2) smooth (S) colonies, which are round, convex, opaque, and possess an even margin; and (3) rough (R) colonies, which are larger, flat, irregular in shape, with rhizoid or filamentous borders (Dubos, 1945). The most common variation was S/R, which could be due to capsular differences (as it is in pneumococci, see Section 9.2), but could also be due to differences in surface chemistry such as electrical charge or hydrophobicity, often associated with antigenic differences. The terms S and R were first used to describe variation in enteric bacteria by Arkwright (1921). In Arkwright's study, S forms were seen to make good stable emulsions when suspended in saline, whereas R forms agglutinated and the cells settled to the bottom of the culture. This agglutination phenomenon was unrelated to antibody-induced agglutination, which could also occur in S forms under the influence of specific antiserum. In most cases, S types could throw off R types, but R types tended to remain stable. An important property that particularly fascinated bacteriologists was that S type cultures were often *virulent* whereas R types were not.

Although it is clear now that the S/R conversion was generally a mutational event, many bacteriologists viewed the change in a much more confused way. The most outspoken proponent of a nongenetic explanation of bacterial dissociation was Philip Hadley, who believed that this phenomenon was a central aspect of the life cycle of the organism (Hadley, 1928). According to Hadley, each bacterial species possesses not one "normal" colony form, but a variety, each of which has to be recognized if one is to define the species. Each of these colony forms was related to a stage in the life cycle of individual cells that comprised the colony structure. Hadley argued vigorously against the so-called "monomorphism" school of Ferdinand Cohn and Robert Koch (see Chapter 3), which insisted on the constancy of bacterial form. "...monomorphism...still clings like a barnacle to modern bacteriology. According to its dictates, whatever departs from the 'normal' must be regarded as an 'involution form,' a degeneration form, a mutant, or a contamination...Are we, then, forced to the conclusion that cultures of the S and R forms are not normal cultures?" (Hadley, 1928).

Arkwright, who had coined the terms smooth and rough, supported this interpretation. He stated that the observed bacterial variations should not be called mutation because "the term mutant...has been given a special meaning by botanists and applied to cases of alleged changes in the germ plasm or chromosomes. As bacteria are not known to possess any mechanism resembling chromosomes the term is now generally considered inappropriate" (Arkwright, 1930).

4.6 TRAINING

A bacteriological concept related to dissociation was *training*. The difference between dissociation and training was that dissociation seemed to occur without any obvious external influence, whereas training involved a specific environmental component. In addition, training differed from adaptation in that the latter was a sudden or discontinuous phenomenon, whereas training resulted in a gradual change. For example, the organism *Propionibacterium pentosaceum* required the vitamin thiamine, but it was possible to "adapt" a culture to grow in the absence of this vitamin by periodic transfer of the culture in gradually decreasing concentrations of thiamine, or by addition of a large inoculum to a thiamine-free medium.

Sir Paul Fildes

The concept of training was apparently first introduced by the eminent British bacteriologist Sir Paul Fildes (for review, see Dubos, 1945; Gale, 1947). In his book on bacterial nutrition, B.C.J.G. Knight, an associate of Fildes, discussed the concept of training in detail (Knight, 1936). One well-studied case concerned the tryptophan requirement of *Bacterium typhosum* (*Salmonella typhosa*). Natural isolates of *Salmonella typhosa* usually required tryptophan for growth, but tryptophan-independent cultures could be isolated. To bring about "training," a culture was grown in a synthetic medium containing ammonium as the principal nitrogen source, together with growth-saturating amounts of tryptophan. Then, transfers were made to media with gradually decreasing concentrations of tryptophan, until eventually a culture was obtained that could grow in the complete absence of the amino acid (Fildes, Gladstone, and Knight, 1933). Note that we are dealing here with *acquisition* of a function, and since mutations were generally considered to result in a loss of function, the idea of mutation perhaps did not seem appropriate. Fildes and co-workers rejected mutation as an explanation for this phenomenon because mutations were known to be rare events and training appeared to be an event that occurred reproducibly. According to Knight (1936):

> A second explanation of the phenomenon of training is that the change of exacting into non-exacting strains is a direct response to the chemical stimulus of the changed nutrient conditions. Enzymic changes of this type are known, which take place even without cell multiplication...The suggestion is, therefore, that the new enzymes are produced as a direct reaction to the chemical stimulus of the new nutrients, in the absence of the normal nutrients.

This obviously Lamarckian explanation completely begged the question, but it did serve to focus, albeit obliquely, on the possible role of heredity in enzyme formation. By this time, Karström's important paper on "adaptive enzyme" formation had appeared (see Section 10.2), and many bacteriologists used Karström's interpretation, unable to sort out changes that occurred in *whole* populations (such as enzyme adaptation) from changes that occurred in small subsets of the population.

Years later, after the Luria/Delbrück paper was published (see below, Section 4.8), Sir Paul Fildes, probably as a result of correspondence with Luria (see later), published a nice study which showed that his phenomenon of "training" was a result of mutation followed by selection (Fildes and Whitaker, 1948). It was shown that a small fraction of tryptophan-positive cells (about 10^{-7} per ml) were present in tryptophan-negative cultures, and that these trp^+ cells arose independently of the tryptophan concentration. When tryptophan was present in the medium, trp^+ cells had no selective advantage, but if the concentration of tryptophan was lowered to suboptimal amounts, the mutants were favored. If the inoculum was low, trp^+ mutants would be absent, but as the population density increased, they would appear, generally at a threshold concentration of 10^7 cells/ml. Thus, in the complete absence of tryptophan, no growth would occur, but if suboptimal levels were present, the inoculum could grow, and if it reached $>10^7$, rare mutants would be present. If such a culture were transferred, the fraction of mutants would be higher in the second round, and lower concentrations of tryptophan could be used. Thus, by successive transfers to lower concentrations of tryptophan, a trp^+ culture was eventually obtained. Without knowing what was going on, it might be concluded that the trp^- strain had been "trained" to trp^+, but it was just that the ability of trp^+ cells to establish themselves depended on the concentration of tryptophan. If a large amount of tryptophan was present, the parent cells outgrew the mutants, and only if the concentration of tryptophan was limiting did mutants grow preferentially over parents. A partial deprivation of tryptophan favored the *detection* of trp^+ mutants, although it did not favor their occurrence. By making viable counts, Fildes and Whitaker (1948) could study the growth of both trp^+ and trp^- cells even before there was visible growth, and in this way the replacement of parent by mutant could be examined. Fildes, like many bacteriologists, had been misled originally by examining cultures only after *visible* turbidity was present.

Fildes and Whitaker (1948) concluded that training was simply "a cumbersome method for selecting genetic mutants."

Fildes and Whitaker do not cite the paper of Luria's student Curcho (1948), published the same year. Curcho showed in a much simpler manner, using the Luria/Delbrück fluctuation test (see Section 4.8), that acquisition of tryptophan independence in *Salmonella typhosa* had a nutritional origin. According to a personal communication from S.E. Luria (May 18, 1989), he had sent a copy of Curcho's results to Fildes, who then returned to work on the problem. Curcho's approach using the fluctuation test not only provided a more straightforward solution to the problem of training, but also permitted calculation of the mutation rate for the *trp* gene.

Sir Cyril Hinshelwood

Another type of training or adaptation became controversial through the activities of the distinguished physical chemist Sir Cyril Hinshelwood. Hinshelwood believed that adaptation of bacteria to growth at high concentrations of inhibitory chemicals could be explained in terms of alternations in reaction rates of the bacterial populations (Hinshelwood, 1946). If a culture was grown in the presence of an inhibitor, the concentration of the substrate for the enzyme that the inhibitor affected would build up, reversing the action of the inhibitor. By making successful transfers to progressively higher concentrations of the inhibitor, the culture could be "trained" to tolerate this inhibitor. It often happens that because of the personality or influence of a proponent, ideas are not ignored, although they should be. Hinshelwood was such a powerful figure in British science that his ideas received deference when they should have received short shrift. He was a physical chemist who had done outstanding work in the field of chemical kinetics. Elected to the Royal Society in 1929 at the early age of 31, he was knighted in 1948, and elected President of the Royal Society in 1951. In 1956 he was awarded the

Nobel Prize in chemistry for his work on chemical kinetics. Among other things, Hinshelwood attempted to apply his mathematical skills and kinetic analyses to bacterial growth. He published a book that was widely read and frequently misunderstood (Hinshelwood, 1946). Hinshelwood believed that variations in reaction rates and kinetics alone could explain change in the bacterial cell. Even as late as 1953, when the Luria/Delbrück study on mutation had become part of genetic orthodoxy, Hinshelwood was invited to participate in an important symposium on adaptation in microorganisms sponsored by the British Society for General Microbiology (Dean and Hinshelwood, 1953).

Hinshelwood's work provides an excellent example of the dangers of an uncritical application of mathematics to biology. Discussing Hinshelwood's work, Luria has stated: "I have often noticed...that biologists are readily intimidated by a bit of mathematics laid before them by chemists or physicists. It was one of the blessings of my too short stay among physicists to be immunized against mathematical humbug" (Luria, 1984).

We can understand now that in most cases, some mutational event led to the acquisition of the new character, but Hinshelwood seemed not to be able to think of a bacterial culture as a mixture of genetically distinct cells. It is interesting that his work on training was published several years *after* the central work of Luria and Delbrück discussed below. However, there may have been a positive side to Hinshelwood's work, since it helped to place into clear perspective the distinction between what Lederberg (1951) called the "neo-Darwinian" and "Lamarckian" concepts of bacterial evolution and to stimulate the development of critical experiments. Among other important studies, the work of Lederberg and Lederberg on indirect selection was motivated in part by a desire to prove Hinshelwood wrong. The Lederbergs' work led to the development of an important technical advance, replica plating (see Section 4.10).

4.7 PHAGE AS AN ELEMENT IN BACTERIAL VARIABILITY

The discovery and genetic significance of bacteriophage is discussed in Chapters 6 and 7. In the early days of phage research, phage was sometimes viewed as an element, even a cause, of bacterial variability. Thus, when Gildemeister (1917) first isolated phage, soon after its discovery by Twort and d'Herelle, he was studying colony morphology of bacteria that had been "freshly" isolated from pathological material. Some colonies that developed were mottled in appearance (he called these *Flatterformen*, flickering forms), due to the presence of phage. The focus soon was on the phage itself, but the discovery resulted from the observations on colony morphology that had a potentially genetic significance. Burnet and McKie (1929), studying a lysogenic *Salmonella*, noted that rough strains liberated more phage than smooth cultures. Later, Burnet and Lush (1936) isolated not only host cells resistant to phage infection, but also phage mutants able to attack host resistants.

Among the most interesting papers of this period was one by Yang and White (1934) on bacteriophage sensitivity of rough and smooth strains of *Vibrio cholerae*. Beginning with what they called "ultrapure" strains, Yang and White reported that they had isolated extremely rough variants that were invariably resistant to phage A, even though they had *never* come in contact with phage. Yang and White (1934) concluded: "On the whole we are inclined to believe that resistance to A phage is not a modification induced by phage action but that resistant elements are present in the ultrapure culture and survive lysis." It was just fortuitous that rough variants were resistant to phage. Yang and White recognized this when they noted:

> The most forcible evidence in favour of a selection theory would be a direct demonstration of resistant units already present in the ultrapure culture. The difficulties in the way of such a demonstration are obvious; the case is not far removed from that of the proverbial needle in the haystack (Yang and White, 1934).

The potential utility of phage for genetics research was often discussed in the 1920s and 1930s, but phage research came of age with Max Delbrück and Salvador Luria. Although the paper by Yang and White (1934) clearly foreshadowed the Luria/Delbrück work, it was not cited either by Burnet or Luria/Delbrück.

4.8 FLUCTUATION TEST: THE PAPER OF LURIA AND DELBRÜCK

The paper of Luria and Delbrück can be said to mark the beginning of modern bacterial genetics. It is instructive that it was published not in a bacteriology journal but in *Genetics*, the journal where "classical" genetics research was published (Luria and Delbrück, 1943). Luria has described the background to how this seminal study was carried out (Luria, 1984).[2]

Originally working independently, Luria and Delbrück came together at the Cold Spring Harbor Laboratory during World War II, initiating a collaboration that was to last for many years (see Chapter 6 for biographical details). The connection between phage and bacterial genetics was made in attempting to explain the origin of bacterial cultures resistant to phage attack (so-called "secondary cultures"). As put by Luria and Delbrück (1943):

> When a pure bacterial culture is attacked by a bacterial virus, the culture will clear after a few hours due to destruction of the sensitive cells by the virus. However, after further incubation for a few hours...the culture will often become turbid again, due to growth of a bacterial variant which is resistant to the action of the virus. This variant can be isolated and freed from the virus and will in many cases retain its resistance to the action of the virus even if subcultured through many generations in the absence of the virus...The resistant bacterial variants appear readily in cultures grown from a single cell. They were, therefore, certainly not present when the culture was started. Their resistance is generally rather specific ...D'Herelle and many other investigators believed that the virus by direct action induced the resistant variants. Gratia, Burnet, and others, on the other hand, believed that the resistant bacterial variants are produced by mutation in the culture prior to the addition of virus. The virus merely brings the variants into prominence by eliminating all sensitive bacteria. Neither of these views seems to have been rigorously proved in any single instance...It may seem peculiar that this simple and important question should not have been settled long ago, but a close analysis of the problem...will show that a decision can only be reached by a more subtle quantitative study than has hitherto been applied... (Luria and Delbrück, 1943).

[2]The fluctuation test is described on pages 74–79 of Luria (1984). However, Luria's memory was faulty. He implies that he began his work as a response to Hinshelwood, yet the latter's work did not become an issue until his book was published in 1946, 3 years after the Luria/Delbrück paper was published (see Section 4.6).

The Theory of the Fluctuation Test

The two opposing hypotheses to explain the presence of phage-resistant host bacteria were (1) mutation followed by selection and (2) acquired hereditary immunity. With the first hypothesis, mutation to resistance could occur at any time prior to the addition of the virus, so that the culture would contain both wild-type cells and "clones of resistant bacteria." In any culture these clones would be of various sizes, depending on when during growth of the culture the mutation occurred. If mutation occurred early after inoculation, then a large number of resistant cells (later called a "jackpot") would be present in the population, whereas smaller numbers would be present if the mutation occurred late in the growth phase (Fig. 4.2, right). It was assumed that in the absence of phage, resistant and sensitive cells grew at equal rates.

With the second hypothesis, acquired hereditary immunity, there would be a small finite probability for any bacterium to survive an attack by the virus. Survival of infection would confer immunity on both the individual and its offspring, but the probability of survival would not run in clones. Thus, if a bacterium survived an attack, there would be no reason to infer that its close relatives, other than its descendants, would also survive the attack.

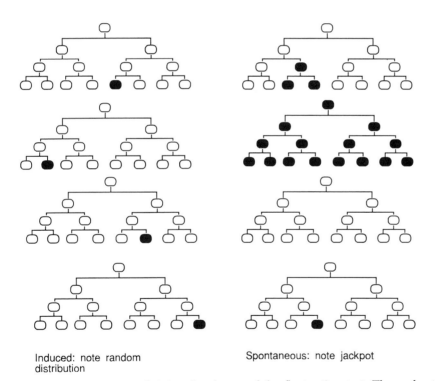

Induced: note random distribution

Spontaneous: note jackpot

Figure 4.2 Diagram explaining the theory of the fluctuation test. The early stages in the development of several small cultures are shown. Resistant bacteria are dark. (*Left*) Anticipated results if resistant bacteria are induced by contact with phage. In this case, the proportion of resistant cells in the population should be more or less the same in each culture. (*Right*) Anticipated results if resistance arises randomly by mutation. Families of resistant siblings should develop, and depending on when the mutation occurred, the sizes of these resistant clones should vary widely.

This last statement contains the essential difference between the two hypotheses. On the mutation hypothesis, the mutation to resistance may occur any time prior to the addition of virus. The culture therefore will contain "clones of resistant bacteria" of various sizes, whereas on the hypothesis of acquired immunity the bacteria which survive an attack by the virus will be a random sample of the culture...on the hypothesis of resistance due to mutation, the proportion of resistant bacteria should increase with time, in a growing culture, as new mutants constantly add to their ranks. [With the acquired hereditary immunity hypothesis,] a constant proportion of resistants may be expected...as long as the physiological conditions of the culture do not change (Luria and Delbrück 1943).

In attempting to assess the proportion of resistant bacteria, Luria[3] made numerous tests on the proportion of resistant bacteria in cultures, but experienced great variations of the proportions from day to day.

Eventually, it was realized that these fluctuations were not due to any uncontrolled conditions of our experiments, but that, on the contrary, large fluctuations were a necessary consequence of the mutation hypothesis and that the quantitative study of the fluctuations may serve to test the hypothesis.

The fluctuation in the numbers of bacteria resistant to bacteriophage α (later called T1) was markedly higher than could be accounted for by sampling errors, a result that was in conflict with the expectations from the hypothesis of acquired immunity (Table 4.2). The distribution of resistants follows a Poisson distribution, as predicted by theory.

Luria (1984) has described how the idea for the fluctuation test came to him while he was attending a faculty dance at the Bloomington, Indiana country club.

During a pause in the music I found myself...watching a colleague putting dimes into [a slot machine]. Though losing most of the time he occasionally got a return. Not a gambler myself, I was teasing him about his inevitable losses, when he suddenly hit the jackpot...gave me a dirty look, and walked away. Right then I began giving some thought to the actual numerology of slot machines; in so doing it dawned on me that slot machines and bacterial mutations may have something to teach each other...Consider first a completely unprogrammed slot machine with a small pool of dimes in it. When one puts in a dime, the machine returns money at random, mostly zero, or one, or more rarely two, exceptionally more...The returns are very different from those of a programmed machine with its excess of zeros and a certain number of jackpots...Realizing the analogy between slot-machine returns and clusters of mutant bacteria was an exciting moment...Next morning I went early to my laboratory...I set up the experimental test of my idea—several series of identical cultures of bacteria, each started with very few bacteria (Luria, 1984).[4]

It should be noted that Luria was "prepared" for this study by earlier statistical work he had done on bacteriophage. He had published a paper on the particulate nature of the "lytic unit" in bacteriophage that made extensive use of the Poisson distribution (Luria, 1939). Thus, although the mathematical analysis in the Luria/Delbrück paper was done by Delbrück, Luria had a firm grasp of the statistical realities before he began the experimental work.

---

[3]Although the paper was co-authored by Luria and Delbrück, it is clear from Luria (1984) that he did all the experimental work, Delbrück providing primarily the mathematical analysis.

[4]Although the slot machine account undoubtedly has some basis, it should be noted that in the original paper, written in 1943, Luria implies that it was *after* great variations in resistance to phage were found that the significance of these fluctuations was realized (Luria and Delbrück, 1943).

Table 4.2   Distribution in Numbers of Phage-resistant
Bacteria in Series of Similar Cultures

Number of resistant bacteria	Number of cultures
0	29
1	17
2	4
3	3
4	3
5	2
6–10	5
11–20	6
21–50	7
51–100	5
101–200	2
201–500	4
501–1000	0

Number of cultures: 87
Volume of cultures: 0.2 ml
Volume of samples assayed: 0.2 ml

Average per sample	28.6
Variance	6431
Average per culture	28.6
Bacteria per culture	2.4×10^8
Mutation rate	
method 1	0.32×10^{-8}
method 2	2.37×10^{-8}

(Data from Luria and Delbrück, 1943.) See text for discussion of the
two methods of calculating mutation rate.

The idea of the fluctuation test is so simple that students often have difficulty grasping it, thinking there should be something more complicated involved. Like many simple ideas, it is obvious once understood. Because Luria and Delbrück spent summers with the strong genetics group at Cold Spring Harbor Laboratory (see Chapter 6), knowledge of the fluctuation test spread quickly through the genetics community and became part of the core of bacterial genetics. Demerec and Fano (1945) of the Cold Spring Harbor group (see Section 2.8) extended the analysis to other bacteriophages, Demerec and Latarjet (1946) and Witkin (1946) used the technique to study radiation resistance, and Demerec (1948) and Luria himself studied resistance to antimicrobial agents (Luria, 1946; Oakberg and Luria, 1947).

Not only did the Luria/Delbrück experiment provide strong support for the phenomenon of mutation in bacteria, it also provided a means for quantitatively calculating mutation rates. The theory for calculating mutation rates was developed by Luria and Delbrück (1943) and Delbrück (1945) and later modified and improved by Lea and Coulson (1949). It is necessary to make a clear distinction between the *fraction* of mutant bacteria in a culture and the mutation *rate*. The fraction of mutants is a function of two factors, the number of mutants occurring and the time during the growth period at which they occur. (It is assumed that mutant and wild-type cells grow at the same rate.) A mutant that appears several generations back from the time of sampling will have developed into a sizeable clone. One can express the mutation rate as the probability that a bacterium will mutate either over a given *time* period or *per cell division*. Since growth rate is a function of environmental conditions, for most purposes it is preferable to calculate mutation rate per cell division.

> ...mutation rates may be estimated from the experiments by two essentially different methods. The first method makes use of the fact that the number of mutations in a series of similar cultures should be distributed in accordance with Poisson's law; the average number of mutations per culture is calculated from the proportion of cultures containing no resistant bacteria at the moment of the test...The second method makes use of the average number of resistant bacteria per culture...the values of the mutation rate obtained by the second method are all higher than the value found by the first method...big clones do not affect the mutation rate calculated by the first method, but they do affect the results of the second method... (Luria and Delbrück, 1943).

The calculation of the mutation rate from actual data requires a numerical approach, since the differential equations that describe the process cannot be explicitly solved. Luria and Delbrück (1943) solved the equation numerically and gave a chart that could be used to calculate mutation rate. Lea and Coulson (1949) modified the equations and presented tables that could be used to calculate mutation rate (a computer can now be used to carry out the numerical analysis).

Luria and Delbrück found that the mutation rate of *Escherichia coli* from sensitivity to resistance to phage was about 10^{-8} mutations per bacterium per division cycle. The realization of this low mutation rate was itself of great significance, since it emphasized the difficulty of isolating mutants and studying the mutation process in bacteria unless selective conditions were available. As Luria and Delbrück pointed out, the study of the mutation process in this case was only possible because of the method (phage resistance) available for detecting the (rare) mutational event.

One problem with calculating mutation rate from the fluctuation test is that a large number of cultures must be set up, and the value obtained is influenced by rare, very early mutations. It is interesting that although all of this work was done many years ago, the methods used then are still among the best for calculating mutation rates (Hayes, 1964; Freifelder, 1987).

A modification of Luria's method for estimating mutation rates makes use of the *chemostat*, a continuous culture device invented by Aaron Novick and Leo Szilard (and independently by Jacques Monod). With the chemostat, one maintains a steady-state population of bacteria and assesses the fraction of mutants at various times. With constant conditions, the fraction of mutants increases linearly (assuming the mutants grow at the same rate as the parent), and the slope of the line gives the mutation rate directly. In an important paper, Novick and Szilard (1951) presented data for both spontaneous and chemically induced mutation. The spontaneous mutation rates obtained were similar to those obtained by Luria and Delbrück.

Parenthetically, the "clonal thinking" that is behind the Luria/Delbrück fluctuation test would ultimately pervade modern cell biology, leading to important insights into such central fields as immunology, human genetics, and oncology.

Fortunately, the bacteriophage that Luria and Delbrück used, T1, is a virulent phage. If they had used a temperate phage, such as *lambda*, they would have found that the phage induced resistance and hence that phage immunity was acquired rather than hereditary (see Section 7.4)

One of the most interesting uses of the fluctuation test was that of Francois Jacob and Elie Wollman, during their study of the origin of Hfr strains (see Section 5.10). Jacob and Wollman developed the hypothesis that Hfr arose spontaneously by mutation of F^+ and proved this by showing that the distribution followed the Luria/Delbrück model. This work not only clarified greatly the mating system in bacteria, but also led to the isolation of numerous Hfr strains that were then used to develop the circular chromosome model for *Escherichia coli*. The legacy of Luria and Delbrück is truly vast!

A modern reexamination of the Luria/Delbrück experiment by Cairns, Overbaugh, and Miller (1988) has caused some excitement (Stahl, 1988) because it seems to show that not all mutational events are "pre-adaptive." Using selective methods that are not lethal as was the T1 infection used by Luria and Delbrück, Cairns, Overbaugh, and Miller (1988) found composite distributions, part random and part apparently "induced" by the selective agent (in this case, lactose). Although the mechanism of this effect is under discussion (Davis, 1989; Lederberg, 1989), it would not be detected under any conditions where the selecting agent is either growth inhibitory or lethal. Therefore, it would not have been manifest in the Luria/Delbrück experiment.

Newcombe's Spreading Experiment

Following up on Luria and Delbrück, the Canadian bacterial geneticist Howard B. Newcombe developed a more direct approach to demonstrating the spontaneous origin of phage-resistant mutants (Newcombe, 1949). Large inocula of bacteria were spread on plates and incubated until a limited population increase had taken place (a few hours). Then, some of the plates were re-spread with a sterile glass rod, so that the progeny in microcolonies that had developed were separated; control plates were not re-spread. Bacteriophage T1 was sprayed on both sets of plates, and incubation continued overnight. According to the adaptation hypothesis, all of the bacteria present on the plates at the time of the re-spreading should be phage-sensitive, since they had not yet come in contact with the phage. Spreading would thus serve only to redistribute members of a homogenous population, and no striking difference in number of phage-resistant colonies should be observed. According to the hypothesis of spontaneous mutation, both susceptible and resistant cells would be present at the end of the initial growth period, and spreading should thus *increase* the number of resistant colonies. Higher counts would thus be expected from the spread than from the unspread plates. This was, of course, what was found, the differences being as much as 50-fold higher. This simple experiment, easily verifiable in any student laboratory, confirmed the conclusion of Luria and Delbrück that phage-resistant mutants arise by spontaneous change prior to contact with phage.

4.9 OTHER KINDS OF BACTERIAL MUTATIONS

The Luria/Delbrück paper provided the stimulus for a large amount of work on the bacterial mutation process. Some of the mutation studies arose because of the demands of the war effort during World War II. Radiation became an important subject in itself, and the discovery of penicillin and its subsequent industrialization led to a massive screening program for higher-yielding mutants of the penicillin-producing mold (see the discussion of Demerec in Section 2.8). At this same time, the seminal work of Beadle and Tatum on *Neurospora* genetics was unfolding, leading to the concept of "one-gene/one enzyme" (see Section 5.1). Mutants provided an essential experimental tool of Beadle and Tatum, and nutritional mutations were of particular interest. Soon, researchers were attempting to isolate a variety of mutants in bacteria.

At the time of Luria's 1947 review (Luria, 1947) a wide variety of mutants had been isolated in *Escherichia coli* and other bacteria. Among the most important were the *nutritional mutants,* since these provided the material for the discovery of the bacterial conjugation process. Roepke, Libby, and Small (1944) isolated a number of *Escherichia coli* mutants requiring such growth factors as nicotinamide, thiamine, methionine, cystine, lysine, threonine, and tryptophan. Tatum and collaborators (Gray and Tatum, 1944; Tatum, 1945, 1946) used X-irradiation to induce nutritional mutants of *Escherichia coli* K-12. Edward Tatum's role turned out to be critical and unique, since his mutants in strain K-12 became the basis of Lederberg's recombination studies (see Chapter 5).

The isolation of nutritional mutants in this early period had to be done by what are now called "brute force" methods, since the penicillin-selection method (see Section 4.12) was not yet available. Tatum (1946) summarized the early data on the frequency of nutritional mutants in irradiated cultures. A few mutant types were obtained when a few hundred to a few thousand colonies were checked. Since the induced mutation rate might have been 10^{-4} to 10^{-5}, and since hundreds of mutant loci could be picked up, the frequency of nutritional mutants reported by Tatum is reasonable; it is thus not surprising that large numbers of nutritional mutants soon became available for genetic analysis.

Nutritional mutants proved to have a variety of uses. They were a great aid in studying metabolic pathways (Davis, 1950), since they formed highly specific perturbations of complex biochemical pathways. This use of nutritional mutants had already been initiated in *Neurospora* by Beadle and Tatum (see Section 5.1), but *Escherichia coli* offered much more favorable material for enzymology. In addition, nutritional mutants of *Escherichia coli* had numerous genetic applications. In particular, reversion rates could be readily studied, since large populations from mutant cultures could be plated on a synthetic culture medium lacking the required nutrient. To designate a mutant that regained the ability to grow in a medium not containing any growth factor, Ryan and Lederberg (1946) introduced the term "prototroph," in analogy with the term "auxotroph" used to indicate an organism that had a nutritional requirement. It was quickly recognized that nutritional mutations were often lethals if the organism was plated on a medium lacking the essential nutrient.

An important idea that came out of a study of nutritional mutants in *Neurospora* was the realization that some mutants exhibited only incomplete genetic blocks, a phenomenon that was called "leakage" (Bonner, 1951). Certain nutritional mutants of *Escherichia coli* were also found to be "leaky," and the term gradually became applied to genes not involving nutritional pathways. Thus, in the *r*II locus of phage T4, which has nothing to do with intermediary metabolism, Benzer isolated mutants that he called leaky (see Section 6.10). Now the term leaky is applied to any mutation that fails to shut off completely the gene (Hayes, 1964).

Mutations for sugar utilization, such as that exhibited by *Escherichia coli mutabile*, were also studied. Antigenic variations, such as the S/R transformation, were also recognized to be due to mutation (Braun, 1953).

Although bacterial mutants had a variety of uses, a true understanding of their nature could only be accomplished by traditional genetic means. As Luria (1947) emphasized, a detailed analysis of mutants required genetic analysis by crossing tests. Without crossing, it is impossible to know if a mutant that exhibits multiple deficiencies is pleiotropic or merely represents a peculiarity in a biochemical pathway. For instance, development of a particular kind of phage resistance always led to the appearance of a proline requirement, yet reversion to proline independence could occur without loss of phage resistance. It was often difficult to explain mutations of these kinds on the basis of the one-gene/one-enzyme theory of gene action. When bacterial crossing became available as a result of Lederberg's work (see Chapter 5), analysis of such questions was quickly undertaken.

4.10 REPLICA PLATING AND INDIRECT SELECTION OF BACTERIAL MUTANTS

Although the Luria/Delbrück fluctuation test and the Newcombe spreading experiment provided a reasonably convincing demonstration that mutation to phage resistance occurred independently of the selective agent, both procedures were perforce indirect. Hinshelwood's book, which discussed extensively the adaptation to inhibitory agents (see Section 4.6), was published after the Luria/Delbrück work (Hinshelwood, 1946) and seemed not to have been influenced by it in the slightest (Newcombe, 1949). Even at the 1953 Society for General Microbiology symposium on adaptation in microorganisms, heated discussions regarding pre- and post-adaptation were carried out. However, the controversy drew to a close after Lederberg and Lederberg (1952) described a procedure, *replica plating*, which permitted isolation of drug-resistant mutants in the *complete absence* of the drug. Although a few recalcitrant individuals still employed Lamarckian interpretations, the results of the replica-plating procedure were difficult to refute. This technique not only settled the question of the origin of drug resistance, but also provided a general technique for studying the genetics of large populations of bacteria that proved to have extremely wide utility. Even today, replica plating, or its modern variants, finds an important place in the bacterial genetics laboratory.

The Replica-plating Technique

In the replica-plating technique, sheets of velveteen cloth are used to transfer impressions of colonies from one plate to another. The fabric is equivalent to a large number of tiny inoculating needles held in fixed array. From a master plate containing a number of colonies, impressions can be transferred to sterile plates containing various kinds of culture media. The presence of colonies on the incubated secondary plates permits an assessment of the phenotypes of the original colonies. Genetic traits that can be readily classified by replica plating include antibiotic-sensitivity, responses to phage (as in phage typing), fermentation characters, nutritional requirements, or any characteristic for which a selective or indicator agar medium can be devised. The replica-plating procedure proved extremely useful in the isolation of auxotrophic mutants and became a stock-in-trade of the bacterial geneticist. Since it could also be used to transfer phage plaques, it also became of value for phage geneticists.

Lederberg (1989) has provided some background on the steps that led to the development of the replica-plating procedure:

> Faced with...problems of phenotypic scoring of an abundance of colonies, Szilard and Novick remarked to me that they had been using multi wired inoculators, even a wire brush for a primitive kind of what I later called replica plating. This was not very satisfactory owing to the poor resolution available with that material. This was in February 1951. For a couple of years prior to that point, possibly the result of a serendipitous accident, it had been common practice in our laboratory to make impressions of the dark and light colonies on eosin methylene blue agar plates by simply pressing a piece of paper onto the agar surface and then mounting this under cellulose tape in our laboratory notebooks.[5]...We had the idea of trying to use such prints as inocula for fresh plates but there was far too much smearing on the paper to allow this to be useful (as was also found by N. Visconti at the Cold Spring Harbor Laboratory).
>
> Meanwhile Howard Newcombe published a report in 1949 which...was a direct translation of the clonal expansion of the individual mutational stem cells. Sometime thereafter I elaborated on his experiment by making a single streak of growth rather than a two-dimensional lawn. I would then move a spreader perpendicular to the stroke, and each clone would then be represented by a line of resistant colonies along the direction of the second stroke. These findings engendered still more intense preoccupation with the imagery of what was happening to mutant clones buried within the bacterial population. If only now there were some constructive method to sample those clones prior to exposing them to the selective agent!
>
> Perhaps the multi point sampling technology of Novick and Szilard could be applied to this broader problem as well as to the tedium of colony scoring? How to improve upon the poor resolution and handling properties of the wire brush? Ed Tatum had taught me to use a beakerful of sterilized tooth picks, one by one, for colony picking; that saved the time needed to flame a platinum loop between picks. The brush was conceptually an ordered array of toothpicks. What might be a functional equivalent?
>
> Paper was unsatisfactory: its lateral capillarity and its compression of the colonies distorted and broke up the original growth pattern. It occurred to me that some fabric with a vertical pile would be an analog of paper on one hand, the wire brush on the other; and I soon collected a wide variety of fabric remnants to put them to empirical tests. The predictable myth that I invaded my wife's wardrobe for this purpose is pure fantasy ...I did ask Dr. Esther Lederberg for her competent assistance in the execution of the experiments, which soon demonstrated that one could rapidly enrich the proportion of resistant mutants by fishing the growth at the point on the original plate where a replica demonstrated the presence of a resistant clone...Our thinking about this as sibling selection was a good joke for Jim Crow and the rest of us in the Genetics Department at Wisconsin. It provided one more illustration of how the traditions of genetic research which might be applied on one hand to dairy husbandry or egg production had such universal validity that they would also bear on bacterial genetics (Lederberg, 1989).

---

[5]There are some samples of these filter-paper prints in Esther Lederberg's Ph.D. thesis in the Memorial Library of the University of Wisconsin-Madison.

Use of Replica Plating in Indirect Selection of Resistant Mutants

Lederberg and Lederberg (1952) used the replica-plating method to isolate resistant mutants without the bacteria ever coming in contact with the selective agent. Both phage resistance and antibiotic resistance were studied. Master plates were prepared by inoculation with high cell densities so that after incubation confluent growth was obtained. Impressions were then made onto plates containing high concentrations of the inhibitory agent. After incubation of these secondary plates, isolated colonies were obtained, reflecting resistant mutants on the master plates. Transfers from the congruent sites on the master plates into broth were made, and the broth cultures were used to repeat the experiment with a more dilute inoculum. After about four stages of such indirect selection with progressively reduced inoculum, isolated colonies on the master plates were obtained, all of which were resistant clones. The clones obtained by indirect selection were never exposed to the inhibitory agent at any time and thus there was no way that the agent could have induced the resistance.

The ramifications of this simple experiment throughout bacterial genetics were numerous. Soon the replica-plating procedure had become a routine method in genetics laboratories and had found its way into most textbooks of either microbiology or genetics.

4.11 RADIATION AND OTHER MUTAGENS

Ever since Muller's classic work on the induction of mutations in *Drosophila* by X-rays (Carlson, 1981), radiation had been used as a tool for probing the nature of the gene. Although most work had been done with X-rays, Demerec (1946) had shown that ultraviolet radiation was also mutagenic in bacteria, greatly increasing the rate of mutation to phage T1 resistance. Alexander Hollaender, at that time at the National Institutes of Health,[6] had shown that the most mutagenic wavelengths of ultraviolet were those at 260 nm, the maximum absorption of nucleic acids (Hollaender and Emmons, 1941; Hollaender, 1945). Kinetics of radiation-induced mutation had been used to conclude that a single "hit" was sufficient for the production of a mutation. Since enzyme molecules were scattered throughout the cell, it seemed unlikely that radiation was acting on them. Rather, it was visualized that a specialized center within the cell must be inactivated by radiation; the most likely such center would be the gene. This evidence was used to support the idea that bacteria had genes, but it would not be until the procedures for crossing bacteria were developed that direct evidence for genes in bacteria could be obtained.

Phenotypic Lag

Demerec and Latarjet (1946) studied the induced mutation process quantitatively and showed that some mutations were only expressed after a lag of one

---

[6]Hollaender was later to be the head of the Biology Division at the Oak Ridge National Laboratory. He played an important role in promoting the development of bacterial genetics through sponsorship of research and symposia.

or more cell divisions had taken place. In the case of phage resistance, it seemed likely that this lag before the mutation was expressed occurred because the phage receptor molecules on the surface of the mutant had to be eliminated (by dilution during division) before resistance would be manifest. In the case of biochemical mutations, phenotypic lag was due either to the segregation of nuclei from multinucleate cells (Witkin, 1951), or to the delay in expression of auxotrophy until preformed enzyme molecules present in the treated cells had been diluted out. Phenotypic lag had practical importance for the use of mutagenesis to isolate mutants. After radiation treatment, a short period of incubation was desirable in order for the mutational event in the genotype to be expressed in an altered phenotype.

Chemical Mutagenesis

Although chemical mutagenesis was discovered during World War II, the work could not be published until the end of the war because the agents used, nitrogen and sulfur mustards, were agents of chemical warfare. Chemical mutagenesis became one of the important topics of discussion at the 1946 symposium of the Cold Spring Harbor Laboratory, the first symposium on microbial genetics to be held after World War II. Demerec in particular carried out extensive studies on chemical mutagenesis of bacteria in the early post-World-War-II period (Demerec and Hanson, 1951). An early review of the work on chemical mutagenesis was given by Auerbach at the 1951 Cold Spring Harbor symposium (Auerbach, 1951). Chemical mutagenesis became an important tool during studies by Benzer and others on the fine structure of the gene (see Section 6.10), and it eventually brought about detailed biochemical understanding of the mutational process (Drake, 1989).

4.12 PENICILLIN SELECTION OF NUTRITIONAL MUTANTS

The use of auxotrophs in bacterial mating studies placed a premium on quick and efficient isolation of auxotrophic mutants. Although a number of "tricks" were developed, the most important was the penicillin-selection process, discovered independently by Davis (1948) and by Lederberg and Zinder (1948).

Studies on the mode of action of penicillin had shown that this antibiotic was bactericidal only to growing cells: Nongrowing cells retained their viability when treated with high concentrations of penicillin (Davis, 1948). If an auxotroph is incubated with penicillin in a synthetic medium in the absence of its required growth factor, it will not grow, whereas the parent prototroph grows and is killed. Although *Escherichia coli* is relatively resistant to penicillin, it can be treated if moderately high concentrations, 100–300 units/ml, are used.

In the application of the method to mutant isolation, the wild type is treated with a mutagen and allowed to incubate in a nutritionally complete medium (for instance, nutrient broth) for a period of time long enough to permit segregation of mutant nuclei and to overcome phenotypic lag. Then the culture is washed to remove nutrients, starved briefly to eliminate intracellular pools, and suspended in a synthetic medium containing penicillin. During a further incubation, the wild-type cells grow and are killed, whereas the auxotrophic mutants do not grow and survive. (Due to inefficiency in action of penicillin, not all wild types are killed.) Thus, the penicillin treatment effectively *enriches* the culture for auxotrophs. The culture is now plated onto a complete agar medium on which the auxotrophs can grow. Each colony is then tested for its ability to grow on minimal medium. Those colonies that cannot grow are presumptive auxotrophs and can then be tested for their particular growth factor requirements (Lederberg, 1950). The utility of the penicillin method is shown by the fact that through 1987, the Lederberg/Zinder paper was cited in over 175 publications and the Davis paper in over 200 publications (Anonymous, 1987).

Subsequently, the penicillin-enrichment method was coupled with the Lederbergs' replica-plating method to markedly increase the efficiency of mutant detection. The colonies on the complete agar medium are replica-plated to minimal agar. Those colonies that do not grow are presumably mutants. Numerous "tricks" to quickly characterize nutritional mutants are given by Lederberg (1950).[7]

Isolation of auxotrophic mutants in the post-World-War-II period greatly benefitted from the increasing availability of synthetic growth factors from chemical companies, especially in the United States. With the rapid expansion of research in biochemistry, a number of companies were organized to synthesize and supply the needed pure chemicals. It therefore became a relatively easy proposition for the bacterial geneticist to isolate a wide variety of auxotrophs, a crucial activity in extending knowledge of the genetic map.

4.13 CONCLUSION

This chapter has covered a period of about 50 years, during which the idea of mutation in bacteria was adopted and rejected numerous times. Bacteriologists, as a group, seemed to have had difficulty separating population from cellular events, while the geneticists were too busy studying linkage and mapping in *Drosophila* to care about bacteria. It was not until the work of Luria and Delbrück that the nature of the variation process in bacteria was clarified. Because Luria and Delbrück were associated with the genetics group at Cold Spring Harbor, their work quickly became accepted. Demerec, a *Drosophila* geneticist and the Director of the Cold Spring Harbor Laboratory (see Section 2.8), turned to bacteria and carried out extensive and important studies on the mutation process. Ultimately, after genetic mapping procedures had been developed in bacteria, this work would lead to a clear idea of gene structure in bacteria (see Chapter 5).

An important point that should be kept in mind, both in this and in subsequent chapters, is that terms such as "mutation" have acquired different meanings as the years progressed. Although the most general definition of mutation is any sudden inheritable change, phenomena as diverse as point mutations, deletions, inversions, transpositions, and changes in chromosome number can all fall under the category of mutation. Thus, the de Vries mutation (which was probably due primarily to changes in ploidy) was something quite different from what Morgan and Muller called mutation. As

---

[7]This valuable review paper is also laced with additional practical comments by an important reviewer, B.D. Davis.

the chemistry of the gene became established, the term acquired a still different meaning. It is extremely important, when reading work published in earlier periods, not to encumber the author's terms with inappropriate meanings.

In the post-World-War-II period, the work on mutation in bacteria was to diverge into two separate directions. In one direction, detailed studies on the mutation process were carried out with the aim of understanding how a mutation itself arose. These studies were initially formal and based on kinetic experiments but acquired a chemical basis after the structure of DNA was announced by Watson and Crick. It was not until the 1960s that the nature of the mutation process was clarified in detail.

In the other direction, mutagenesis studies became concerned with the isolation of large numbers of mutants for genetic and biochemical analysis. The procedures for mutagenesis and mutant selection served mainly as tools. Those who used mutagenesis in this way had little concern about the nature of the mutation process, so long as it worked and gave them large numbers of useful mutants.

In 1947, Luria considered the then current state of bacterial genetics. After a thorough review of the literature, most of which (except for his own work) was inconclusive, Luria turned to the work of Lederberg and Tatum, which had just been reported in preliminary form.

> The discovery of biochemical mutations in bacteria with production of specific growth factor deficiencies permitted Lederberg and Tatum to demonstrate by a brilliant technique the recombination of characters in mixed cultures of different mutants. These studies, still in the preliminary stage, appear to be among the most fundamental advances in the whole history of bacteriological science (Luria, 1947).

Prophetic words, indeed! We consider the work of Lederberg in the next chapter.

REFERENCES

Anonymous. 1987. This week's citation classic. *Current Contents* 30: 16–17.

Arkwright, J.A. 1921. Variation in bacteria in relation to agglutination both by salts and by specific serum. *Journal of Pathology and Bacteriology* 24: 36–60.

Arkwright, J.A. 1930. Variation. pp. 311–374 in *A System in Bacteriology in Relation to Medicine*. His Majesty's Stationery Office, London.

Auerbach, C. 1951. Problems in chemical mutagenesis. *Cold Spring Harbor Symposia on Quantitative Biology* 16: 199–213.

Beijerinck, M.W. 1900. On different forms of hereditary variation of microbes. *Proceedings of the Section of Sciences, Kon. Akademie van Wetenschappen, Amsterdam*. 3: 352–365. Reprinted in pp. 37–47 *Beijerinck, M.W. Collected Works*, volume 4. Martinus Nijhof Publishers, Delft, 1921.

Beijerinck, M.W. 1910. Variability in *Bacillus prodigiosus*. *Proceedings of the Section of Sciences, Kon. Akademie van Wetenschappen, Amsterdam* 12: 640–649.

Beijerinck, M.W. 1912. Mutation bei Mikroben. *Folia Mikrobiologica* 1: 1–97.

Bonner, D.M. 1951. Gene-enzyme relationships in *Neurospora*. *Cold Spring Harbor Symposia on Quantitative Biology* 16: 143–157.

Braun, W. 1953. *Bacterial Genetics*. Saunders, Philadelphia. 238 pp.

Burnet, F.M. and Lush, D. 1936. Induced lysogenicity and mutation of bacteriophage within lysogenic bacteria. *Australian Journal of Experimental Biology and Medical Science* 14: 27–38.

Burnet, F.M. and McKie, M. 1929. Observations on a permanently lysogenic strain of

B. enteritidis Gaertner. Australian Journal of Experimental Biology and Medical Science 6: 277–284.

Burri, R. 1910. Über scheinbar plötzliche Neuerwerbung eines bestimmten Gärungsvermögens durch Bakterien der Coligruppe. *Zentralblatt für Bakteriologie, II Abt.* 28: 321–345.

Cairns, J., Overbaugh, J., and Miller, S. 1988. The origin of mutants. *Nature* 335: 142–145.

Carlson, E.A. 1981. *Genes, Radiation, and Society.* Cornell University Press, Ithaca.

Cole, L.J. and Wright, W.H. 1916. Application of the pure-line concept to bacteria. *Journal of Infectious Diseases* 19: 209–221.

Curcho, M.G. 1948. Mutation to tryptophan independence in *Eberthella typhosa. Journal of Bacteriology* 56: 374–375.

Davis, B.D. 1948. Isolation of biochemically deficient mutants of bacteria by penicillin. *Journal of the American Chemical Society* 70: 4267.

Davis, B.D. 1950. Studies on nutritionally deficient bacterial mutants isolated by means of penicillin. *Experientia* 6: 41–50.

Davis, B.D. 1989. Transcriptional bias: A non-Lamarckian mechanism for substrate-induced mutations. *Proceedings of the National Academy of Sciences* 86: 5005–5009.

Dean, A.C.R. and Hinshelwood, C. 1953. Observations on bacterial adaptation. *Symposium of the Society for General Microbiology* 3: 21–45.

Delbrück, M. 1945. Spontaneous mutations of bacteria. *Annals of the Missouri Botanical Garden* 32: 223–233.

Demerec, M. 1946. Induced mutations and possible mechanisms of the transmission of heredity in *Escherichia coli. Proceedings of the National Academy of Sciences* 32: 36–46.

Demerec, M. 1948. Origin of bacterial resistance to antibiotics. *Journal of Bacteriology* 56: 63–74.

Demerec, M. and Fano, U. 1945. Bacteriophage-resistant mutants in *Escherichia coli. Genetics* 30: 119–136.

Demerec, M. and Hanson, J. 1951. Mutagenic action of manganous chloride. *Cold Spring Harbor Symposia on Quantitative Biology* 16: 215–228.

Demerec, M. and Latarjet, R. 1946. Mutations in bacteria induced by radiations. *Cold Spring Harbor Symposia on Quantitative Biology* 11: 38–50.

Drake, J.W. 1989. Mechanisms of mutagenesis. *Environmental and Molecular Mutagenesis* 14 (Supplement 16): 11–15.

Dubos, R.J. 1945. *The Bacterial Cell.* Harvard University Press, Cambridge. 460 pp.

Fildes, P. and Whitaker, K. 1948. "Training" or mutation of bacteria. *British Journal of Experimental Pathology* 29: 240–248.

Fildes, P., Gladstone, G.P., and Knight, B.C.J.G. 1933. The nitrogen and vitamin requirements of B. typhosus. *British Journal of Experimental Pathology* 14: 189–196.

Freifelder, D. 1987. *Microbial Genetics.* Jones and Bartlett, Boston.

Gale, E.F. 1947. *The Chemical Activities of Bacteria.* University Tutorial Press, London. 199 pp.

Gildemeister, E. 1917. Weitere Mitteilungen über Variabilitätserscheinungen bei Bakterien die bereits bei ihrer Isolierung aus dem Organismus zu beobachten sind. *Zentralblatt für Bakteriologie, I Abteilung, Original* 79: 49–62.

Gray, C.H. and Tatum, E.L. 1944. X-ray induced growth factor requirements in bacteria. *Proceedings of the National Academy of Sciences* 30: 404–410.

Hadley, P. 1928. The dissociative aspects of bacterial behavior. pp. 84–101 in Jordan, E.O. and I.S. Falk (ed.), *The Newer Knowledge of Bacteriology and Immunology.* University of Chicago Press, Chicago.

Hayes, W. 1964. *The Genetics of Bacteria and Their Viruses.* John Wiley, New York.

Henderson-Smith, J. 1913. On the organisms of the typhoid-colon group and their differentiation. *Zentralblatt für Bakteriologie, I Abt.* 68: 151–165.

Hinshelwood, C.N. 1946. *The Chemical Kinetics of the Bacterial Cell.* The Clarendon Press, Oxford. 284 pp.

Hollaender, A. 1945. The mechanism of radiation effects and the use of radiation for the production of mutations with improved fermentation. *Annals of the Missouri Botanical Garden* 32: 165–259.

Hollaender, A. and Emmons, C.W. 1941. Wavelength dependence of mutation production in the ultraviolet with special emphasis on fungi. *Cold Spring Harbor Symposia on Quantitative Biology* 9: 179–186.

Knight, B.C.J.G. 1936. *Bacterial Nutrition*. His Majesty's Stationery Office, London. 182 pp.

Lea, D.E. and Coulson, C.A. 1949. The distribution of the numbers of mutants in bacterial populations. *Journal of Genetics* 49: 264–285.

Lederberg, J. 1950. Isolation and characterization of biochemical mutants of bacteria. *Methods in Medical Research* 3: 5–22.

Lederberg, J. 1951. *Papers in Microbial Genetics*. University of Wisconsin Press, Madison.

Lederberg, J. 1989. Replica plating and indirect selection of bacterial mutants: isolation of preadaptive mutants in bacteria by sib selection. *Genetics* 121: 395–399.

Lederberg, J. and Lederberg, E.M. 1952. Replica plating and indirect selection of bacterial mutants. *Journal of Bacteriology* 63: 399–406.

Lederberg, J. and Zinder, N.D. 1948. Concentration of biochemical mutants of bacteria with penicillin. *Journal of the American Chemical Society* 70: 4267–4268.

Lewis, I.M. 1934. Bacterial variation with special reference to behavior of some mutabile strains of colon bacteria in synthetic media. *Journal of Bacteriology* 28: 619–639.

Luria, S.E. 1939. Sur l'unité lytique du bactériophage. *Comptes rendus des séances de la Société de biologie* 130: 904–907.

Luria, S.E. 1946. Spontaneous bacterial mutations to resistance to antibacterial agents. *Cold Spring Harbor Symposia on Quantitative Biology* 11: 130–137.

Luria, S.E. 1947. Recent advances in bacterial genetics. *Bacteriological Reviews* 11: 1–40.

Luria, S.E. 1984. *A Slot Machine, A Broken Test Tube: An Autobiography*. Harper and Row, New York.

Luria, S.E. and Delbrück, M. 1943. Mutations of bacteria from virus sensitivity to virus resistance. *Genetics.* 28: 491–511.

Massini, R. 1907. Über einen in biologischer Beziehung interessanten Kolistamm (*Bacterium coli mutabile*). Ein Beitrag zur Variation bei Bakterien. *Archiv für Hygiene* 61: 250–292.

Monod, J. and Audureau, A. 1946. Mutation et adaptation enzymatique chez *Escherichia coli-mutabile*. *Annales de l'Institut Pasteur* 72: 868–878.

Neisser, I.M. 1906. Ein Fall von Mutation nach de Vries bei Bakterien und andere Demonstrationen. *Zentralblatt für Bakteriologie, I. Abt. (Referate)* 38: 98–102.

Newcombe, H.B. 1949. Origin of bacterial variants. *Nature* 164: 150.

Novick, A. and Szilard, L. 1951. Experiments on spontaneous and chemical induced mutations of bacteria growing in the chemostat. *Cold Spring Harbor Symposia on Quantitative Biology* 16: 337–343.

Oakberg, E.F. and Luria, S.E. 1947. Mutations to sulfonamide resistance in *Staphylococcus aureus*. Genetics 32: 249–261.

Roepke, R.R., Libby, R.L., and Small, M.H. 1944. Mutation or variation of *Escherichia coli* with respect to growth requirements. *Journal of Bacteriology* 48: 401–412.

Ryan, F.J. and Lederberg, J. 1946. Reverse-mutation and adaptation in leucine-less *Neurospora*. *Proceedings of the National Academy of Sciences* 32: 163–173.

Simon, M.I. and Silverman, M. 1983. Recombinational regulation of gene expression in bacteria. pp. 211–227 in Beckwith, J., Davies, J., and Gallant, J.A. (ed.), *Gene Function in Prokaryotes*. Cold Spring Harbor Laboratory, Cold Spring Harbor, New York.

Stahl, F.W. 1988. A unicorn in the garden. *Nature* 335: 112–113.

Stomps, T.J. 1954. On the rediscovery of Mendel's work by Hugo de Vries. *Journal of Heredity* 45: 293–294.

Tatum, E.L. 1945. X-ray induced mutant strains of *Escherichia coli*. *Proceedings of the National Academy of Sciences* 31: 215–219.

Tatum, E.L. 1946. Induced biochemical mutations in bacteria. *Cold Spring Harbor Symposia on Quantitative Biology* 11: 278–284.

van Iterson, G., den Dooren de Jong, L., and Kluyver, A.J. 1983. *Martinus Beijerinck. His Life and Work*. Science Tech Publishers, Madison, Wisconsin.

Witkin, E.M. 1946. Inherited differences in sensitivity to radiation in *Escherichia coli*. *Proceedings of the National Academy of Sciences* 32: 59–68.

Witkin, E.M. 1951. Nuclear segregation and the delayed appearance of induced mutants in *Escherichia coli*. *Cold Spring Harbor Symposia on Quantitative Biology* 16: 357–372.

Yang, Y.N. and White, F.B. 1934. Rough variation in *V. cholerae* and its relation to resistance to cholera-phage (type A). *Journal of Pathology and Bacteriology* 38: 187–200.

Yudkin, J. 1953. Origin of acquired drug resistance in bacteria. *Nature* 171: 541–546.

5
MATING

Just before the discovery of mating in bacteria, one of the leaders in the field of biochemical genetics wrote the following words:

> The genetic definition of a gene implies sexual reproduction. It is only through segregation and recombination of genes during meiosis and fusion of gametes that the gene exhibits its unitary property. In bacteria, for example, in which cell reproduction is vegetative, there are presumably units functionally homologous with the genes of higher organisms, but there is no means by which these can be identified by the techniques of classical genetics (Beadle, 1945).

Within a year, the crossing techniques that Beadle had found lacking would be available, and exploration of the field of bacterial genetics could seriously begin.

Because of the widespread existence of cytoplasmic inheritance in algae, protozoa, and higher plants (Sapp, 1987), a common hypothesis at the time was that the genes of bacteria were more analogous to the extrachromosomal factors of higher organisms than to nuclear genes. Prior to 1946, sexual phenomena had been sought in bacteria but had not been reproducibly found. Despite numerous papers purporting to describe life cycles in bacteria (for reviews, see Löhnis, 1921; Bisset, 1950; Hutchinson and Stempen, 1954), no bacterial system was available in which genetic analysis could be carried out. Unsuccessful attempts to cross bacteria had been made by Sherman and Wing (1937) and Gowen and Lincoln (1942). The main reason these early attempts were unsuccessful was because the genetic markers used (primarily fermentative or antigenic) did not permit the detection of rare recombinants.

5.1 THE *NEUROSPORA* BACKGROUND

Research on the genetics of fungi, especially *Neurospora*, provided an important link between classical and bacterial genetics. Life cycles in fungi were well known, and the important idea of homothallic and heterothallic mating systems, first arising from the early work of the geneticist Blakeslee (1904), was well established. An important area of research in the 1930s involved studies on the nature of gene action, best characterized as physiological or biochemical genetics. One of the most ardent proponents of biochemical genetics was George W. Beadle, who had begun work on the genetic control of eye pigment in *Drosophila* but turned to *Neurospora* as a more favorable experimental organism (Beadle, 1945; Kay, 1989).

Beadle began work on biochemical genetics as a junior staff member in Morgan's department at CalTech. Boris Ephrussi from Paris was at CalTech in 1933 as a visiting scientist, and he and Beadle began to work together. They were interested in the problem of how genes affected eye color in *Drosophila* (Beadle, 1963). The concept of a biochemical link to genetics had been suggested many years ago by Garrod (see Section 2.6) but had not been followed up by geneticists. (Beadle was to subsequently reintroduce Garrod's concept of inborn errors of metabolism, with far-reaching consequences for human genetics.) Carrying out complicated grafting experiments, first at CalTech and subsequently in Paris, Beadle and Ephrussi obtained evidence that a metabolic sequence, each step controlled by a separate gene, led from an initial precursor to the final eye pigment. Returning from Paris, Beadle joined Stanford as a junior faculty member, where he and the bacteriologist/biochemist Edward L. Tatum tried unsuccessfully to identify the eye-color substances. Because of the difficulty of doing biochemical work on an organism as complex as *Drosophila*, Beadle and Tatum turned to a simpler organism, the bread mold *Neurospora crassa*.

Although in the 1930s nothing was known about the chemistry of the gene, a common idea was that genes acted by controlling specific chemical reactions in cells. Beadle and Tatum reasoned that if a one-to-one relation existed between genes and specific reactions, it should be possible to select mutants concerned with particular reactions. Although Beadle provided the important link between classical genetics and *Neurospora*, it was Tatum who provided the biochemical background (Tatum, 1959; Lederberg, 1979, 1990). Tatum had received his Ph.D. at the University of Wisconsin working on the nutritional requirements of propionic-acid bacteria. After a postdoctoral period in The Netherlands, he accepted a position, in 1937, as a biochemist in Beadle's department at Stanford University. Tatum was familiar with the fungi from his contacts with the mycologist Nils Fries in The Netherlands and was aware that a number of them had been cultivated on synthetic media. By the late 1930s, the nutrition of organisms was beginning to be understood in detail (see Section 3.8), and some synthetic vitamins and amino acids were available commercially. Beadle was already familiar with genetic studies on *Neurospora* because the distinguished American mycologist B.O. Dodge, the discoverer of the *Neurospora* sexual phase, had introduced the system to Morgan and his group at CalTech when Beadle was there. Also, at Morgan's prompting, crossing techniques for *Neurospora* had been worked out by graduate student Carl C. Lindegren.

Beadle and Tatum began their *Neurospora* work in the spring of 1941 and quickly made important discoveries. *Neurospora* was a heterothallic ascomycete fungus, the two mating types being morphologically and physiologically identical but self-incompatible. Because its life cycle and mating system were under experimental control, it could be readily hybridized. One of the advantages of *Neurospora* was that all products of meiosis were available for genetic analysis. Meiosis occurs in such a way that the spindle is aligned with the length of the axis of the ascus. The four strands (tetrads) of meiosis are thus arranged so that two nuclei in one half of the ascus contain the chromosomes derived from one of the two homologs, and the other two nuclei contain the chromosomes from the other homolog. After a mitotic division, eight spores are formed, arranged in groups of four. If the two homologs differ by a single gene, the four ascospores at one end are of one genetic constitution and the other four are of the opposite constitution. The ascospores are thus seen in an ordered sequence that reflects the events that occurred during meiosis. The ascospores that were the products of meiosis were removed from the ascus with a micromanipulator and allowed to germinate, and the resulting cultures were characterized phenotypically. Thus, genetic mapping was easy and direct. The nutrition of *Neurospora* was simple: It grew on a synthetic medium with a sole organic carbon source, requiring only the single vitamin biotin as a growth factor.

By modern standards, the mutant isolation procedure was cumbersome. The asexual spores and hyphae were multinucleate, so one could not isolate mutants (usually assumed to be recessive) except via the uninucleate sexual spores. Thus, *Neurospora* asexual spores were exposed to X rays or ultraviolet radiation to induce mutations, the irradiated spores were crossed with strains of the opposite mating type, and single ascospore strains were established on complete medium (containing all likely growth factors). Cumbersome and time-consuming micromanipulation was therefore necessary to prepare cultures to be tested for mutants. The strains obtained were then characterized nutritionally. Growth-factor-dependent strains could be analyzed genetically through additional crosses, and their biochemistry could be studied. Because all eight ascospores from a single meiotic event could be obtained, detailed analysis of crossing-over and linkage could be made.

Strains of *Neurospora* differing from the original wild type were obtained that failed to grow in the absence of various vitamins: thiamine, pyridoxine, *p*-aminobenzoic acid, pantothenic acid, inositol, nicotinic acid, and choline. Other strains were obtained that required one of the amino acids arginine, lysine, leucine, valine, methionine, tryptophan, proline, and threonine. Genetic analysis showed that each mutant differed from the wild type by a single gene. By studying the biochemical pathways of certain of these strains, Beadle, Tatum, and their co-workers (most noteworthy, David Bonner and Norman Horowitz) showed that the mutants lacked the ability to carry out particular reaction steps. On the other hand, mutations of single *nonallelic* genes that gave rise to the same nutritional requirements represented genetic blocks of *different* biosynthetic reactions. In addition, mutant strains grown on an exogenous supply of end product would *accumulate* the intermediate immediately *prior* to the genetic block, provided this intermediate was not otherwise metabolized. Thus, genotypically different mutants provided a means of accumulating different intermediates (Bonner, 1946). This made possible the detailed study of biochemical pathways, an extremely important development in the early post-World-War-II era. Enzyme analyses showed, in some cases, that particular enzymes were lacking in the mutants. These results led to the *one-gene/one-enzyme hypothesis* (Beadle, 1945). However, the hypothesis did not state in so many words that the gene contained *all* the information for the enzyme. Indeed, it was considered that the gene was also protein (or nucleoprotein), and one model was that the gene imposed, directly or indirectly, a specific configuration on the enzyme (Bonner, 1946).

According to Beadle (1945), the gene was viewed as the ''master molecule or template in directing the final conformation of the protein molecule as it is put together from its component parts.'' This idea was in keeping with Linus Pauling's instructive theory of antibody formation, which hypothesized that the antigen controlled the *folding* of the protein polypeptide, but not its amino acid sequence (see Section 10.14).

It is difficult now to appreciate the great significance of the one-gene/one-enzyme hypothesis. At the time the hypothesis was advanced, almost nothing was known about the chemical nature of genes. The crucial role of *amino acid sequence* for the folding and function of proteins was barely suspected. For instance, in the discussion of David Bonner's paper on the "one-to-one" hypothesis at the 1946 Cold Spring Harbor Symposium on Quantitative Biology, Max Delbrück presented a rather critical attack on the whole idea, using the philosophical views of Karl Popper on the nature of scientific proof. Delbrück insisted that the experiments of Beadle and Tatum were designed to show the role of genes in enzyme specificity and hence would be unlikely to find evidence *against* the hypothesis. Delbrück stated: "...in order to make a fair appraisal of the present status of the thesis of a one-to-one correlation between genes and species of enzymes, it is necessary to begin with a discussion of methods by which the thesis could be *disproved*. If such methods are not readily available, then the mass of 'compatible' evidence carries no weight whatsoever..." (discussion in Bonner, 1946).

A key point of the one-gene/one-enzyme hypothesis, however, was that it focused immediately on a type of research that was very "do-able": how enzyme proteins are affected by mutation. Already at this time, temperature-sensitive and pH-sensitive mutations were known and recognized to provide significant information. Bonner (1946) noted that such mutations might give important information about how the gene exerts its control over enzyme properties. Despite such insight, it would be more than 15 years before the genetics community would turn to temperature conditional mutations as a tool for deciphering gene action. In 1946, geneticists did not actually believe that a study of enzymes would be all that revealing. Certainly Delbrück himself did little to encourage the study of enzymes (see Chapters 6 and 7). According to a personal communication from B.D. Davis (August 1989), who attended the 1946 Cold Spring Harbor symposium: "After the [meeting], Beadle said he, Norm Horowitz, and I were the only ones still believing in the core of truth." It is understandable, however, that geneticists in particular might have accepted the one-gene/one-enzyme hypothesis slowly. At this time, the structure of enzymes was poorly understood. Even more serious, single mutations in regulatory genes can simultaneously abolish synthesis of several enzymes, making interpretation in 1946 extremely difficult.

One of the most significant outcomes of the *Neurospora* work for future developments in bacterial genetics was the recognition that nutritional mutants were *conditional* lethals (Tatum and Beadle, 1945). In the absence of the required growth factor, the mutant could not grow, but it could be maintained in culture indefinitely as long as the growth factor was present. Thus, the mutant could be readily selected against, an important consideration in the work of Lederberg and Tatum on mating in bacteria (to be discussed below). Of importance to the enzymologist, the inactivation of specific genes was equivalent to the chemical poisoning of specific enzymes, except that the genes were highly specific, whereas enzyme poisons were often rather non-specific.

Gene *complementation*, another discovery from the *Neurospora* work that became of relevance to bacteria later (leading ultimately to Seymour Benzer's *cis-trans* test and the whole *operon* concept; see Chapters 6 and 10), was first seen when two mutant genes were put together in the same cell. In *Neurospora* hyphae, the cells are always multinucleate, and when crosses are made, the mingling of dissimilar nuclei results in the formation of heterocaryons that can persist for variable lengths of time before diploid formation occurs. Beadle and Coonradt (1944) showed that gene complementation could take place within heterocaryons, and concluded that they were similar to diploid heterozygotes. Thus, even though a persistent diploid stage did not exist in *Neurospora*, it was possible to characterize mutants as "dominant" and "recessive." Heterocaryons provided the first real approach in a microorganism for distinguishing the gene as a physiological unit from the gene as a unit of crossing-over or mutation.

Even during World War II, the Beadle/Tatum laboratory at Stanford attracted considerable attention (see Kay [1989] for an interesting account of how Beadle managed to obtain research funds for what was essentially basic research). Among the visitors was Francis J. Ryan of Columbia University, who spent a sabbatical year there. Ryan did some early *Neurospora* work with Beadle and Tatum and then brought the system back with him to Columbia. Ryan is most noteworthy in the present context as the teacher of Joshua Lederberg, but he made major contributions to microbial genetics at many levels for a number of years before his early death (Atwood, Schneider, and Ryan, 1951; Sager and Ryan, 1961).

5.2 TATUM'S WORK ON *ESCHERICHIA COLI*

In addition to the *Neurospora* work that he did together with Beadle, Tatum independently carried out mutational studies on bacteria. The interest here was to extend the *Neurospora* work to another organism, and Tatum's bacteriological background naturally drew him to *Escherichia coli*. During 1944 and 1945, Tatum carried out a series of studies on the isolation of growth-factor-requiring mutants of *Escherichia coli* and other bacteria (Gray and Tatum, 1944; Tatum, 1945). Fortuitously, the strain of *Escherichia coli* used was the fertile K-12, a strain that had been used in the student course at Stanford for a number of years. (A brief description of the origin of *Escherichia coli* K-12 is given in Lederberg 1950d.) As Lederberg would subsequently show, most strains of *Escherichia coli* isolated from nature were unable to mate.

In his second *Escherichia coli* paper, Tatum (1945) reported on the first double mutants (Table 5.1). The fact that two or more biochemically distinct requirements were obtained in the same mutant was considered significant in terms of bacterial heredity. Since crossing was not yet possible, these double mutants provided the only evidence for heredity in bacteria. "This is presumptive evidence for the existence of genes in bacteria..." (Tatum, 1945). However, the most significant thing about these mutants is that they were to figure prominently in Lederberg's first mating studies (see below). Although Tatum did not know it, during the isolation of strain 679-680 (see Table 5.1), the F plasmid present in K-12 was lost, an important consequence for the later discovery of sex compatibility in *Escherichia coli* (see Section 5.6).

Table 5.1   Isolation of Double Mutant
Strains of *Escherichia coli*

Strain number	Substances required
58–161	methionine, biotin
58–278	phenylalanine, biotin
58–309	cystine, biotin
58–336	isoleucine, biotin
58–580	thiamine, biotin
58–593	thiamine, biotin
58–610	thiamine, biotin
58–741	histidine, biotin
58–2651	proline, biotin
679–183	proline, threonine
679–440	proline, threonine
679–662	glutamic acid, threonine
679–680	leucine, threonine

The double mutants were isolated from strain 58, which required biotin, and strain 679, which required threonine. It was later found (Cavalli, Lederberg, and Lederberg, 1953) that the original parent, strain K-12, contained an F factor that was lost during the isolation of mutant 679–680. (Data from Tatum, 1945.)

5.3 LEDERBERG'S EARLY WORK

Some of the background to the discovery of mating in *Escherichia coli* has been given by Lederberg (1986, 1987), and Zuckerman and Lederberg (1986). As noted above, Francis Ryan had worked with Beadle and Tatum and had brought the *Neurospora* system back to Columbia University. As an undergraduate student at Columbia University, Lederberg worked in Ryan's laboratory, first as a dishwasher, later doing actual research. Lederberg received his Bachelor of Arts degree in 1944, during the middle of World War II, and entered a U.S. Navy medical school program. Attending medical school at Columbia University (College of Physicians and Surgeons), he found time to continue research in Ryan's laboratory. The announcement of the discovery in 1944 that the transforming principle of pneumococcus was DNA (see Chapter 9) piqued Lederberg's interest, and he attempted a similar transformation by DNA in *Neurospora*. Using a *leucineless* mutant of *Neurospora*, Lederberg attempted to bring about transformation. However, he quickly discovered that the *leucineless* mutant spontaneously reverted to the wild-type condition. Ryan had previously noted that this strain, when cultured on minimal medium, would gradually initiate growth, a phenomenon he then called *adaptation*. Lederberg showed that the acquisition of the ability to grow on leucine was due to a reverse mutation, and genetic crosses showed that the revertant was allelic with the mutant. Lederberg and Ryan called these revertants *prototrophs*, a term that was to become widely used in bacterial genetics. This *Neurospora* work resulted in Lederberg's first scientific publication (Ryan and Lederberg, 1946). However, the main significance of this work, for subsequent developments, was that it introduced the *prototrophic recovery technique*, a procedure that was to be so important in the discovery of mating in bacteria.

In 1945, Lederberg decided to try to develop a genetic system in a bacterium. He conceived of the use of nutritional mutants as a means of searching for mating in bacteria. With the prototrophic recovery approach in mind, he devised a possible experiment. "The basic protocol...entailed the use of a pair of nutritional mutants, say A^+B^- and A^-B^+. If crossing occurred, one could plate out billions of cells in a selective medium...one should be able to find even a single A^+B^+ recombinant. This experimental design was encouraged by Beadle and Coonradt's report of nutritional symbiosis in *Neurospora* heterocaryons" (Lederberg, 1987).

Lederberg therefore isolated nutritional mutants of *Escherichia coli* and carried out appropriate crosses. A *methionineless* strain was selected by serial passage through basal medium containing sulfanilamide and methionine, a procedure that had been reported by others to permit isolation of such a mutant. A *prolineless* strain was isolated as a spontaneous mutant. The early crossing experiments, done with a strain of *Escherichia coli*, 6522, that was not fertile, were unsuccessful.[1]

During the summer of 1945, Tatum moved from Stanford to Yale University, where he was to set up a new program in microbiology in the Botany Department. Ryan suggested that Lederberg write to Tatum and request his mutant strains and also a possible collaboration. As noted earlier (see above), Tatum had isolated *double mutants* of *Escherichia coli*. Such double mutants were of special value since their reversion to prototrophy should be extremely low. Another value perceived by Lederberg for obtaining Tatum's strains was the following: In the event that mating types or sterility factors were present in *Escherichia coli*, mating might not occur. Self-incompatibility was common in fungi, and one of the best ways of obtaining distinct mating types was to use independent isolates. This idea was expressed in Lederberg's first letter to Tatum: "It should...be advantageous to use stocks of heterogeneous origin in the event that there exist mating types, sterility factors, etc."[2] This idea, of course, proved to be true. However, Lederberg had no way of knowing that Tatum's *Escherichia coli* K-12 would be one of the rare strains containing an F plasmid and that this plasmid had been lost in the threonine⁻ leucine⁻ double mutant (see Table 5.1).

Lederberg's application to Tatum was accepted.[3] Taking a leave from Columbia medical school, he moved to New Haven in late March 1946 and immediately began the recombination experiments. Initially, there was no intent that Lederberg should be studying for a Ph.D. degree (this was awarded after the fact; Lederberg, 1947a). Using Tatum's double mutants in K-12, Lederberg carried out initial recombination experiments. Fortunately, the genes in the double mutants Tatum had isolated, biotin-methionine (B^-M^-) and threonine-proline (T^-P^-) were linked closely enough so that their simultaneous use in prototroph recovery experiments did not un-

---

[1]I have in my possession a copy of a data book page sent to me by Lederberg dated 18 September 1945. Crosses were made between proline and methionine mutants, on minimal medium. No colonies were obtained in several replicates after incubation for 24 and 48 hours. Controls for cross-feeding (syntrophism) and reversion were also negative. A day after this negative experiment, Lederberg wrote Tatum proposing a collaboration.

[2]Personal communication from J. Lederberg of a letter he sent to E.L. Tatum 19 September 1945. This sentence was deleted from the version of the letter published in Lederberg (1987).

[3]Tatum had probably conceived independently of the use of nutritional mutants in a search for bacterial mating (Lederberg, 1986).

necessarily complicate genetic analysis. In addition, mutants resistant to bacteriophage T1 were isolated in some of the mutant stocks to provide an additional (unselected) genetic marker. According to Lederberg (1987), it took about six weeks from the time the first crossing experiments were set up in mid-April 1946 until well-controlled, positive results were obtained.

The first published results of these experiments were presented at the July 1946 symposium at Cold Spring Harbor (Lederberg and Tatum, 1946a). The data (Table 5.2) showed that only when two mutants were mixed were prototrophs obtained: When single strains were plated, only the parental type was obtained. Even more significant, when prototrophs (recombinants) were tested for T1 resistance (the unselected marker), it was found that some were sensitive, others were resistant, but the frequency depended on which parent carried the resistant marker. Cautiously, Lederberg and Tatum made the following conclusions from these results (see also, Lederberg and Tatum, 1946b):

1. The bacteria were haploid, as shown by the fact that T1 resistance and sensitivity segregated in the selected recombinants and that the ratios were reversed in the reciprocal crosses.
2. A diploid stage did not persist, since the recombinant colonies were composed of individuals of only the same genotype.
3. Linkage relationships existed, since T1 resistance tended to be more closely linked with T^+L^+ than with M^+B^+.

Subsequent studies (discussed below) led to the development of the first linkage map of *Escherichia coli* within a year of the initial discovery of mating.

Reviewing these results a year later, Luria (1947) marveled at the "brilliant technique" used. Viewed from our present perspective, it might be asked why this simple experiment was considered so brilliant:

1. It represented the first use of conditional mutants to select against the parental types. The use of nutritional requirements not only as genetic markers, but also as *selective* agents was completely new.
2. The mutants used were *double* mutants, so that reversion artifacts were avoided.
3. The use of an *unselected marker* (T1 resistance) permitted a separation of physiological from genetic phenomena.
4. The prototrophic recovery technique had enormous sensitivity, so that even rare events could be studied. Although this high sensitivity was used in this case because the frequency of the mating process was low, eventually this same sensitivity would be used by Seymour Benzer, Milislav Demerec, Francois Jacob, and others to study the fine structure of the gene (see Chapters 6, 8, and 10). Lederberg's research pointed the way.

According to Lederberg (1987), the discussion following his paper at the Cold Spring Harbor symposium was critical and lively.[4] André Lwoff, who had done extensive work on nutrition and cross-feeding (syntrophism) in

---

[4]According to a personal communication from Aaron Novick, it was Max Delbrück in particular whose questions were especially troublesome, but Lederberg effectively countered all of Delbrück's arguments.

Table 5.2 Nutritional Types Isolated from Single and Mixed Cultures

From single and mixed	From mixed only	From single and mixed
A. $B^-M^-P^+T^+$ .... and .... $B^+M^+P^-T^-$		$B^+M^+P^+T^-$ $B^+M^-P^-T^+$
$B^+M^-P^+T^+$ $B^-M^+P^+T^+$	$B^+M^+P^+T^+$*	
B. $B^-M^-P^+T^+R$ .... and .... $B^+M^+P^-T^-$		**
**	$B^+M^+P^+T^+R$* $B^+M^+P^+T^+$*	
C. $B^-M^-P^+T^+$ .... and .... $B^+M^+P^-T^-R$		**
**	$B^+M^+P^+T^+R$* $B^+M^+P^+T^+$*	
D. $B^-M^-P^+T^+R$ .... and .... $B^+M^+P^-T^-R$		**
**	$B^+M^+P^+T^+R$*	
E. $B^-\Phi^-C^-P^+T^+$ .... and .... $B^+\Phi^+C^+P^-T^-$		$B^+\Phi^+C^+P^-T^+$ $B^+\Phi^+C^+P^+T^-$
$B^-\Phi^-C^+P^+T^+$ $B^+\Phi^+C^-P^+T^+$ $B^+\Phi^-C^-P^+T^+$	$B^+\Phi^+C^+P^+T^+$** $B^+\Phi^-C^+P^+T^+$ $B^-\Phi^+C^+P^+T^+$	

* Prototroph.
** See A for biochemical variations.
The letters refer to requirements for essential metabolites as follows:
  B = biotin   M = methionine
  Φ = phenylalanine   P = proline
  C = cystine   T = threonine
  R = Resistance to virus T1.

(Reprinted, with permission, from Lederberg and Tatum, 1946.)

Table 5.3 Relative Proportions of Various Nutritional Cell Types in a Mixed Culture

TYPE	NUMBER OF THIS TYPE ISOLATED*	NUMBER OF PROTOTROPHS	RATIO OF THIS TYPE TO PROTOTROPHS	REMARKS
$B-\phi-C-T+L+B_1+V_1{}^s$	(Parental type. Present in large excess)			
$B+\phi+C+T-L-B_1-V_1{}^r$	(Parental type. Present in large excess)			
$B+\phi+C+T+L+B_1+$	86		1.00	Prototrophs
$B+\phi+C+T+L+B_1-$	36	37	0.97	Thiamineless
$B+\phi+C+T-L+B_1+$	2	31	0.06	Threonineless
$B+\phi+C+T+L-B_1+$	4	55	0.07	Leucineless
$B-\phi+C+T+L+B_1+$	5	56	0.09	Biotinless
$B+\phi-C+T+L+B_1+$	1	52	0.02	Phenylalanineless
$B+\phi+C-T+L+B_1+$	1	19	0.05	Cystineless
$B+\phi+C+T+L-B_1-$	3	16	0.19	Possible single-reversion type
$B-\phi-C+T+L+B_1+$	2	41	0.05	Possible single-reversion type
$B-\phi+C+T+L+B_1-$†	3	28	0.11	
$B-\phi+C+T-L+B_1+$†	(Isolated in a similar experiment)			
$B-\phi+C+T+L-B_1+$†	(Isolated in a similar experiment)			

* These figures do not include results of tests of virus resistance. Of 49 prototrophs tested, 20 (41%) were resistant. Seven out of 20 thiamineless (35%) were resistant.

† It should be noted that these types represent double-requirement recombination types.

(Reprinted, with permission, from Tatum and Lederberg, 1947.)

bacteria (see Section 7.4), was especially concerned that cell aggregation of the two phenotypes could explain the results. However, Lederberg had avoided cross-feeding by carefully purifying and repurifying the prototrophic recombinants. Furthermore, cultures of a recombinant prototroph were irradiated with ultraviolet radiation at such a dosage that the number of colonies was reduced by 10^{-5}. It would be assumed that at this level of killing, any purported aggregates would only have single viable cells remaining, and hence the possibility of cross-feeding would be reduced. Yet, several hundred colonies derived from ultraviolet radiation were tested, and all remained prototrophic. In addition, the segregation of the unselected T1 resistance marker argued against cross-feeding. Many of these controls were mentioned in the initial Cold Spring Harbor paper but were presented in considerable detail in the first complete paper on this subject (Table 5.3) (Tatum and Lederberg, 1947).

Other explanations for the results were more difficult to deal with. Transformation by DNA released spontaneously from parent cells was initially ruled out by showing that prototrophs could not be obtained when cell filtrates were used (Lederberg and Tatum, 1946b). Later, Lederberg obtained some purified DNase from Maclyn McCarty and showed that the enzyme, which destroyed the pneumococcus transforming principle (see Section 9.7), had no effect on the number of prototrophs obtained.[5] Attempts to discern

[5] Transformation in *Escherichia coli* had been reported the same year but could not subsequently be repeated (see Boivin and Vendreley, 1946 and Section 9.9).

the putative zygote microscopically were unsuccessful, but because of the low frequency of recombination, mating pairs would not be expected.

Interest in the discovery of bacterial mating was widespread and Lederberg received numerous requests for cultures (Lederberg, 1987). Within a year, the phenomenon had been confirmed in other laboratories, and research on the process was actively under way. Publications from Lederberg and Tatum in 1947 presented experimental details and further controls and experiments (Lederberg, 1947b; Tatum and Lederberg, 1947).

The unlikelihood of reversion was emphasized in numerous papers (Lederberg and Tatum, 1953) (Fig. 5.1). The mating procedure was carried out by mixing high densities of washed cells and plating on minimal agar medium to which various supplements were added. The procedure required, therefore, that the mating take place on the agar plate. This precluded the study of the kinetics of the mating process or of any details of cytological interactions. Lederberg's interest was, primarily, in the use of the procedure for genetics, and he soon presented the first genetic map of *Escherichia coli* (Fig. 5.2) (Lederberg, 1947b).

> In constructing a map, and calculating distances, it has been taken for granted that there is in *E. coli* a system of linear linkage, such as has been demonstrated quite conclusively in *Drosophila*...What direct evidence may one bring to bear on this question?
>
> The method which one is forced to employ in hybridizing this bacterium introduces certain complications. The classical proof of linearity is based on the additive character of distances...between loci occurring within the same linkage group. The determination of map distances is based upon a comparison between parental and new combinations of linked genes, as determined in the progeny of zygotes selected at random. In *E. coli*, on the other hand, one is limited to the recovery of that recombination class in which there has necessarily been an interchange between certain biochemical loci...For this reason, it is not possible to obtain a direct measure of the absolute distance between factors which are located within this critical region, and any argument in favor of linearity which is based

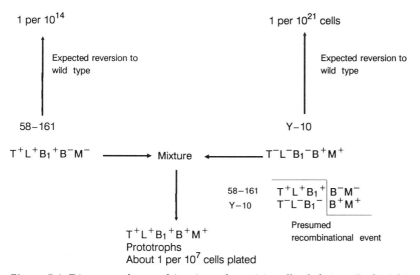

Figure 5.1 Diagram of recombination of nutritionally deficient *Escherichia coli* K-12. The reversion rates for the parental strains are calculated from the reversion rates for individual characters ($\sim 1 \times 10^7$ cells). (Based on Lederberg and Tatum, 1953.)

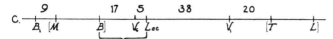

Figure 5.2 The first linkage map for *Escherichia coli* K12. The map distances given are remarkably close to those established by more modern techniques. (Reprinted, with permission, from Lederberg, 1947b.)

on the segregations of such factors may have the flavor of circular reasoning. It would be preferable to study the segregations of factors which are assigned to loci distal to the biochemical factors whose recombination is the basis of the detection of sexual offspring. The stocks with which this might be accomplished are not yet available, but it is hoped that they will be for future work.

That there does exist some sort of linkage system is made highly credible by the results of the "reverse crosses"...(Lederberg, 1947b).

In this 1947 paper, Lederberg interpreted the mating system as one in which *Escherichia coli* was haploid, with a single chromosome. Because all of the stocks had been isolated using a single strain, it was assumed, following the fungal analogy, that *Escherichia coli* was homothallic. "The evidence suggests that this bacterium has a life-cycle comparable to *Zygosaccharomyces*: the vegetative cells are haploid (but not necessarily uninucleate); fertilization is homothallic or unrestricted genetically; the putative zygote undergoes immediate reduction without any intervening mitosis. The number of gametophytes issuing from a single zygote is not definitely known" (Lederberg, 1948). By analogy with the mating process in higher organisms, it was assumed that mating resulted in the formation of a diploid by two single cells and that meiosis occurred without cell division, leading to the formation of haploid segregants.

When further stocks became available, and the genetic map was extended, the interpretations became progressively more complex. It was not until the discovery of F^+ and Hfr strains, and the study of the kinetics of the mating process, that the organization of the *Escherichia coli* map could be constructed and classical zygote formation could be eliminated (see Section 5.11).

According to a personal communication (February, 1989) from Lederberg, his model of K-12 mating was influenced by his knowledge of the conjugation system of the protozoan *Paramecium*, which by this time had been well established by Tracy Sonneborn (see Sonneborn, 1937; Sapp, 1987). In *Paramecium*, diploid cells of opposite mating type come together, undergo meiosis, and exchange haploid nuclei across a cytoplasmic bridge. The diploid state is restored by nuclear fusion. There is no cell fusion, and each of the conjugating cells retains its cellular identity throughout the mating process. "I always used the term conjugation, and *Paramecium* was the model I had in mind for interpreting the process: Sonneborn's work."[6]

However, the *Paramecium* conjugation work soon became entwined in a complex system of presumed cytoplasmic inheritance (Sonneborn, 1943), so that the simple conjugation analogy for K-12 got lost. The references in Lederberg's papers to Sonneborn's work are to the phenomenon of cytoplasmic inheritance of the *kappa* factor, rather than to the conventional mating system (see also Hayes, 1953a, for an attempt to analogize the F factor to *kappa*).[7]

[6] Personal communication from J. Lederberg, February, 1989.
[7] "Having brought bacteria into the fold of genetics, Lederberg had an overwhelming desire to maximize the unity." B.D. Davis, personal communication, August, 1989.

5.4 THE PRE-HAYES ERA OF BACTERIAL GENETICS

It is convenient to divide research on bacterial genetics into two phases: pre-Hayes and post-Hayes. In the pre-Hayes period, mating in bacteria was envisioned as a conventional sex process, perhaps modified by aspects of "relative sexuality," but nevertheless a standard haploid/diploid/meiosis mechanism. After Hayes, it was known that bacteria were *not* just small cells, but constituted a completely different kind of cell. Although the terms *procaryote* and *eucaryote* were apparently not introduced into the English language until 1962 (Stanier and van Niel, 1962), it gradually became clear during the 1950s that something was different about bacteria.

Over the five years between 1947, when the mating process was first accepted, and 1952, when Hayes's first work was published, a number of important advances were made, despite the fact that the overall mechanism of mating remained a mystery. Since the efficiency of mating was extremely low, Lederberg decided that it would be difficult to study the process at the cellular or physiological level. He therefore concentrated on formal genetic studies. He developed new efficient methods for isolating nutritional mutants, including the powerful penicillin selection method and the replica-plating method discussed in Chapter 4 (Lederberg and Zinder, 1948; Lederberg, 1950a; Lederberg and Lederberg, 1952). New genetic markers were isolated, of which the most important were those for β-galactosidase (lactase) (Lederberg, 1950b) (see Section 10.6) and antibiotic resistance (Lederberg, 1950c). The isolation of a streptomycin-resistant mutant was to prove important, since it permitted the development of an efficient procedure for screening strains for ability to mate (Lederberg, 1950c, 1951a) and also led to the important discovery by Hayes concerning polarity in the mating process (see Section 5.5).

During 1950 and 1951, interpretation of the mating process still followed the classical line, and some unusual fermentation-positive recombinants that segregated fermentation-negative clones were interpreted as heterozygous diploids (Lederberg, 1950d). Single-cell isolations with a micromanipulator of presumed diploids were done in an attempt to study the segregation process (Zelle and Lederberg, 1951), but what these studies actually showed is still not clear. Postulates of nondisjunction clearly attempted to relate confusing matings to known phenomena in higher organisms (Lederberg, 1950d). This whole work culminated in a paper that developed a formal interpretation of *Escherichia coli* genetics in terms of a branched four-armed chromosome (Lederberg, Lederberg, Zinder, and Lively, 1951). This paper became famous in the genetics community because of its length and inordinate complexity (see also Section 8.1).

An important control experiment performed during this period was that of Davis (1950), which showed that cell-to-cell contact was essential for the recombination process to occur. As we have discussed earlier, one explanation for the recombination process was some sort of transformation, in which lysing cells would release DNA that would transform other cells. Lederberg had attempted to rule this out by using cell extracts and DNase, but there had always been a doubt that an unstable transforming principle or one resistant to DNase might be involved. To analyze the process, Davis constructed a U-tube with a bacterial-impassible fritted glass filter at the bottom. When two mating strains were inoculated on opposite sides of the filter and the medium was flushed back and forth between the two sides, no recombinants were obtained. This simple control, which gave negative results in the present case, was to prove very useful to Zinder and Lederberg several years later when the process of phage-mediated genetic exchange (transduction) was discovered in *Salmonella typhimurium* (see Section 8.1). One may speculate on how the whole field of bacterial genetics might have developed if Lederberg had discovered transduction *before* discovering mating!

During this same period, Lederberg also carried out work on gene-enzyme relationships in *Escherichia coli*, making important early studies on the enzyme β-galactosidase (see Section 10.6). This work probably started because of the numerous lac^- mutants that Esther Lederberg had isolated (E.Z. Lederberg, 1950).[8] J. Lederberg introduced the use of the synthetic chromogenic substrate *o*-nitro-phenyl-β-D-galactoside for the assay of the enzyme, a substrate that was to prove extremely valuable in studies on enzyme induction.

Other studies that were to prove important in focusing attention on peculiarities of the mating process were those dealing with the prevalence of *Escherichia coli* strains capable of mating. Cavalli and Heslot (1949) showed that one of seven strains of *Escherichia coli* from the British culture collection was able to mate with a Lederberg tester strain.[9]

Using the streptomycin-selection procedure that he had developed (Lederberg, 1950c), Lederberg (1951b) tested a large number of clinical isolates of *Escherichia coli*. Since the clinical isolates were all streptomycin-sensitive but prototrophic, mating with streptomycin-resistant nutritional mutants of K-12 could be tested by plating on minimal medium containing streptomycin (Lederberg, 1951a). In this way, a large number of isolates could be tested without the laborious procedure of first isolating nutritional mutants. From about 140 isolates, 9 were found that were capable of mating with K-12.

5.5 THE WORK OF WILLIAM HAYES

William Hayes was an Irish medical bacteriologist who had been doing clinical bacteriology in London in the early post-World-War-II period. From studies on antigenic phase variation in *Salmonella*, he became interested in bacterial genetics (Hayes, 1966). He decided that before he began *Salmonella* studies he should become acquainted with genetic phenomena in *Escherichia coli*. While attending a course in bacterial chemistry at Cambridge, he met Cavalli-Sforza, who gave him the basic K-12 strains to work with. Hayes decided to study not simply the products, but also the kinetics of the mating process and developed a procedure using streptomycin resistance to follow it (essentially Lederberg's streptomycin-selection procedure; see above). He

---

[8]Esther Lederberg, née Zimmer, was born in New York City in 1922 and received her B.A. from Hunter College in 1942. She worked with Alexander Hollaender at the National Institutes of Health. She received her M.A. from Stanford University in 1946 and then moved to Yale. During summers she also worked on radiation biology at Cold Spring Harbor with Alexander Hollaender and Milislav Demerec. At Yale, she worked with Norman Giles on *Neurospora*. She received her Ph.D. in genetics from the University of Wisconsin, working with J. Lederberg, although her official thesis advisor was the chairman of the Genetics Department, R.A. Brink.

[9]Luca Cavalli (later Cavalli-Sforza) was an Italian population geneticist who at that time was in R.A. Fisher's department at the University of Cambridge. He was later to do considerable collaborative work with Lederberg. He also isolated the first strain of *Escherichia coli* capable of mating at high frequency, which he called Hfr (Cavalli-Sforza, 1950).

Table 5.4 Streptomycin Inhibition of Recombination: Demonstration of the Unidirectional Transfer of Genetic Material

Strain	W677 SM^R	WM677 SM^S	
58–161 SM^R	recombinants	no recombinants	SM present
58–161 SM^S	recombinants*	no recombinants	SM present

58–161: Biotin (B)$^-$, methionine (M)$^-$; later found to be F$^+$. W677; Leucine (L)$^-$, threonine (T)$^-$, thiamine (B1)$^-$; later found to be F$^-$. In this experiment, B$^+$M$^+$L$^+$T$^+$ prototrophic recombinants were selected. The asterisk indicates unexpected recombinants. Streptomycin was present in all combinations at 200 μg/ml. (Based on Hayes, 1952a.)

spread mixtures of a streptomycin-resistant mutant of one parent (strain A) and a streptomycin-sensitive strain of the other parent (strain B) on minimal agar plates and then at intervals added a lethal concentration of streptomycin to the plates. The idea was that no colonies should appear until resistant prototrophs began to arise. The initial experiments worked beautifully: There were no colonies if streptomycin was added prior to two hours after mating, but then their number increased with time. Just to be certain, Hayes carried out the same experiment with the resistant pattern *reversed*, strain A being sensitive and strain B being resistant. The results were astounding: The presence of streptomycin during mating had little effect (Table 5.4). About the same number of recombinant colonies emerged from all the samples, even when streptomycin was added immediately after plating the mixture.

> I discussed these results with [my closest colleague] Denny Mitchison and I think it was he who first suggested that one of the parents, A, might be acting as a gene donor and the other, B, as a recipient. This was confirmed by treating each sensitive parent with streptomycin to a survival of less than 10^{-6} colony-forming cells, and then mating with an untreated suspension of the other. The upshot was that the crosses in which strain B had been treated were invariably sterile while treated A suspensions always generated recombinants although their numbers might be markedly reduced as compared with normal crosses. It followed that parent B was the recipient or "female" whose continued viability was essential for the whole process of recombination and segregation, while the A donor (or "male") cell was dispensable once genetic transfer had been effected. I suggested that the male cell extruded a surface "gamete" which was taken up by the female cell on contact, so that blocking male protein synthesis by streptomycin did little to inhibit its fertility.[10]

Hayes's interpretation was based at least in part on the then accepted facts about the *Paramecium* conjugation system, which involved transient coupling of cells of opposite mating type and transfer of nuclei and cytoplasmic particles via a conjugation bridge (see above and Hayes, 1953a). It was fortunate that streptomycin, unlike penicillin, did not cause lysis; it left the killed cells intact and functional in DNA transmissal.

Hayes began his first work on K-12 mating in July 1950, and the work described above was published as a note in *Nature* in early 1952 (Hayes, 1952a). Another paper, using ultraviolet radiation instead of streptomycin

---

[10]William Hayes, personal communication. The material quoted is from a mini-autobiography prepared for the Royal Society.

and giving essentially the same results, was published later that same year (Hayes, 1952b). Hayes also presented papers on the work at the Society for General Microbiology meeting in London in June 1952 and at an international symposium on microbial genetics held in Pallanza, Italy in September. An attendee at both of these meetings was James D. Watson, who later wrote of the Pallanza meeting:

> Bill's appearance was the sleeper of the three day gathering; before his talk no one except Cavalli-Sforza knew he existed. As soon as he had finished his unassuming report, however, everyone in the audience knew that a bombshell had exploded in the world of Joshua Lederberg! (Watson, 1968).

Watson, who at this time had not yet completed his work with Crick on the structure of DNA, went on to collaborate with Hayes on a primarily theoretical paper that attempted to explain genetic crosses in *Escherichia coli* in terms of three linkage groups (Watson and Hayes, 1953).

Recalling these times later, Elie Wollman has written:

> A new era in the study of bacterial recombination dawned in 1952 when the first indications of sexual polarity and sexual differentiation were uncovered. In the spring of that year I visited William Hayes for the first time in his laboratory at the Postgraduate Medical School in London. His working conditions were then so modest that they made our musty attic in the Pasteur Institute look almost luxurious by comparison. I was particularly impressed by his tiny petri plates, 3 to 4 centimeters in diameter and cut out from the bottom of vials, and by the watchmaker's eye-lens with which he counted the minute colonies of recombinants appearing on these plates. Shortly after this visit, I gave an account of the new developments in recombination in bacteria, and of the genetic basis of lysogeny, at the first international conference on "Bacteriophage" held at Royaumont. Max Delbrück, who was present at Royaumont and who had been all along somewhat suspicious of genetic recombination in bacteria, listened with interest to my description of Hayes' work. Though still far from convinced that there was anything to this sexual polarity business at all, he decided to invite Hayes to give a paper at the following Cold Spring Harbor Symposium on Viruses (Wollman, 1966).

After the Cold Spring Harbor symposium, Delbrück invited Hayes to spend some time at CalTech, thus initiating a long-term friendship.

Lederberg had been quite aware of the limitations of the prototroph-recovery technique. As he had frequently stated, not all the products of recombination were being obtained, so that one could not check on the regularity of the segregation process. Hayes's data, showing that the two partners in the cross were not equivalent, opened up the door for all the subsequent studies on the nature of the bacterial mating process. Among the most important was the discovery of the F particle and the determination of the nature of the Hfr strains.

5.6 AN INFECTIVE FACTOR CONTROLLING SEX COMPATIBILITY

The ability of *Escherichia coli* K-12 to mate is determined by the presence of a conjugative plasmid called F. Strains that possess this F factor plasmid are genetically males and are designated F^+. The F factor codes for the F pili and the *tra* genes that control the transfer of part of the male genome to the female (see Section 5.18). The F plasmid is found in some, but not all, wild-type *Escherichia coli* and was present in the original K-12 strain that was isolated at

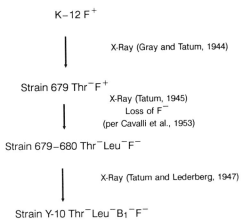

Figure 5.3 Pedigree of the F$^-$ strain (Y-10) isolated from the F$^+$ K-12. (Assembled from Gray and Tatum, 1944; Tatum, 1945; Tatum and Lederberg, 1947.)

Stanford University. When Tatum isolated nutritional mutants of this strain, they were also F$^+$, but the F factor was lost in certain strains during further mutagenesis (see Table 5.1). One well-studied strain that had lost the F factor was strain Y-10, which was threonine$^-$, leucine$^-$, and thiamine$^-$. This strain had been isolated from K-12 in three steps (Fig. 5.3). It was during the step between threonine$^-$ and threonine$^-$leucine$^-$ that the F factor was lost.

Escherichia coli K-12 cells that lack the F factor generally behave as females and are called F$^-$. Also, F$^+$ and Hfr strains can be temporarily converted to F$^-$ by use of appropriate culture conditions (such strains are called *phenocopies*). Note, however, that not all strains lacking the F factor are F$^-$, since it is essential that the female be compatible to the F$^+$ in order to receive the F factor or other genes. (Compatibility also depends on the nature of the conjugative plasmid. Some especially promiscuous plasmids are able to transfer themselves through a wide variety of Gram-negative bacteria, even ones unrelated to *Escherichia coli*. Also, many plasmids are not conjugative at all.)

In addition to coding for maleness, the F factor also codes for its own ability to replicate independently of the chromosome and for its transfer to other cells. F$^-$ cells that receive the F factor as a result of mating thus become F$^+$ and are able themselves to transfer F. The F factor therefore behaves as an *infectious particle*. This infectiousness of F was the property that first led to its discovery by Hayes (1953b) and independently by the Lederbergs and Cavalli (Lederberg, Cavalli, and Lederberg, 1952). This work, carried out in the wake of Hayes's discovery that one partner of the mating pair behaved as a donor and the other as a recipient, served to clarify considerably the nature of the mating process. However, it was not until a few years later that the relationship of the F factor to the transfer of chromosomal genes was clarified through the work of Wollman and Jacob (see below).

Evidence of Self-incompatibility

It should be noted that in his original (1945) studies, Lederberg had considered the possibility of sexual incompatibility, but since all of the isolates used had been derived from the single K-12 wild type, he assumed that *Escherichia coli* possessed a homothallic mating system (discussed above). The first evidence of a compatibility difference was obtained independently by Hayes (1953b) and by Cavalli, Lederberg, and Lederberg (1953). In the first Lederberg/Cavalli studies, a single-colony isolate, W-1607, was found to be completely infertile under conditions where all other strains showed fertility. Subsequently, it was discovered that strain Y-10, discussed above (see Fig. 5.3), was also infertile, and since the lineage of Y-10 was known, it could be concluded that Y-10 had "lost" something that one of its forerunners had. It was then shown that sex compatibility in *Escherichia coli* could be defined in the following manner:

$F^- \times F^-$ was sterile;

$F^+ \times F^-$ was fertile;

$F^+ \times F^+$ was also fertile, although at a lower efficiency.

Furthermore, prototrophic recombinants selected in an F^- background were uniformly F^+. Finally, even if streptomycin-resistant females were selected on streptomycin agar, they retained all genetic markers of the original F^- parent but became F^+. Thus, the F particle transferred itself even in the absence of chromosomal gene transfer. This transfer of F was remarkably efficient. Within one-hour incubation of an $F^+ \times F^-$ mixture, 10% of the original female cells had become F^+. On the other hand, cell-free preparations from F^+ cells were unable to convert F^- to F^+; cell-to-cell contact was necessary.

Although the infectious nature of F might seem a radical idea, it should be noted that at this same time Norton Zinder in Lederberg's laboratory was studying phage-mediated gene transfer, the process that was to be called *transduction* (see Section 8.1). Furthermore, genetic change mediated by free DNA in pneumococcus (*transformation*) was also viewed as an infective process at this time (see Section 9.10). The infectiousness of the fertility factor was discussed at the 1953 Cold Spring Harbor symposium on viruses by Hayes (1953a) and by Zinder (1953).

As noted earlier, Cavalli-Sforza (1950) had isolated a strain capable of mating at very high frequency. This strain, called Hfr (later HfrC), had the F^+ phenotype since it mated with F^- but not with F^+. Even more mysterious, progeny of Hfr $\times$ F^- matings did not become F^+ but *retained* the F^- character. Lederberg, Cavalli, and Lederberg (1952) postulated that Hfr stocks may contain the F^+ agent in a masked or bound form (an idea not too far from that ultimately deduced by Jacob and Wollman, as discussed below).

Although these studies of Lederberg, Cavalli, and Lederberg (1952) were extremely interesting, there was much they did not explain. The physical basis of the transfer of the F factor was unclear. Since phage-mediated transduction had been discovered this same year (see Section 8.1), it was natural to look for a virus, especially considering the infectious nature of F, but no evidence of such a particle was found.

In attempting to relate the F phenomenon to Hayes's discovery of donors and recipients, Lederberg, Cavalli, and Lederberg (1952) continued to seek an analogy with sexual phenomena in higher organisms:

The F^+/F^- relationships suggest speculations on heterogametic differentiation. Evidence supporting a physiological differentiation of F^+ and F^- parents is pro-

Table 5.5 Evidence for Gain and Loss of an Infectious Particle That Confers Fertility

Mating	Number of prototroph colonies	Interpretation
58–161/sp × W677/sp	0	loss of F
58–161/sp × W677	0	loss is in 58–161
58–161 × W677/sp	109	loss is in 58–161
58–161 × W677	71	control
58–161 F$^+$ × W677 (F$^+$ was constructed from 58–161/sp)	105	F is infectious

The 58–161 F$^+$ strain was constructed in the following way: 58–161/sp/SM^R was made resistant to sodium azide, then mixed with 58–161 Az^S. After incubation and streaking, 25 Az^R colonies were picked and tested for ability to mate with W677. In two separate experiments, 8/25 and 10/25 recovered colonies yielded prototrophic colonies and had, therefore, become F$^+$. (Data from Hayes, 1953b.)

vided by Hayes' experiments which imply that streptomycin-"killed" cells can function as F$^+$ but not as F$^-$...The effects of this differentiation on the segregation mechanism are still *sub judice*, but analogies are not difficult to find in the higher organisms, *e.g.*, in the selective elimination of paternal sex chromosomes in *Sciara coprophila*... (Lederberg, Cavalli, and Lederberg, 1952).

Hayes's own discovery of sex incompatibility occurred independently of that of the Lederbergs and Cavalli (Cavalli, Lederberg, and Lederberg, 1953; Hayes, 1953b; these papers were published back-to-back). Hayes has described his work:

...Dr. Clive Spicer, who had worked with the Lederbergs and with whom I had been discussing my [work], told me that he had a pair of parental K12 strains, similar to my A and B parents, which had lost their capacity to produce recombinants on storage. I had been attempting without success to isolate a male strain that had lost its postulated vector by looking for A colonies that were no longer fertile with the B female...Spicer...kindly gave me his strains which I crossed with my fertile ones. It turned out that it was indeed the male strain that was defective. The crucial experiment was to see whether its fertility could be infectively restored by contact with a normal male. Accordingly I labelled the defective A...strain with two independent mutations...and then grew it overnight in mixed culture with my fertile A strain...Before I knew the result I wrote to Luca Cavalli...telling him what I had done and saying that if the experiment worked he would have to accept the fact of infectious transfer. The experiment did work; 25% of colonies of the re-isolated [defective] strain A were now normally fertile.

However, when I got Cavalli's reply to my letter he said that he already knew what the result of my experiment would be since he and the Lederbergs had done basically the same experiment three weeks earlier ...Thus these two quite different modes of experimentation and thinking converged in the coincidental discovery of the first transmissible plasmid, the F factor.[11]

Although Hayes's results (Table 5.5) were quite similar to those of the Lederbergs and Cavalli, the interpretation of these results was quite different. As indicated above, Lederberg had attempted to fit the results into a more-or-less conventional mating system as might be found in higher organisms. In the Lederberg view, the F factor brought about diploid formation in some manner, and the fact that only a restricted range of genetic markers appeared in the prototrophs was due to aberrant segregation events (see also Nelson and Lederberg, 1954). In Hayes's view, there was a "one-way transfer

[11]William Hayes. Personal communication of mini-autobiography written for the Royal Society.

of a restricted group of genes from the F$^+$ to the F$^-$, followed by crossing-over and meiosis limited to the immigrant genes and their alleles, so that the genotype of the progeny remains basically that of the F$^-$ parent..." (Hayes, 1953b). (These opposing views would come to be called *post-zygotic* and *pre-zygotic elimination*, respectively.)

The concept of one-way transfer arose from the streptomycin experiments discussed earlier (Hayes, 1952a), which showed that F$^+$ cells killed by streptomycin could still transfer genes to the F$^-$. Although Hayes admitted that Lederberg's idea of relative sexuality could explain the results, he stated that "it is simpler to suppose that fertility depends on the presence of a gene carrier in F$^+$ cells and its absence from F$^-$ cells" (Hayes, 1953b).

None of the work at this time, however, explained how the transfer of F brought about gene transfer. Since F transfer was very efficient, but gene transfer was not, something was missing from the models. This something would not be found until studies on the origin of Hfr strains were carried out by Wollman and Jacob (see below).

5.7 HFR STRAINS AND THE NATURE OF THE MATING PROCESS

As noted earlier, the first Hfr strain (HfrC) had been isolated in 1950, but its significance had not been realized (Cavalli-Sforza, 1950). Following the discovery of the F factor, Cavalli and the Lederbergs showed that this Hfr strain behaved like a male but did not transfer the F factor, since recombinants from Hfr $\times$ F$^-$ crosses were always F$^-$. Furthermore, the Hfr state in HfrC was unstable, and revertants showed all the character of F$^+$ strains. It was thus postulated that Hfr strains harbored the F factor in a masked or bound form (Lederberg, Cavalli, and Lederberg, 1952).

During his studies, Hayes (1953a) also isolated an Hfr strain as a spontaneous occurrence, but his Hfr strain (subsequently called HfrH)[12] did not show any evidence of reverting to F$^+$. Hayes showed that in his Hfr only a restricted group of genes was transferred. He postulated that the F agent "had been replaced by a new agent, HF, which has a different type of association with the genetic structure of the cell" (Hayes, 1953a). According to Hayes, "...the main importance of HfrH was that I gave it to Elie Wollman and Francois Jacob of the Institut Pasteur, Paris, with whom I had already established a close liaison, in whose hands it played a key role in the many experiments of their brilliant series that revealed the true nature of *E. coli* sexuality."[13]

Wollman and Jacob

Elie Wollman and Francois Jacob had approached their studies of bacterial mating from quite a different direction than either Hayes or Lederberg. Elie Wollman was the son of Eugéne and Elisabeth Wollman, two distinguished bacteriophage workers at the Pasteur Institute who had done pioneering research on lysogeny (see Section 7.3). A student of Lwoff, Elie Wollman had spent the years 1948–1950 in Delbrück's laboratory at CalTech working with

[12]This strain became the basic strain in the development of the genetic map of *Escherichia coli*.
[13]Hayes, mini-autobiography for the Royal Society.

Gunther Stent on adsorption of bacteriophage T4. (Some of this personal history is given in Wollman, 1966.) He returned to Paris at the time of Lwoff's revolutionary work on lysogeny (see Section 7.4) and initiated studies on the genetics of the prophage. The Lederbergs had just discovered that *Escherichia coli* K-12 was lysogenic for a phage they called *lambda*, and since K-12 could be studied genetically, the genetics of prophage became a feasible topic of investigation. Wollman was able to show that the prophage could behave as a genetic marker and that it was linked to the *gal* locus (see Section 7.5).

Francois Jacob entered Lwoff's laboratory at the Institut Pasteur a few years after Wollman (Jacob, 1988). He had done his Ph.D. degree on the phenomenon of lysogeny and especially on the nature of the prophage of the temperate virus (see Section 7.5). He had also studied *colicinogeny*, the phenomenon of production by certain strains of substances lethal to closely related strains (see Section 5.9). This phenomenon had been considered to be related in some way to the phenomenon of lysogenicity. Wollman and Jacob began a collaboration on a study of the genetics of lysogeny. Soon, they had discovered the phenomenon of zygotic induction, which made possible a study of the kinetics of the DNA transfer process from Hfr to F⁻ (see below).

Interrupted Mating

Hayes had already shown that in HfrH only a restricted group of genes was transferred to the F⁻. He and Watson had suggested, in light of their "three chromosome model," that the F factor became associated with one of the three chromosomes (Watson and Hayes, 1953). Wollman and Jacob then studied the kinetics of transfer of genes from HfrH to F⁻ and showed that the longer one incubated the mixture of donor and recipient cells, the more genes were transferred, and that the transfer of genes occurred in an *ordered* manner (Wollman and Jacob, 1955). This major discovery, which provided the essential clue to the nature of the *Escherichia coli* mating system, led to the important *interrupted mating experiments*, in which a high-speed blendor was used to separate mating pairs after different periods of time (Fig. 5.4).[14] By separating mating pairs at various times, certain genes were found in recombinants earlier than others. By comparing the order of entry as observed in the blendor experiments to the genetic map, it could be shown that the time of transfer correlated with the position on the map. Wollman and Jacob (1955) suggested that one could actually use the *time of transfer* as a means of expressing the genetic map of *Escherichia coli*, a suggestion that proved to be correct. Interrupted mating subsequently became the basis for the whole system of genetic mapping in *Escherichia coli*.

Following Hayes, Wollman and Jacob (1955) made the following conclusion:

1) The segment of the Hfr chromosome subject to high frequency recombination is an oriented segment R→O, the order of transmission of characters [to recombinants] being a function of their distance to the unknown origin O. The probability that a given character is present in recombinants is the smaller, the farther the character is from the O origin.
2) There may be genetic recombination even when a small segment of the Hfr

[14]The high-speed blendor had been first used by T.F. Anderson to strip phage particles from bacterial cells and played a major role in the Hershey/Chase experiment (see Section 6.13).

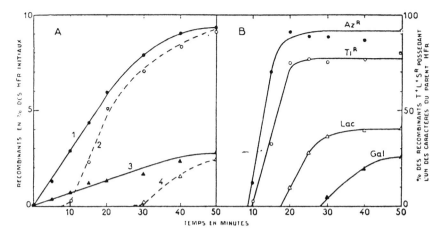

Figure 5.4 The first published interrupted mating experiment. A mating of strep-tomycin-sensitive Hfr (10^7/ml) × streptomycin-resistant F^- (5×10^8/ml) was carried out. Samples were removed at various times and diluted, and a fraction was subjected to mechanical blending, the other remaining as control. A portion of each fraction was plated on streptomycin-containing medium to determine the frequency of recombin-ants. (*A*) Time course of appearance of recombinants of $T^+L^+S^R$ (curve 1, before blending; curve 2, after blending) and gal$^+S^R$ (curve 3, before treatment; curve 4, after treatment). (*B*) Analysis of genetic recombinants obtained after each treatment. From each plating, 120 colonies were examined to determine the presence of unselected markers derived from the Hfr. The graphs give the time of appearance of the unselected markers. (Reprinted, with permission, from Wollman and Jacob, 1955.)

> parent chromosome is transferred to an F^- bacterium, which agrees with the view of Hayes [on partial transfer]. As to the actual mechanism of integration of the transferred genetic material, recombination in *E. coli* K-12 may resemble transduc-tion [that is, the integration of small chromosomal segments] (translated from Wollman and Jacob, 1955).

Thus, Wollman and Jacob suggested that their data could be explained by a model in which the segment of the chromosome that was transferred from the Hfr was limited to the segment O (origin)—R (point at which the F factor attached). Of course, one of the main points of this model was that the transfer from the donor to the recipient did not occur all at once, but in stages, and could be interrupted by the simple physical process of agitation. Furthermore, it had to be assumed that the interruption of the mating process brought about by agitation did not prevent the genetic fragment that had already been transferred from becoming integrated into the recipient's genome.

The idea that mating resembled transduction and that the agent that was involved in the transfer, the F factor, resembled in some way a virus was to be made even more concrete with the development of the episome concept (see later).

The Wollman/Jacob interpretation was criticized by Lederberg in a letter to *Science* in response to a news report of this work (Lederberg, 1955a). The argument revolved around whether one was dealing with a prezygotic or a postzygotic segregation phenomenon. In keeping with the attempt to retain an analogy with higher organisms, Lederberg advanced some alternate ex-planations:

Unfortunately, the true constitution of the fertilized cell cannot be inferred with certainty from data on haploid segregants, owing to peculiarities of the meiotic mechanism in *Escherichia coli* K-12. In the cited experiments, fertilization might have been incomplete, or it might have been complete with later disturbances of chromosome pairing to account for the segregation effects. Furthermore, if fertilization were fractional, the gradient of recovery of various loci might be due, as proposed, to the preferential orientation of gametic chromosomes, or to dependence on one locus (a centromere?) for the completion of synapsis, crossing-over and segregation. These limitations of inference are equally applicable to undisturbed matings and have provoked a diversity of hypotheses...all the experimental data are equally consistent with a second hypothesis that fertilization is regularly complete, but is coincident with chromosome breakage at specific, predetermined points on the F^+ chromosome, so that certain segments are deleted after meiosis (Lederberg, 1955a). (See also Lederberg, 1955b, for a more detailed discussion of these ideas.)

Lederberg's alternate interpretations were soon shown by Hayes and the Pasteur group to be incorrect. By 1957, Lederberg had fully accepted the Wollman/Jacob model, as shown by his Harvey Lecture of that year (Lederberg, 1957).

5.8 STAGES IN THE MATING PROCESS

The availability of the Hfr strain made possible for the first time the study of the physiology of the mating process (Jacob and Wollman, 1955). Using the interrupted mating technique and a genetic analysis of recombinants, three stages of the process could be discerned: *pairing*, which resulted in association of male and female cells; *transfer*, in which an ordered transfer of genes from the Hfr to the F^- occurred; and *integration*, in which some or all of the transferred genetic material underwent recombination with the F^- genome. The kinetics of the first step, pairing, could be studied by diluting mating mixtures at various intervals, thus reducing the population densities to the point where the probability of impact of male and female cells was low (Hayes, 1953a). The second step, transfer, could be measured by the interrupted mating technique.[15]

The distinction between transfer and the third step, integration, could be made by means of *zygotic induction*, a phenomenon discovered by Wollman and Jacob: If the prophage for *lambda* entered an F^- cell that was sensitive to *lambda*, the prophage became induced and the recipient was lysed (see Section 7.5). This phenomenon provided a critical demonstration that the Lederberg interpretation of postzygotic elimination (see above) was incorrect. If Lederberg's idea had been correct, the entire genome would always pass from donor to recipient and zygotic induction would *always* occur.

Because zygotic induction does not require integration, it was possible to use this phenomenon to measure when the transfer process took place. By subtracting the frequency of recombination from the frequency of cells producing *lambda*, the extent of transfer could be calculated.

Low doses of ultraviolet radiation strongly affected the integration of transferred genes, without affecting greatly the transfer process. Jacob and Wollman (1955) noted that UV also had a similar effect on recombination in

---

[15]The famous PaJaMo experiment (see Section 10.9) provided evidence that there was no significant *cytoplasmic* transfer during mating.

bacteriophage *lambda*. They concluded, therefore, that there were close analogies between the recombination process in bacteria and the process of phage replication. This analogy between bacteria and phage was to provide considerable stimulus for thinking of the mating process in a new light. No longer could the process be considered a conventional sexual process that had been modified because of certain "peculiarities" of the bacterial system.

5.9 COLICINOGENY

The discovery in the mid-1950s of the transfer of colicinogeny by cell-to-cell contact was to play an interesting role in the interpretation of the mechanism of mating. *Colicins* are antibiotic substances produced by certain *Escherichia coli* strains that cause the death of other strains (Reeves, 1972). The Belgian scientist Pierre Frédéricq had shown that when col$^+$ bacteria were mixed with col$^-$ bacteria, the col$^+$ character was transmitted from the col$^+$ to the col$^-$ bacteria at high frequency (Frédéricq, 1954). As noted by Jacob and Wollman, the *col* factor, like the F factor, was transferred at high frequency, and no linkage appeared to exist between this factor and the other genetic characters of the F$^+$ bacteria.[16] Jacob and Wollman thus concluded that the F factor and the *col* factor were not located on the same structure as the other genetic determinants of bacteria. This idea eventually led to the concept of the *episome* and *plasmid* (see below and Section 7.7), but for the moment it is mainly of interest because it focused attention on how the F factor might be related to the Hfr phenotype.

5.10 THE CONVERSION OF F$^+$ TO HFR

Given the fact that F behaved as an independent factor, how then was it involved in *genetic* transfer?

> It could be supposed that, when F$^+$ and F$^-$ bacteria mate, there is only a low probability for each F$^+$ bacterium to transfer a segment of its genetic material to the recipient bacterium. Alternatively, it could be supposed that F$^+$ bacteria, as such, are unable to transfer any of their genetic material upon conjugation but that, in any F$^+$ population, there is a small fraction of individuals which can do so at high frequency. According to the former hypothesis, the population of F$^+$ bacteria would be *homogeneous* in its ability to give rise to genetic recombinants. According to the latter hypothesis, on the contrary, the population of F$^+$ bacteria would be *heterogeneous*, and would contain a small proportion of Hfr mutants. It should be recalled in this context that the two Hfr strains first isolated, that of Cavalli and that of Hayes, arose from the same strain of F$^+$ bacteria (Jacob and Wollman, 1961).

Jacob and Wollman recognized that if the latter hypothesis, heterogeneity of F$^+$ bacteria, was correct, then the origin of Hfr from F$^+$ was analogous to a mutational event and could be investigated by the methods that had already been used to study the mutation process in bacteria, the fluctuation test and indirect selection (see Chapter 4). Using the procedure of Luria and Delbrück (1943), Jacob and Wollman (1956) grew F$^+$ bacteria from small inocula in 50 separate tubes and then analyzed the resulting cultures for ability to transfer genes at high frequency (hence Hfr) to an F$^-$ tester strain (Table 5.6). The

---

[16]It is now known that the colicin-producing ability studied by Frédéricq is controlled by a different transmissible plasmid, the Col I plasmid (Falkow, 1975).

Table 5.6   Use of the Fluctuation Test to Demonstrate the Mutational Origin of Hfr from F$^+$

	F$^+$ bacteria per tube $\times 4 \times 10^8$	Samples from independent cultures: Recombinants T$^+$L$^+$S^R per 0.15 ml in each tube	Samples from the same culture: Recombinants T$^+$L$^+$S^R per 0.15 ml from flask
Minimum number of bacteria per tube	55	1	10
Maximum number per tube	82	116	23
Average	67.90	15.30	16.33
Variance	45.77	351.54	13
Chi square	33.3	1105	12
Probability	>0.1	<0.01	0.6

A culture with genotype F$^+$M$^-$S^S grown in limiting methionine was inoculated at 250 bacteria/ml in 50 tubes, 1.2 ml per tube. Another portion was kept in a single flask. After growth, samples were mixed with an excess of F$^-$ bacteria of genotype T$^-$L$^-$S^R. After 1 hr, the contents of the tubes were spread on a selective medium and scored for T$^+$L$^+$S^R recombinants. Note that the variance is much higher with the 50 separate tubes than with 50 samples from the same tube. This is the result that would be predicted from the hypothesis that F$^+$ → Hfr is due to a rare mutation. (Data from Jacob and Wollman, 1956.)

control consisted of an analysis of 50 samples taken from the *same* culture. The statistical variance was considerably higher with the experimental tubes than with the control. It was concluded, therefore, that the Hfr character arose as a result of a mutational process.[17] Jacob and Wollman (1956) also used the indirect selection procedure of Lederberg and Lederberg (1952) to isolate Hfr mutants from F$^+$. They prepared lawns of F$^+$ cells on complete agar medium and replica-plated onto a selective medium seeded with F$^-$ bacteria. They picked from the original plate regions corresponding to the positions of recombinants on the replicas, and after several rounds, isolated Hfr cultures.

Curing F$^+$ Bacteria

Another important discovery was that F$^+$ bacteria could be converted to F$^-$ by growth in the presence of acridine dyes (Hirota, 1956), although Hfr bacteria were unaffected (Hirota, 1960). This phenomenon, called *curing*, made the F particle seem even closer to phage, since acridines were known to cure virus infection but had no effect on chromosomally bound prophage. Curing provided additional support for the hypothesis that the sex factor F was not integrated into the bacterial chromosome.

Implications

The results of these studies on the origin of Hfr bacteria had far-reaching implications:

1. The mutational origin of Hfr from F$^+$ showed that the F factor somehow

---

[17]Note that there are a number of different Hfr strains represented in the data of the table. The number of Hfr strains to be expected would depend on the number of homologous sites in the chromosome at which F can integrate.

entered a different state. Because all recombinants in Hfr × F⁻ crosses remained F⁻ (unless selected for remote markers), it was concluded that the F⁺ → Hfr transition consisted of the integration of the F factor into the chromosome at its distal end.

2. The difficulties in analyzing crosses between F⁺ and F⁻ could now be understood. A population of F⁺ bacteria actually contained a majority of bacteria that were themselves *incapable* of forming genetic recombinants, together with a small proportion of different Hfr mutants. In some cases, genetic recombination in F⁺ populations was due to the rare Hfr cells present, which could explain the *low frequency* of chromosome transfer in F⁺ even when the frequency of transfer of F itself was high.

3. Using the mutational approach, Hfr bacteria could now be isolated at will. Depending on which character was used in selection, different types of Hfr could be isolated, all of which transmitted efficiently the characters used for selecting recombinants. It was found that these different Hfr strains varied in the order in which their genetic characters were transferred during conjugation. This led directly to the concept of the circular chromosome, as well as to the development of a detailed genetic map (see later).

4. F⁺ bacteria could be converted at will into F⁻ by use of acridines.

Acceptance of the fragmentary nature of gene transfer from Hfr to F⁻ was probably made easier by knowledge of the growing body of facts about transduction, which had been well established since the mid-1950s (see Section 8.3).

Once Hfr strains were available, it became possible to study the conjugation process by microscopy. In addition to electron microscopy, Anderson (1958) carried out a careful pedigree analysis of single recombinants using a micromanipulator, exploiting Hfr and F⁻ strains that were distinguishable microscopically. The micromanipulation studies that had been done earlier by Zelle and Lederberg (1951) had not been readily interpretable. With a high-frequency conjugation process, it now became possible to remove single exconjugant cells and obtain families of clones by successive isolation of cells after division. By this procedure, it was possible to confirm *directly* the hypothesis of one-way transfer from Hfr to F⁻. Segregation data of F⁻ recombinants also confirmed the fragmentary nature of the process.

In an attempt to unify the various genetic recombination processes in bacteria, Jacob and Wollman coined the terms "meromixis" and "merozygote." *Meromixis* referred to any process of *partial* genetic transfer, and the product of meromixis was an incomplete zygote, a *merozygote* (Wollman, Jacob, and Hayes, 1956). Actually, the term meromixis had almost the same meaning as the term *transduction* when the latter was first defined, but the term transduction lost its broad meaning almost immediately after it was coined (see Chapter 8).

5.11 THE CHROMOSOMAL ORGANIZATION OF HFR BACTERIA

In the standard conjugation experiment of Jacob and Wollman, an F⁻ strain that was mutant for various nutritional markers and was streptomycin-resistant was mated with a wild-type Hfr that was streptomycin-sensitive. By

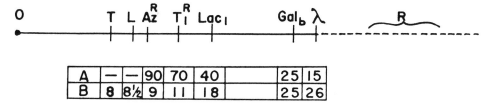

Figure 5.5 Correlation of time of entry with genetic linkage. (*Top*) Genetic map of the segment of the HfrH chromosome that entered initially. Recombinants of genotype $T^+L^+S^R$ were selected and then tested for the presence of the other markers (unselected). The location of the different characters is shown in *A* as the percentage of $T^+L^+S^R$ recombinants that had inherited the different Hfr alleles. In *B* is shown the time at which individual Hfr characters start penetrating into F^- recipient cells in an interrupted mating experiment. The correlation between the two kinds of maps is excellent. (Reprinted, with permission, from Wollman, Jacob, and Hayes, 1956.)

plating the mixture on streptomycin, the Hfr donor was eliminated. (This was the procedure introduced by Lederberg [1951a].) The parent F^- recipient was eliminated by plating on a medium lacking one of its required nutrients. The colonies that developed could then be characterized for other unselected nutritional markers by replica plating onto various other plates. In the genetic analysis, a single F^- strain could be used, and the results thus depended primarily on the order of transfer of chromosomal genes of the particular Hfr. Preferably, the selection should employ a nutrient whose corresponding gene was transferred *early*.

Analysis of the conjugation process between various Hfr strains and F^- using interrupted mating led to the visualization of the linkage group as a linear structure defined by its two extremities, the origin and the terminus (Wollman, Jacob, and Hayes, 1956; Jacob and Wollman, 1961). The map determined by the Lederberg procedure (frequency of unselected markers) correlated with the map determined by the time of entry (Fig. 5.5), but the time of entry map was easier to obtain. The origin was that portion of the linkage group that entered first.[18]

Each Hfr type was characterized by the genes it transmitted at high frequency and by the subsequent order of transmission (Table 5.7). However, if the linkage groups of different Hfr types were compared, it was apparent that in all Hfr types, the linkage relationships between characters were constant and uniform. Whichever characters were transferred first, and whatever order the transfer of these characters, an unequivocal order could be established in which the different genetic characters of *Escherichia coli* K-12 existed. By comparison of various Hfr strains, it appeared to Jacob and Wollman that the genetic map could be represented formally as a circle, but the order of transfer could be in *either* direction from some defined site that was characteristic for each independently isolated Hfr.

Although circular genetic maps were subsequently developed for various bacteriophages (see Section 6.11), the idea of genetic circularity was first advanced by Jacob and Wollman for *Escherichia coli* K-12. In their own words:

[18]The total map length in *Escherichia coli* K-12 is 100 minutes. Since *Escherichia coli* doubles in 30–60 minutes, depending on conditions, it takes much longer than one doubling time for complete transfer of the *Escherichia coli* genome.

Table 5.7 Linkage Groups of Different Independently Isolated Hfr Types

Types of Hfr	ORDER OF TRANSFER OF GENETIC CHARACTERS																		
Hfr H	O	T	L	Az	T_1	Pro	Lac	Ad	Gal	Try	H	S-G	Sm	Mal	Xyl	Mtl	Isol	M	B_1
1	O	L	T	B_1	M	Isol	Mtl	Xyl	Mal	Sm	S-G	H	Try	Gal	Ad	Lac	Pro	T_1	Az
2	O	Pro	T_1	Az	L	T	B_1	M	Isol	Mtl	Xyl	Mal	Sm	S-G	H	Try	Gal	Ad	Lac
3	O	Ad	Lac	Pro	T_1	Az	L	T	B_1	M	Isol	Mtl	Xyl	Mal	Sm	S-G	H	Try	Gal
4	O	B_1	M	Isol	Mtl	Xyl	Mal	Sm	S-G	H	Try	Gal	Ad	Lac	Pro	T_1	Az	L	T'
5	O	M	B_1	T	L	Az	T_1	Pro	Lac	Ad	Gal	Try	H	S-G	Sm	Mal	Xyl	Mtl	Isol
6	O	Isol	M	B_1	T	L	Az	T_1	Pro	Lac	Ad	Gal	Try	H	S-G	Sm	Mal	Xyl	Mtl
7	O	T_1	Az	L	T	B_1	M	Isol	Mtl	Xyl	Mal	Sm	S-G	H	Try	Gal	Ad	Lac	Pro
AB 311	O	H	Try	Gal	Ad	Lac	Pro	T_1	Az	L	T	B_1	M	Isol	Mtl	Xyl	Mal	Sm	S-G
AB 312	O	Sm	Mal	Xyl	Mtl	Isol	M	B_1	T	L	Az	T_1	Pro	Lac	Ad	Gal	Try	H	S-G
AB 313	O	Mtl	Xyl	Mal	Sm	S-G	H	Try	Gal	Ad	Lac	Pro	T_1	Az	L	T	B_1	M	Isol

(Reprinted, with permission, from Jacob and Wollman, 1961.)

> Although the linkage group of each specific Hfr type must be represented as a linear structure defined by its [point of origin], there is no reason to choose the specific linear structure of any one strain as the basis for constructing a general chromosomal map of *Escherichia coli* K12. One is led, therefore, to place all the known genetic characters on a closed curve, a circle...From this representation, the linkage group of any particular Hfr type may be easily deduced by opening the closed curve at a defined point and inserting the origin O at one of the two extremities thus formed. This process establishes unequivocally both the nature of the genetic segment transferred with high frequency, and the order of transfer of the genetic characters located on this segment (Jacob and Wollman, 1961).

Jacob and Wollman were careful at this time to avoid stating categorically that the *Escherichia coli* chromosome was circular. They stated only that the genetic map could formally be represented as a circle. (This caution was warranted. Bacteriophage T4 has a circular genetic map even though its DNA is linear. The circularity here is due to the fact that in different phage particles the DNA molecules end at different positions, the phage DNA thus being circularly permuted. This is discussed in Section 6.11.) However, from the fact that Hfr arose initially from F$^+$, they made the (erroneous) proposal that

> ...the chromosome of an F$^+$ bacterium and that of an Hfr cell differ in configuration, and that the mutation from F$^+$ to Hfr results in a change in the structure of the chromosome (Jacob and Wollman, 1961).

However, Jacob and Wollman soon abandoned the hypothesis that only the F$^+$ cell had a circular chromosome when they showed that if two Hfr cells of the *same* type were mated (using a physiological "trick" to convert one to F$^-$), the map obtained was also circular. Thus, the linearity of the Hfr chromosome was apparent rather than real, applying only during the transfer process. The Hfr also has a closed loop DNA, but this circle *opens* during transfer to the F$^-$ (Hayes, 1964).

In their original illustrations of the F factor, Jacob and Wollman depicted it as a *linear* structure, even when extrachromosomal, whereas the chromosome was depicted as circular (Jacob and Wollman, 1961). However, Alan Campbell proposed a model for integration of F that involved a circular structure in analogy to *lambda* (Campbell, 1962). Then Cairns's autoradiography work (see Section 3.9) provided physical evidence of the circular chromosome. At the 1963 Cold Spring Harbor symposium, where Cairns (1963) presented his work to a large audience, Jacob, Brenner, and Cuzin (1963) presented their *replicon* model of chromosome replication in bacteria, a model that almost *required* circularity of chromosomal and F factor DNA. This model attempted to account for the known features of DNA replication in both bacteria and viruses. Although it was later to be found that many phage never replicated in a cyclized form, the replicon model focused attention on those genetic elements, such as the F factor, which replicated independently of the chromosome. Eventually, this would lead to an understanding of the physical nature of plasmids. It is now well established that many procaryotic genomes are circular (see Section 5.18), but at the time that Jacob and Wollman hypothesized the circular chromosome in *Escherichia coli*, circular DNA (as opposed to a circular genome) was unknown. In their early writing, Jacob and Wollman were careful to maintain the distinction between genetics and biochemistry:

> F⁺ bacteria have a sex factor which is not integrated into the structure of their chromosome. Their chromosome would therefore be (effectively or potentially) circular... (translated from Wollman and Jacob, 1959).

Although many details of the mating process remained to be worked out, the studies of Jacob and Wollman established a new paradigm for bacterial genetics. No longer could the mating process in bacteria be considered a normal, albeit somewhat "peculiar," example of conventional sexual reproduction. The bacterial process was *strikingly different* from the process in higher organisms.

5.12 F PRIME FACTORS

Genetic analysis in bacteria generally depends on detection of rare events, and bacterial geneticists are thus attuned to the recognition of small differences. One such difference first recognized by Edward A. Adelberg became the beginning of an important new understanding of the F factor. It can also be considered to be the foundation of genetic studies with plasmids.

Initially, the F factor was viewed as merely a carrier of genetic elements involved only in its own replication and transfer. Then Adelberg and Burns (1960) discovered an aberrant F factor that had undergone genetic exchange with a portion of the chromosome. Jacob and Adelberg (1959) had first isolated a modified F⁺ that had incorporated the chromosomal *lac* genes. This variant had arisen from an Hfr in which the F factor had become incorporated close to the *lac* locus. When F spontaneously detached, it excised *lac* with it and hence transferred the chromosomally derived *lac* gene, and *lac* alone, at high frequency. Such modified F particles came to be called F *prime* (Hayes, 1964). Another F variant was isolated in which the proline locus had been incorporated. Jacob and Adelberg (1959) thus concluded that any segment of the bacterial chromosome and the F factor could associate together and form a unit of replication and transmission. They recognized that this process of genetic transfer was comparable to that of transduction.[19]

The idea that F *prime* arose by recombination with the bacterial chromosome led to a facile method for isolating a large number of different F *primes*. The procedure was to interrupt mating *early*, but to select for a marker known to be transferred *late*. Many recombinants selected in such a way contained a modified F, and this could be checked by mating such recombinants in a second round of conjugation and showing that they converted F⁻ to F⁺. In this way, F *prime* plasmids could be isolated that include genes from virtually the entire *Escherichia coli* chromosome. (This approach was essentially in vivo cloning, since a defined portion of the genome was isolated and at-

[19]The background to the work on F prime factors has been given to me by Elie Wollman (personal communication, 1989). "Adelberg had brought back to Berkeley some of our Hfr strains. I spent the year 1958–59 in Berkeley—finishing the writing of our book [Wollman and Jacob, 1959]. Once Ed Adelberg came to me telling me that one of the Hfr strains had changed: the frequency of recombinants was less than the expected, but all were donors of intermediate frequency. I suggested that, by comparison with HFT phage the sex factor had left its site accompanied by neighboring genetic fragments. This was verified experimentally. Lwoff, who had come to visit, brought the news back to Francois Jacob who immediately used it for making partial Lac diploids. This is the history of F prime factors."

tached to a vector—the F particle. As discussed in Chapter 11, this approach foreshadowed the development of in vitro cloning methods in the 1970s and 1980s.)

Jacob and Wollman (1961) called the transfer of host genes at high frequency by the F (sex) factor *sexduction* (sometimes called *F-duction*), in analogy with transduction by bacteriophage. The intent was to express the infectious nature of the process and to indicate that the sex factor was doing something more than transferring itself. However, the term sexduction never became popular and has disappeared from the bacterial genetics literature.

In addition to their importance for understanding the nature of F itself, F *primes* have found wide use as tools for studying the nature of gene action. For instance, F *prime lac* was used by Jacob and Wollman to construct partial diploids for complementation (*cis-trans*) tests (see Section 10.10). It was also possible, using the *lac* genes as a marker, to follow the transfer of the F factor to species other than *Escherichia coli*, making feasible the isolation of the F plasmid and its physical characterization (see below).

5.13 GENETIC HOMOLOGY BETWEEN *ESCHERICHIA COLI* AND *SALMONELLA*

The close taxonomic relationship between *Escherichia* and *Salmonella* prompted early attempts to cross species of these two types (Baron, Spilman, and Carey, 1959). It was used to introduce *Escherichia coli* genes into *Salmonella* (Zinder, 1960a). Because *Salmonella* could also be mapped by transduction (see Chapter 8), it was possible to develop rather quickly a detailed map of this organism. It is now known that the map of *Salmonella typhimurium* is almost identical to that of *Escherichia coli* except for a large segment, about 20 minutes long, which is inverted. Also, the use of F *prime lac* and other F *primes* permitted the transfer of genes to other Gram-negative bacteria.

5.14 MALE-SPECIFIC BACTERIOPHAGES

It was natural to look for physiological differences that could distinguish male from female cells of *Escherichia coli*. Zinder (1960b) discovered a color indicator reaction that was specific for Hfr and F^+ cultures, but the most fascinating discovery was a group of bacteriophages that were specific for male cells (Loeb, 1960; Dettori, Maccacaro, and Piccinin, 1961). Initially, the most interesting thing about these male-specific phages was that they contained RNA instead of DNA, the first bacteriophages to be so found. Zinder (1980) has described the history of the discovery and early research on these phages in some detail. Such phages proved of considerable utility in the early days of messenger RNA, since they provided one of the sources of RNA for developing systems of cell-free protein synthesis. (Another group of male-specific phages, including fd and M13, contain single-stranded DNA. There are also female-specific phages, such as phage T7.)

In addition to their general interest as RNA phages, the male-specific phages provided an important insight into the mechanism of pairing and chromosome transfer during the mating process. Following up on some earlier observations, Brinton, Gemski, and Carnahan (1964) showed that these RNA phages attached to a specific kind of pilus (the sex pilus) on the surface of male cells. The phage particles could be used as a specific reagent in electron microscopy to visualize this pilus and to observe its behavior during genetic crosses. It could be shown that F pili were necessary for chromosome transfer and played the initial role in making contact between the male and female cells.

5.15 THE *REC⁻* PHENOTYPE

Although details of the mechanism of recombination are beyond the scope of this book, a phenotype that provided important insights should be mentioned here. The *rec⁻* phenotype exhibits markedly reduced ability to undergo genetic recombination as well as very high sensitivity to ultraviolet radiation. The first *rec⁻* was isolated by Clark and Margulies (1965) from mutagenized F⁻ cells. Three separate genes were found, labeled *recA*, *recB*, and *recC*. The *rec* genes are known to play a role in repair of DNA damage, and the recA protein binds to single-stranded DNA and promotes homologous pairing. The isolation of *rec⁻* mutants has helped to provide important insights into the molecular mechanism of recombination.

5.16 AN IMPORTANT SYNTHESIS

As shown by the fluctuation test, the mutational origin of Hfr from F⁺ was clearly a *single* event, but an event that had multiple consequences. Jacob and Wollman visualized this event as the attachment of the F factor at a given point on the circular chromosome and the opening up of this structure. Hfr bacteria were thought to have the F factor integrated into their genome at a location distal from the point of origin. The F factor was therefore an element that could exist in *two* alternate states: autonomous and integrated. In the autonomous state, the F factor could control its own replication and transfer, but did not have a significant effect on gene transfer. In the integrated state, the F factor was no longer infectious, but still was able to bring about conjugation and chromosome transfer. In this respect, the F factor resembled the *prophage*, which could exist in either an integrated or autonomous state.

Jacob and Wollman (1958) called such elements that could exist in either an integrated or an autonomous state *episomes* (see Section 7.7). The term *episome* subsequently fell into disuse, being replaced by the more general term *plasmid* (Hayes, 1969), but the term episome was important in focusing attention on the nature and peculiarities of genetic elements in bacteria. Soon, understanding of the physical state of DNA in bacteria would clarify the nature of the bacterial chromosome and the mechanism of genetic recombination. Bacterial genetics became, not an end in itself, but a *tool*. Within a short period of time, this work led to the development of the operon model of Jacob and Monod (see Section 10.10).

5.17 RESISTANCE TRANSFER FACTORS

Soon after the discovery of the F factor and the development of the episome concept, a new type of transmissible genetic factor was discovered, the resistance transfer factor, generally abbreviated RTF or R factor. The R factor coded for multiple drug resistance and was thus not only of theoretical but also of practical importance. R factors were first discovered in Japan among strains of antibiotic-resistant *Shigella* (Watanabe, 1963).[20] The first indication that these strains were unusual was the observation that in particular *Shigella* epidemics, some strains isolated from patients were sensitive to all available antibiotics, whereas other strains isolated in the same epidemic showed multiple antibiotic resistance (generally to the antibiotics streptomycin, chloramphenicol, and tetracycline). Furthermore, the administration of a single antibiotic such as chloramphenicol led to the development in the patient of a multiply resistant strain. Since chloramphenicol should not be a selective agent for all antibiotic resistance genes, this observation could hardly be attributed to mutation. One hypothesis to explain this peculiar result was that multiply resistant *Escherichia coli* present in the intestinal tract had transferred this resistance to the *Shigella*. The ability of the multiply resistant *Escherichia coli* to transfer resistance to *Shigella* was quickly confirmed by experimentation.

All attempts to transfer multiple antibiotic resistance by cell-free filtrates were unsuccessful. Accordingly, it was concluded that cell-to-cell contact (mating) was necessary for the transfer to occur. However, transfer occurred independently of presence or absence of the F factor. Although multiply resistant strains behaved quite differently from the natural isolates and could be isolated by mutation by exposing sensitive strains to single drugs one after another, such multiply resistant strains could not transfer any of their antibiotic resistance characters by mixed cultivation. In addition, treatment with acridine dyes made multiply resistant cells simultaneously sensitive to all the antibiotics, indicating that the resistance factors were in an autonomous (curable) state. Because no other measurable properties of the donor were transferred, it was concluded that multiple drug resistance was a kind of infective heredity.

The behavior of R factors fit neatly into the episome concept of Jacob and Wollman, and R factors were interpreted as episomes very soon after their discovery (Watanabe and Fukasawa, 1961). In an influential review, Campbell (1962) had drawn a broad picture of the episome, encompassing within it not only the entities first studied by Jacob and Wollman: F factor, colicin, and *lambda* phage, but also R factors. The R factors clearly existed in an autonomous state, and it was soon shown that they could also become integrated into the chromosome. Thus, R factors were true episomes.

Space does not permit a further discussion of these interesting genetic elements, which have been studied so extensively both from a genetic and epidemiological point of view. Thorough reviews can be found in symposia during the 1960s and 1970s (see, e.g., Meynell, 1973; Falkow, 1975).

5.18 PHYSICAL EVIDENCE FOR STRUCTURE OF PLASMIDS

As the genetic work was proceeding in the late 1950s, work was also under way on the detailed structure of DNA. One of the most important ideas to come out of this work was the discovery that the DNA base compositions of different organisms were different (see Section 9.15). DNA base composition

---

[20]Much of the early work on resistance transfer factors was published in the Japanese language. The Watanabe (1963) review provided the first overview of the phenomenon in English and will be used extensively in the present section.

influenced the buoyant density in CsCl gradients in the ultracentrifuge, and because of this, molecules with different DNA base compositions could be distinguished in mixtures. Thus, the technique of density gradient centrifugation provided a unique method for detecting genetic transfer between bacteria whose DNA base compositions differed.

Among the enteric bacteria, the base composition of *Escherichia coli* and *Salmonella* were 50% GC, whereas that of *Serratia marcescens* was 58% GC. Because it was possible to transfer the F factor from *Escherichia coli* or *Salmonella* to *Serratia*, it was possible to use density gradient centrifugation to obtain physical evidence of the transfer. Using an F factor containing *lac*, Marmur, Rownd, Falkow, Baron, Schildkraut, and Doty (1961) were able to detect the F factor DNA after it had been transferred to *Serratia*. In the *Serratia* carrying F, a small band of DNA corresponding to the *Escherichia coli* buoyant density was present in addition to the large band due to *Serratia* DNA. The fact that the small band was present showed clearly that the F factor had not become integrated into the *Serratia* genome, even though the *lac* genes were expressed. This study not only confirmed that genetic exchange could occur between organisms of different genera, but also showed clearly that the episomal element was DNA. It also raised the possibility that the isolation and purification of the F factor might be possible, since it appeared to exist in a free state in the recipient cells. This work was followed up extensively by Falkow and colleagues (Falkow, Wohlhieter, Citarella, and Baron, 1964).

As evidence continued to build for the circularity of genetic elements, it was natural to look for evidence of circularity in episomal and plasmid DNA. The Kleinschmidt technique (see Section 3.9) permitted visualization of DNA molecules at high resolution, and as research on the physical properties of DNA continued, isolation of undegraded DNA of various kinds could be done with more confidence. The first plasmid shown to be circular was the *Col* E1 colicinogenic factor isolated from *Proteus mirabilis* (Roth and Helinski, 1967). The plasmid DNA was purified on methylated albumin kieselguhr columns and examined in the electron microscope with the Kleinschmidt technique. Two forms of circular DNA were seen, a highly supercoiled form and an open circular form. When the plasmid DNA was treated with high pH followed by rapid neutralization, it retained its configuration, showing that the ends of the molecules were covalently linked and thus arranged in the circles. A few months later, Helinski's laboratory also reported that an F factor of *Proteus mirabilis* was also circular (Hickson, Roth, and Helinski, 1967). Further discussion of this area is beyond the scope of this book.

5.19 CONCLUSION

We began this chapter at a point in history when bacterial genetics was virtually a complete mystery, and we end it now at a point where knowledge was so advanced that the merging of genetics and biochemistry was possible. The significant development was not just the *fact* of bacterial mating, or the gradually developing genetic map of *Escherichia coli*, but the *methodology* of bacterial genetics, which by 1960 permitted the discovery of the intimate details of the relationship between DNA and protein. Although the one-gene/one-enzyme concept arose out of work on the fungus *Neurospora*, it would be work on the bacterium *Escherichia coli* that would reveal the precise details of gene/enzyme relationships. We will return to these ideas in some detail in Chapter 10. However, there is more to the story of bacterial genetics than just mating in *Escherichia coli* K-12. We have already alluded to the process of *transduction*, which was being studied almost simultaneously with mating. And we have noted how a synthesis of research on the temperate bacteriophage *lambda* and the F factor led to the development of the episome concept. Research on bacterial viruses has played a major role in the development of our understanding of genetic phenomena in procaryotes, and we thus turn now to a discussion of phages and their genetic properties.

REFERENCES

Adelberg, E.A. and Burns, S.N. 1960. Genetic variation in the sex factor of *Escherichia coli*. *Journal of Bacteriology* 79: 321–330.

Anderson, T.F. 1958. Recombination and segregation in *Escherichia coli*. *Cold Spring Harbor Symposia on Quantitative Biology* 23: 47–58.

Atwood, K.C., Schneider, L.K., and Ryan, F.J. 1951. Selective mechanisms in bacteria. *Cold Spring Harbor Symposia on Quantitative Biology* 16: 345–355.

Baron, L.S., Spilman, W.M., and Carey, W.F. 1959. Hybridization of *Salmonella* species by mating with *Escherichia coli*. *Science* 130: 566–567.

Beadle, G.W. 1945. Biochemical genetics. *Chemical Reviews* 37: 15–96.

Beadle, G.W. 1963. *Genetics and Modern Biology*. American Philosophical Society, Philadelphia.

Beadle, G.W. and Coonradt, V.L. 1944. Heterocaryosis in *Neurospora crassa*. *Genetics* 29: 291–308.

Bisset, K.A. 1950. *The Cytology and Life-History of Bacteria*. Williams and Wilkins, Baltimore. 136 pp.

Blakeslee, A.F. 1904. Sexual reproduction in the *Mucorineae*. *Proceedings of the American Academy of Arts and Sciences* 40: 205–319.

Boivin, A. and Vendreley, R. 1946. Role de l'acide désoxy-ribonucléique hautement polymérisé dans le déterminisme des caractères héréditaires des bactéries. Signification pour la biochemie générale de l'hérédité. *Helvetica Chimica Acta* 29: 1338–1344.

Bonner, D. 1946. Biochemical mutations in *Neurospora*. *Cold Spring Harbor Symposia on Quantitative Biology* 11: 14–24.

Brinton, C.C., Gemski, P., and Carnahan, J. 1964. A new type of bacterial pilus genetically controlled by the fertility factor of *Escherichia coli* K 12 and its role in chromosome transfer. *Proceedings of the National Academy of Sciences* 52: 776–783.

Cairns, J. 1963. The bacterial chromosome and its manner of replication as seen by autoradiography. *Journal of Molecular Biology* 6: 208–213.

Campbell, A.M. 1962. Episomes. *Advances in Genetics* 11: 101–145.

Cavalli, L.L. and Heslot, H. 1949. Recombination in bacteria: outcrossing *Escherichia coli* K 12. *Nature* 164: 1057–1058.

Cavalli, L.L., Lederberg, J., and Lederberg, E.M. 1953. An infective factor controlling sex compatibility in *Bacterium coli*. *Journal of General Microbiology* 8: 89–103.

Cavalli-Sforza, L.L. 1950. La sessualità nei batteri. *Bolletino Istituto Sieroterapico Milanese* 29: 281–289.

Clark, A.J. and Margulies, A.D. 1965. Isolation and characterization of recombination deficient mutants of *Escherichia coli* K12. *Proceedings of the National Academy of Sciences* 53: 451.

Davis, B.D. 1950. Nonfiltrability of the agents of genetic recombination in *Escherichia coli*. *Journal of Bacteriology* 60: 507–508.

Dettori, R., Maccacaro, G.A., and Piccinin, G.L. 1961. Sex-specific bacteriophages of *Escherichia coli* K 12. *Giornale di Microbiologia* 9: 141–150.

Falkow, S. 1975. *Infectious Multiple Drug Resistance*. Pion Limited, London. 300 pp.

Falkow, S., Wolhieter, J.A., Citarella, R.V., and Baron, L.S. 1964. Transfer of episomic elements to *Proteus*. I. Transfer of F-linked chromosomal determinants. *Journal of Bacteriology* 87: 209–219.

Frédéricq, P. 1954. Transduction génétique des propriétés colinogèes chez *E. coli* et *Shigella sonnei*. *Comptes rendus Société Biologiques* 148: 399–402.

Gowen, J.W. and Lincoln, R.E. 1942. A test for sexual fusion in bacteria. *Journal of Bacteriology* 44: 551–554.

Gray, C.H. and Tatum, E.L. 1944. X-ray induced growth factor requirements in bacteria. *Proceedings of the National Academy of Sciences* 30: 404–410.

Hayes, W. 1952a. Recombination in *Bact. coli* K-12: unidirectional transfer of genetic material. *Nature* 169: 118–119.

Hayes, W. 1952b. Genetic recombination in *Bact. coli* K12: analysis of the stimulating effect of ultra-violet light. *Nature* 169: 1017–1018.

Hayes, W. 1953a. The mechanism of genetic recombination in *Escherichia coli*. *Cold Spring Harbor Symposia on Quantitative Biology* 18: 75–93.

Hayes, W. 1953b. Observations on a transmissible agent determining sexual differentiation in *Bacterium coli*. *Journal of General Microbiology* 8: 72–88.

Hayes, W. 1964. *The Genetics of Bacteria and their Viruses*. John Wiley, New York. 740 pp.

Hayes, W. 1966. Sexual differentiation in bacteria. pp. 201–215 in Cairns, J., Stent, G.S., and Watson, J.D. (ed.), *Phage and the Origins of Molecular Biology*. Cold Spring Harbor Laboratory of Quantitative Biology, Cold Spring Harbor, New York.

Hayes, W. 1969. Introduction: what *are* episomes and plasmids? pp. 4–8 in Wolstenholme, G.E.W. and M. O'Connor (ed.), *Bacterial Episomes and Plasmids*. Little, Brown and Co., Boston.

Hickson, F.T., Roth, T.F., and Helinski, D.R. 1967. Circular DNA forms of a bacterial sex factor. *Proceedings of the National Academy of Sciences* 58: 1731–1738.

Hirota, Y. 1956. Artificial elimination of the F factor in *Bact. coli* K12. *Nature* 178: 92.

Hirota, Y. 1960. The effect of acridine dyes on mating type factors in *Escherichia coli*. *Proceedings of the National Academy of Sciences* 46: 57–64.

Hutchinson, W.G. and Stempen, H. 1954. Sex in bacteria. Evidence from morphology. pp. 29–41 in *Sex in Microorganisms*. Symposium of the American Association for the Advancement of Science, Washington, D.C.

Jacob, F. 1988. *The Statue Within*. Basic Books, New York. 326 pp.

Jacob, F. and Adelberg, E.A. 1959. Transfert de caractères génétiques par incorporation au facteur sexuel d'*Escherichia coli*. *Comptes rendus des Séances de l'Académie des Sciences* 249: 189–191.

Jacob, F. and Wollman, E.L. 1955. Étapes de la recombinaison génétique chez *Escherichia coli* K 12. *Comptes rendus Académie des Sciences* 240: 2566–2568.

Jacob, F. and Wollman, E.L. 1956. Recombination généticque et mutants de fertilité chez *Escherichia coli*. *Comptes rendus Académie des Sciences* 242: 303–306.

Jacob, F. and Wollman, E.L. 1958. Les épisomes, éléments génétiques ajoutés. *Comptes rendus Académie des Sciences* 247: 154–156.

Jacob, F. and Wollman, E.L. 1961. *Sexuality and the Genetics of Bacteria*. Academic Press, New York. 374 pp..

Jacob, F., Brenner, S., and Cuzin, F. 1963. On the regulation of DNA replication in bacteria. *Cold Spring Harbor Symposia on Quantitative Biology* 28: 329–348.

Kay, L.E. 1989. Selling pure science in wartime: the biochemical genetics of G.W. Beadle. *Journal of the History of Biology* 22: 73–101.

Lederberg, E.Z. 1950. *Genetic Control of Mutability in the Bacterium* Escherichia coli. Ph.D. thesis, University of Wisconsin, Madison.

Lederberg, J. 1947a. *Genetic Recombination in* Escherichia coli. Ph.D. thesis, Yale University, New Haven.

Lederberg, J. 1947b. Gene recombination and linked segregations in *Escherichia coli*. *Genetics* 32: 505–525.

Lederberg, J. 1948. Problems in microbial genetics. *Heredity* 2: 145–198.

Lederberg, J. 1950a. Isolation and characterization of biochemical mutants of bacteria. *Methods in Medical Research* 3: 5–22.

Lederberg, J. 1950b. The β-D-galactosidase of *Escherichia coli*, strain K-12. *Journal of Bacteriology* 60: 381–392.

Lederberg, J. 1950c. The selection of genetic recombinations with bacterial growth inhibitors. *Journal of Bacteriology* 59: 211–215.

Lederberg, J. 1950d. Genetic studies with bacteria. pp. 263–289 in Dunn, L.C. (ed.), *Genetics in the 20th Century*. Macmillan, New York.

Lederberg, J. 1951a. Streptomycin resistance: a genetically recessive mutation. *Journal of Bacteriology* 61: 549–550.

Lederberg, J. 1951b. Prevalence of *Escherichia coli* strains exhibiting genetic recombination. *Science* 114: 68–69.

Lederberg, J. 1955a. Genetic recombination in bacteria. *Science* 122: 920.

Lederberg, J. 1955b. Recombination mechanisms in bacteria. *Journal of Cellular and Comparative Physiology* 45 (Supplement 2): 75–107.

Lederberg, J. 1957. Bacterial reproduction. *Harvey Lectures* 53: 69–82.

Lederberg, J. 1979. Edward Lawrie Tatum. *Annual Review of Genetics* 13: 1–5.

Lederberg, J. 1986. Forty years of genetic recombination in bacteria. *Nature* 324: 627–629.

Lederberg, J. 1987. Genetic recombination in bacteria: a discovery account. *Annual Review of Genetics* 21: 23–46.

Lederberg, J. 1990. Edward Lawrie Tatum. *Biographical Memoirs*. National Academy of Sciences. (In press.)

Lederberg, J. and Lederberg, E.M. 1952. Replica plating and indirect selection of bacterial mutants. *Journal of Bacteriology* 63: 399–406.

Lederberg, J. and Tatum, E.L. 1946a. Novel genotypes in mixed cultures of biochemical mutants of bacteria. *Cold Spring Harbor Symposia on Quantitative Biology* 11: 113–114.

Lederberg, J. and Tatum, E.L. 1946b. Gene recombination in *Escherichia coli*. *Nature* 158: 558.

Lederberg, J. and Tatum, E.L. 1953. Sex in bacteria: genetic studies, 1945–1952. *Science* 118: 169–175.

Lederberg, J. and Zinder, N.D. 1948. Concentration of biochemical mutants of bacteria with penicillin. *Journal of the American Chemical Society* 70: 4267.

Lederberg, J., Cavalli, L.L., and Lederberg, E.M. 1952. Sex compatibility in *Escherichia coli*. *Genetics* 37: 720–730.

Lederberg, J., Lederberg, E.M., Zinder, N.D., and Lively, E.R. 1951. Recombination analysis of bacterial heredity. *Cold Spring Harbor Symposia on Quantitative Biology* 16: 413–441.

Loeb, T. 1960. Isolation of a bacteriophage specific for the F^+ and Hfr mating types of *Escherichia coli* K-12. *Science* 131: 932–933.

Löhnis, F. 1921. Studies upon the life cycles of the bacteria. *Memoirs of the National Academy of Sciences* 16: 1–252.

Luria, S.E. 1947. Recent advances in bacterial genetics. *Bacteriological Reviews* **11:** 1–40.

Luria, S.E. and Delbrück, M. 1943. Mutations of bacteria from virus sensitivity to virus resistance. *Genetics* 28: 491–511.

Marmur, J., Rownd, R., Falkow, S., Baron, L.S., Schildkraut, C., and Doty, P. 1961. The nature of intergeneric episomal infection. *Proceedings of the National Academy of Sciences* 47: 972–979.

Meynell, G.G. 1973. *Bacterial Plasmids*. MIT Press, Cambridge, Massachusetts. 164 pp.

Nelson, T.C. and Lederberg, J. 1954. Postzygotic elimination of genetic factors in *Escherichia coli*. *Proceedings of the National Academy of Science* 40: 415–419.

Reeves, P. 1972. *The Bacteriocins*. Springer-Verlag, New York. 142 pp.

Roth, T.F. and Helinski, D.R. 1967. Evidence for circular DNA forms of a bacterial plasmid. *Proceedings of the National Academy of Sciences* 58: 650–657.

Ryan, F. J. and Lederberg, J. 1946. Reverse-mutation and adaptation in leucineless *Neurospora*. *Proceedings of the National Academy of Sciences* 32: 163–173.

Sager, R. and Ryan, F.J. 1961. *Cell Heredity*. John Wiley, New York. 411 pp.

Sapp, J. 1987. *Beyond the Gene*. Oxford University Press, New York. 266 pp.

Sherman, J.M. and Wing, H.U. 1937. Attempts to reveal sex in bacteria; with some light on fermentative variability in the coli-aerogenes group. *Journal of Bacteriology* 33: 315–321.

Sonneborn, T.M. 1937. Sex, sex inheritance and sex determination in *Paramecium aurelia*. *Proceedings of the National Academy of Sciences* 23: 378–395.

Sonneborn, T.M. 1943. Gene and cytoplasm. *Proceedings of the National Academy of Sciences* 29: 329–343.

Stanier, R.Y. and van Niel, C.B. 1962. The concept of a bacterium. *Archiv für Mikrobiologie* 42: 17–35.

Tatum, E.L. 1945. X-ray induced mutant strains of *Escherichia coli*. *Proceedings of the National Academy of Sciences* 31: 215–219.

Tatum, E.L. 1959. A case history in biological research. *Science* 129: 1711–1719.

Tatum, E.L. and Beadle, G.W. 1945. Biochemical genetics of *Neurospora*. *Annals of the Missouri Botanical Garden* 32: 125–253.

Tatum, E.L. and Lederberg, J. 1947. Gene recombination in the bacterium *Escherichia coli*. *Journal of Bacteriology* 53: 673–684.

Watanabe, T. 1963. Infective heredity of multiple drug resistance in bacteria. *Bacteriological Reviews* 27: 87–115.

Watanabe, T. and Fukasawa, T. 1961. Episome-mediated transfer of drug resistance in *Enterobacteriaceae*. I. Transfer of resistance factors by conjugation. *Journal of Bacteriology* 81: 669–678.

Watson, J.D. 1968. *The Double Helix*. Atheneum, New York.

Watson, J.D. and Hayes, W. 1953. Genetic exchange in *Escherichia coli* K12: evidence for three linkage groups. *Proceedings of the National Academy of Sciences* 39: 416–426.

Wollman, E.L. 1966. Bacterial conjugation. pp. 216–225 in Cairns, J., Stent, G.S., and Watson, J.D. (ed.), *Phage and the Origins of Molecular Biology*. Cold Spring Harbor Laboratory of Quantitative Biology, Cold Spring Harbor, New York.

Wollman, E.L. and Jacob, F. 1955. Sur le mécanisme du transfert de matériel génétique au cours de la recombinaison chez *Escherichia coli* K 12. *Comptes rendus Académie des Sciences* 240: 2449–2451.

Wollman, E.L. and Jacob, F. 1959. *La sexualité des bacteéries*. Masson, Paris. 247 pp.

Wollman, E.L., Jacob, F., and Hayes, W. 1956. Conjugation and genetic recombination in *Escherichia coli* K-12. *Cold Spring Harbor Symposia on Quantitative Biology* 21: 141–162.

Zelle, M.R. and Lederberg, J. 1951. Single-cell isolations of diploid heterozygous *Escherichia coli*. *Journal of Bacteriology* 61: 351–355.

Zinder, N.D. 1953. Infective heredity in bacteria. *Cold Spring Harbor Symposia on Quantitative Biology* 18: 261–269.

Zinder, N.D. 1960a. Hybrids of Escherichia and Salmonella. *Science* 131: 813–815.

Zinder, N.D. 1960b. Sexuality and mating in Salmonella. *Science* 131: 924–926.

Zinder, N.D. 1980. Portraits of viruses: RNA phage. *Intervirology* 13: 257–270.

Zuckerman, H. and Lederberg, J. 1986. Postmature scientific discovery? *Nature* 324: 629–631.

6
PHAGE

The history of the bacterial viruses, bacteriophages, is long and confusing, and much of it is not relevant to an understanding of modern bacterial genetics. Phage research is often divided into two eras, pre-Delbrück and post-Delbrück: "...the continuity of 'modern' phage research really dates only from 1938, when Max Delbrück took up work in this field" (Stent, 1963). However, this may be an inaccurate representation, since phage genetics, the hallmark of the Delbrück era, was actually carried out by F.M. Burnet from the late 1920s through the mid 1930s. Furthermore, although Delbrück's contributions (discussed below) were numerous and meritorious, there is some reason to believe that his powerful influence seriously hindered certain aspects of phage research, such as *lysogeny* (see Chapter 7).

The discovery and early history of phage research, including the controversies concerning priority and interpretation, have been given in considerable detail by Varley (1986). Stent (1963) provides some historical discussion of phage research up until the early 1960s and provides a brief introduction to the early, pre-Delbrück, era. Mathews (1971) discusses the history of modern phage research and presents the literature up to the end of the 1960s.

6.1 DISCOVERY AND EARLY WORK

Although viruses that affected plants and animals had been known since the late 19th century, the discovery of a virus that attacked bacteria was first made by Frederick W. Twort in England in 1915 and by Felix d'Herelle in France in 1917 (Twort, 1915; d'Herelle, 1917). Although there was controversy subsequently as to whether d'Herelle was aware of Twort's paper when he made his initial discovery, it seems likely that the two papers were independent (Duckworth, 1976). Considerations of priority aside, the "nature" of the phage phenomenon was uncertain for many years.

Twort's Work

Frederick Twort was a medical bacteriologist working at a veterinary research institute in London. During the course of work on the evolutionary origin of pathogenic from nonpathogenic bacteria, he made his observations on phage. On agar slants plated with ultrafiltrates of vaccinia virus, Twort observed that colonies of a contaminating micrococcus were undergoing what he termed a "glassy transformation." Twort showed that the material responsible for this phenomenon was transmissible, filterable, invisible in the light microscope, and did not multiply in the absence of living bacteria. In his first paper, Twort advanced three hypotheses to explain these observations: (1) The bacterial "disease" might be a stage in the life cycle of the micrococcus, a stage in which the independent units were extremely small; (2) the material might be a filterable virus, an obligate parasite of the micrococcus; (3) the material might be an enzyme secreted by the bacterium that had the power to induce similar bacteria to produce the same enzyme. One or another of the latter two alternatives, a virus or a self-reproducing enzyme, became the favored theories of the various "schools" of phage research throughout the 1920s. With World War I intervening, Twort abandoned research on the "glassy transformation" phenomenon.

d'Herelle's Work

Felix d'Herelle was a French-Canadian bacteriologist who had been raised in France. Most of his research was done in association with the Institut Pasteur, either in Paris or at one of the many satellite laboratories around the world. d'Herelle first discovered the bacteriophage phenomenon during studies on *Shigella* dysentery in humans. d'Herelle was interested in the immunological phenomena involved in the recovery of patients from dysentery. As part of a study on fecal material from dysentery patients, he filtered bloody stools and added them to cultures of a pathogenic *Shigella*. After incubation, the resulting culture was also filtered, and a drop of the filtrate was added to further broth cultures, as well as to a culture on agar.

> I placed the tube of broth culture and the agar plate in an incubator at 37°. It was the end of the afternoon, in what was then the mortuary, where I had my laboratory. The next morning, on opening the incubator, I experienced one of those rare moments of intense emotion which reward the research worker for all his pains: at the first glance I saw that the broth culture, which the night before had been very turbid, was perfectly clear: all the bacteria had vanished, they had dissolved away like sugar in water. As for the agar spread, it was devoid of all growth and what caused my emotion was that in a flash I had understood: what caused my clear spots was in fact an invisible microbe, a filtrable virus, but a virus parasitic on bacteria (d'Herelle, 1949).

In his first paper, d'Herelle (1917) coined the word *bacteriophage*, signifying an entity that "eats" bacteria. Although the nature of the bacteriophage remained a controversy for many years, d'Herelle's term, proving more felicitous than Twort's "glassy transformation," was retained. Immediately, d'Herelle postulated a connection between phage infection and patient recovery from dysentery. He considered bacteriophage to play a major role in immunity, and he also viewed it as a potential therapeutic agent. Although these ideas have not borne fruit, they certainly motivated large amounts of research on bacteriophage in the first several decades after its discovery.

The Nature of the Bacteriophage

In his early work, d'Herelle (1921) showed that phage was particulate and could be titrated by limiting dilution or by counting plaques (*taches vierges* or *plages* in French). He found that if an active phage was diluted almost to the limit of its activity and equal small amounts of this dilution were added to a series of cultures, some tubes showed typical lysis after incubation and evidence of high phage titer, whereas other tubes showed normal growth with no evidence of the presence of phage. This experiment, which is essentially the *most-probable-number* procedure of bacteriologists, led d'Herelle to conclude (rightly) that the active bacteriophage agent was transmissible and particulate. Another important observation germane to ideas on the nature of phage was that the phage particles could be shown to *absorb* to sensitive, but not to insensitive, bacteria. Finally, the lysis of bacteria by phage required active growth of the bacteria. The phenomenon was thus not a simple enzymatic lysis such as that of the lysozyme (discovered about the same time by Fleming, 1922).

d'Herelle also showed that after phage absorption, there was a latent period, after which there was a sudden rise in the phage titer, increases of from 6- to 60-fold often occurring within 15 minutes. The titer would then remain constant for another 30–45 minutes until another large increase occurred. It is noteworthy that these detailed observations on the discontinuous nature of phage reproduction preceded by many years the seminal work of Ellis and Delbrück on the "one-step growth curve" (see later). Work during the 1920s and 1930s on the physical nature of the phage particle, reviewed by Burnet (1930) and Stent (1963), showed that phage particles were indeed submicroscopic, filterable, of a linear dimension of the order of 0.1 μm, and contained DNA and protein (see Section 6.12).

d'Herelle's interpretation of bacteriophage as a virus was quick to raise controversy. His most prominent opponents were the Belgian immunologists André Gratia and Jules Bordet. Bordet, one of the most distinguished immunologists of his day and a Nobel laureate, had discovered complement and had developed the so-called "unitary hypothesis of antibody," which stated that various immune phenomena such as agglutination, precipitation, and bacteriolysis were all due to various interactions of the antigen with a single kind of antibody molecule. Bordet, objecting especially to d'Herelle's insistence on a role of bacteriophage in immunity, quickly developed an alternative theory for phage:

> According to d'Herelle, the lysis is due to a living being, to a filtering virus. We, on the contrary, believe that the lytic principle originates from the bacteria themselves, which, when touched by this active substance are capable of regenerating it, the factor responsible for the phenomenon being thus unceasingly reproduced—on the condition, however, that the bacteria be still living and provided with the alimentary substances necessary for their growth (Bordet, 1922).

Intracellular or Extracellular Origin of Phage

The idea of Bordet and Gratia that phage had an intracellular origin was, of course, founded in experiment. In his initial work, Bordet had shown that certain bacterial strains could produce bacteriophages active against other, closely related, strains. These strains were, as we now know, *lysogenic* (see Chapter 7). d'Herelle insisted that Bordet's phage-producing cultures were contaminated and that the phage could be eliminated by careful restreaking. Both workers were correct: d'Herelle was dealing with virulent phages, whereas Bordet was dealing with temperate phages. The inability to make this important distinction resulted in extensive confusion and controversy during the first two decades of bacteriophage research. Even Max Delbrück refused to believe the facts about lysogeny (see Section 7.4). In the present chapter, we deal strictly with phenomena involving virulent phage, the area that Delbrück and the so-called "phage group" concentrated on. The whole controversy between the d'Herelle and the Bordet schools is well covered in Burian, Gayon, and Zallen (1988). Lysogeny will be discussed in Chapter 7.

6.2 BACTERIOPHAGE RESEARCH IN THE 1930s

Burnet's Work

Serious genetic studies on phage did not begin until the work of Delbrück, Hershey, and Luria in the 1940s, but F.M. Burnet did some earlier work that served as an important starting point. Although d'Herelle believed that there was only one phage that attacked all bacteria (the so-called "unicity" theory of phage), Burnet showed that "phage" was not a single entity but that a great variety existed, with varying physical and biological properties and with species specificity. Furthermore, if neutralizing antibodies were made against one phage, such antibodies would cross-react with some, but not all, other bacteriophages attacking the same species. Thus, Burnet showed that serological cross-reaction rather than species specificity was the best criterion for establishing the relatedness of phages. Adsorption studies showed that the fixation of phage particles to the bacterial cell was also specific, and Burnet viewed adsorption as analogous to an antigen-antibody reaction. He also showed that phage-resistant variants appeared in phage-infected cultures and that these variants were no longer able to adsorb phage. Moreover, the inability to adsorb phage was due to the loss of a surface receptor, and cell-free extracts of phage-sensitive cells could fix phage particles specifically, whereas those from phage-resistant cells could not. Burnet also did important work on lysogeny (see Chapter 7), and by 1936, when he left phage research for immunology, the stage was well set for Max Delbrück.

Northrop's Work

Another early student of bacteriophage was John H. Northrop, a biochemist at the branch of the Rockefeller Institute in Princeton, New Jersey. Northrop shared the Nobel Prize with James B. Sumner and Wendell Stanley in 1946 for his work on the crystallization of enzymes. He discovered that the enzymes *trypsin* and *chymotrypsin* were formed in the pancreas as precursors (zymogens), which he called *trypsinogen* and *chymotrypsinogen*, respectively, and that the conversion of precursor to active enzyme was the result of a proteolytic attack on the precursor (Northrop, 1937). Since trypsin itself would cause the trypsinogen → trypsin transformation, the reaction was autocatalytic. He looked for a similar model for bacteriophage production.

Northrop was Stanley's senior colleague at Rockefeller/Princeton (Kay, 1986b). Because of Stanley's work on the crystallization of tobacco mosaic virus, a relationship between viruses and enzymes seemed evident to Northrop, and he initiated studies on the multiplication of a staphylococcus phage. Although the phage was not crystallized, it was prepared in highly purified form. Its size was determined by sedimentation and diffusion analysis and it was shown chemically to be a nucleoprotein. Much of this work was done in association with a colleague, A.P. Krueger. From studies on the kinetics of phage formation, Northrop concluded that phage produced itself autocatalytically from a precursor present in normal cells, rejecting d'Herelle's idea that phage was a living organism (Northrop, 1937).

6.3 DELBRÜCK AND THE BEGINNINGS OF MODERN RESEARCH

Max Delbrück became the hub of the so-called "phage group," and his background and life have been the interest of numerous scientists and historians. Much of the "flavor" of the Delbrück era can be obtained from the *Festschrift* that took place on the occasion of Delbrück's sixtieth birthday (Cairns, Stent, and Watson, 1966). A view of his life is given in the book by Fischer (1985) (for an English translation, see Fischer and Lipson, 1988). Kay (1985; 1986a) has presented the background of Delbrück's switch from physics to biology in the 1930s, and his relationship with colleagues at CalTech. A short biography with a useful summary of key dates in Delbrück's life was published by Hayes (1982), and numerous remembrances of Delbrück and his times can be found in Cairns, Stent, and Watson (1966) and in Luria (1984). The Delbrück archives at CalTech are superb, and Fischer has made extensive use of these archives in his biography (as have I in the present book).

Although Delbrück's life and early work are interesting and useful in understanding his place and role in the origins of molecular biology, it would involve unnecessary duplication to repeat the details here. Therefore, only a brief outline of Delbrück's life is given; the following concentrates on his scientific work on phage. It should be noted that only part of the phage work is relevant to the purposes of the present volume, so that numerous areas, well-covered in Stent (1963), are omitted.

Biographical Notes

Max Delbrück was born in Berlin in 1906, the son of a professor of history at the University of Berlin. Attending the University of Göttingen, he first studied astronomy but switched to theoretical physics, working under Max Born and Walter Heitler. He received his doctorate in 1930 with a thesis on quantum aspects of chemical bonding. He spent the fall of 1929 at the Physics Department of the University of Bristol (U.K.), where he concentrated on improving his English. During 1931–1932 he studied with the eminent Niels Bohr in Copenhagen, and with Wolfgang Pauli in Zurich, under sponsorship by the Rockefeller Foundation (U.S.). This was the beginning of an important relationship of Delbrück with the Rockefeller Foundation that would be critical in his subsequent move to the United States.

Niels Bohr had an important influence on Delbrück, and was responsible for Delbrück's initial interest in biology (Kay, 1985). One of Bohr's important concepts in theoretical physics was the principle of complementarity. Although complementarity is more a philosophical than a scientific principle, it had great influence on the development of modern physics, and Bohr felt that complementarity also had implications for experimental biology. According to the complementarity principle, when one is dealing with systems of extreme complexity it is impossible to make experimental observations without in the process modifying the system under study. Bohr felt that because full development of biological principles would require disturbance of the living system, life would not be reducible to concepts of physics. Delbrück was intrigued with the biological implications of complementarity (Delbrück, 1949a), and continued to correspond with Bohr about this matter throughout his career. Delbrück's "black box" approach to phage has often been noted (Cohen, 1968), and Delbrück's reluctance to "do" biochemistry on phage, which required tearing the system apart, possibly arose from his initial fascination with Bohr's complementarity principle (Blaedel, 1988). We can also contrast here the general approach of the physicist with that of the chemist. The physicist designs experiments that lead to an understanding of deep principles, whereas the chemist fusses with details that have a physical basis but can be studied without special recourse to the underlying physical principles. As a physicist, Delbrück was inclined to ignore the chemical details.

Fate may have played a role in Delbrück's turn toward biology, as the following quotation suggests: "...Bohr availed himself of every opportunity to make plain his ideas about complementarity...There is a special story about the lecture which he prepared for the inaugural meeting of the Second International Congress of Light Therapy, which took place in Copenhagen in 1932. On the same day that the meeting took place, the German physicist Max Delbrück, who at that time was associated with the Bohr institute, returned to Copenhagen after a journey. Léon Rosenfeld picked him up at the railway station and by going directly to Christiansborg, where the lecture was to be held, they arrived just in time. 'To claim,' Rosenfeld wrote later, 'that we were fascinated by the lecture, whose title was "Light and Life," would be a romantic overstatement, but it is a fact that when Delbrück read the paper afterwards and pondered over it, he became so enthusiastic about the vision which it revealed over the far-flung expanses of biology that he immediately decided to take up the challenge'" (Blaedel, 1988).

Upon his return to Berlin from Copenhagen in 1932, Delbrück worked for five years at the Kaiser Wilhelm Institute for Chemistry, as an assistant of Lise Meitner, who was collaborating with Otto Hahn on radiation chemistry. In addition to significant work on the physical chemistry of ionizing radiation, Delbrück began biological work. A few years earlier, the American geneticist Hermann Muller had discovered that ionizing radiation was able to induce mutations (see Section 2.8), and in 1932 Muller spent some time in Berlin with the Russian emigré geneticist Nicolai Timoféeff-Ressovsky (Carlson, 1971). Muller and Timoféeff-Ressovsky studied the types of genetic changes in *Drosophila* induced by different wavelengths of radiation, one of the few approaches available in the 1930s for ascertaining the nature of the gene. Delbrück became familiar with Timoféeff-Ressovsky's work and, together with the experimental physicist K.G. Zimmer (Zimmer, 1966), a collaboration was initiated that resulted in a lengthy paper that provided the first direct evidence that a single mutation in a gene was the result of a single ionization event. This was the first work to analyze genes as "targets" of radiation, and it became the foundation of an important field of genetics called *target theory*. In the collaboration, Timoféeff-Ressovsky provided the biological and genetic work, Zimmer the measurements of radiation doses, and Delbrück the theoretical analysis. This paper, published in a rather obscure journal (Timoféeff-Ressovsky, Zimmer, and Delbrück, 1935), provided a quantum model of gene mutation that turned out to be incorrect, but it stimulated considerable interest and can be looked upon as one of the earliest approaches to molecular biology. In this paper, Delbrück calculated the size of a gene as 9 μm, between 50 and 100 times larger than the actual size. (Research using target theory to probe the size of the gene continued through the 1960s in many laboratories, using such experimental systems as phage and transforming DNA. Discussion of this interesting work is beyond the scope of the present book.)

Although the paper by Timoféeff-Ressovsky, Zimmer, and Delbrück (1935) was published in an obscure journal, reprints found their way to a number of people, including the well-known theoretical physicist Erwin Schrödinger. In 1944, Schrödinger published a small book called *What Is Life?* in which he referred in detail to the Delbrück model for the gene (Schrödinger, 1944). Although the model, and most of Schrödinger's book, were quickly outmoded, the exposure that Delbrück received through the book caused a number of physicists to turn to Delbrück when they became interested in biology. In the post-World-War-II period, numerous such scientists found their way to Delbrück's laboratory at CalTech (Fig. 6.1).

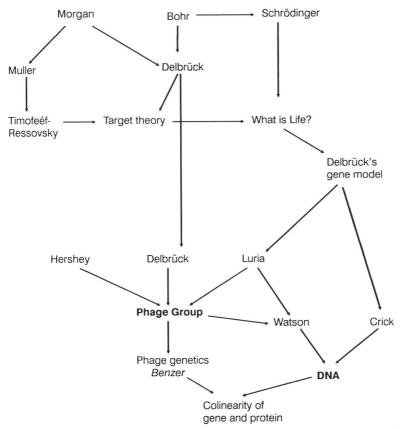

Figure 6.1 Influences on Delbrück and a few of the influences by Delbrück on others. Note that most of the individuals listed became Nobel Prize winners.

CalTech and Delbrück's Initial Phage Work

Following up his interest in genetics, and with financial support from the Rockefeller Foundation, Delbrück came in 1937 to the California Institute of Technology (CalTech). He selected CalTech because of T.H. Morgan, who at that time (see Section 2.7) was head of the Biology Division. Although intending originally to work on *Drosophila*, Delbrück quickly decided that the fruit fly provided too complicated a system, and he turned to bacteriophage. (Years later, Delbrück recalled: "I didn't make much progress in reading these forbidding-looking papers [on *Drosophila* genetics]; every genotype was about a mile long, terrible" [Fischer and Lipson, 1988]). The story of how Delbrück learned of bacteriophage has been well told by Emory Ellis (1966) and by Fischer and Lipson (1988). Ellis had begun to study phage through an interest in cancer research. Since certain cancers were caused by viruses, and since phages were viruses, Ellis chose phage as the simplest system to approach the cancer problem. Isolating phage from sewage, he used d'Herelle's plaque assay method to study phage growth quantitatively. From the beginning, Ellis's interest was in the phage growth process, rather than in the medical or bacteriological aspects of phage that had interested d'Herelle and many of the other early phage workers. Fortunately, the phage that Ellis isolated was virulent, so that problems of lysogeny (see Chapter 7) did not enter.

One-step Growth

As discussed earlier, d'Herelle had already shown the step-like nature of phage growth, but Ellis and Delbrück refined the analysis and put it on a quantitative basis (Ellis and Delbrück, 1939). With the experiment set up in the proper way, it was possible to derive two important parameters from the one-step growth analysis, the latent period and the burst size (Fig. 6.2). It is important to emphasize that the Ellis/Delbrück paper provided the solid basis that was necessary for all subsequent work on phage genetics and biochemistry.

For Delbrück, phage had both genetic and virological aspects. The application of phage to genetics was brought to him first by Ellis, as discussed above, but other discussions of the genetic relevance of viruses were being published at this time, of which the most influential was that of Wendell Stanley (1938). At the 1941 Cold Spring Harbor symposium, which was attended by both Luria and Delbrück, Stanley wrote the following: "There is a striking similarity between the properties that have been found for the viruses...and the properties that have been ascribed to genes...it...seems quite likely that a study of viruses will provide much information pertinent to the question of the nature of genes and of their mode of action" (Stanley and Knight, 1941). Some of Stanley's work and influence is discussed by Cohen (1979, 1986).

The relevance of phage for broader questions of virology became almost a mission for Delbrück. He insisted that bacteriophages should be considered *viruses* that attacked *bacteria*, rather than the something different that might be implied from d'Herelle's term *bacteriophage*. This point was made extensively in the introduction to Delbrück's first review on bacterial viruses

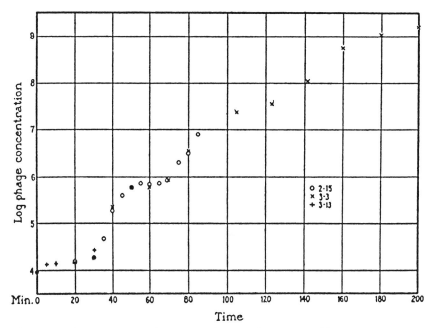

Figure 6.2 One-step growth curves of an uncharacterized bacteriophage against *Escherichia coli*. A diluted phage preparation was mixed with a suspension of bacteria containing 2×10^8 cells per ml, and diluted 1 to 50 after 3 minutes. At this time, 70% of the phage had become adsorbed. The total number of infective centers was determined at intervals from plaque assays. Three experiments, done on different days, are plotted. (Reprinted, with permission, from Ellis and Delbrück, 1939.)

(Delbrück, 1942). "In d'Herelle's view the bacteriophages are small cells, in Bordet's view they are modified bacterial proteins. The issue is one which can only be settled by a clearer understanding of what actually goes on when the bacteriophage is reproduced...The bacteriophages are to be classified with the animal and plant viruses as bacterial viruses."

"Many of us were later involved in a plot to interest the 'real plant and animal virologists' in bacteriophages by calling them 'bacterial viruses'" (Anderson, 1966; see also Delbrück, 1959). This idea became especially important because Stanley (1935) had crystallized tobacco mosaic virus, later shown to be a nucleoprotein (Kay, 1986b). (The term "nucleoprotein" had a different meaning in 1939 than it does today, since the ionic nature of the linkage between protein and nucleic acid was not known and the accepted building block of nucleic acid was the tetranucleotide; see Chapter 9).

Continuing the phage work after Ellis returned to cancer research, Delbrück focused on the nature of the replication process itself, using the one-step growth procedure as a tool. He recognized clearly the relevance of phage for the problem of the reproduction of the gene (Delbrück, 1942). The geneticist Hermann Muller had first suggested, many years ago, that phage might provide important insights into the nature of the gene. In a 1922 paper that was essentially ignored in its time (Carlson, 1971) but was widely quoted later, Muller discussed the newly discovered bacterial viruses, then known as *d'Herelle bodies*: "If these d'Herelle bodies were really genes, fundamentally like our chromosome genes, they would give us an utterly new angle from which to attack the gene problem" (Muller, 1922). (Muller eventually received recognition from molecular biologists for this prescient statement. I heard Seymour Benzer quote the above statement about 1961 in the introduction to a lecture he gave at Indiana University in Bloomington, with Hermann Muller sitting in the front row. See Chapter 1 for a more extensive quotation from Muller's paper.) However, it is important to note that at the time Muller made this statement, the word "gene" had quite a different meaning than it later acquired in the age of molecular biology.

During the years 1937–1942, Delbrück was especially concerned with what he considered to be the erroneous and rather dangerous views of John Northrop and A.P. Krueger regarding the nature of phage replication. On the basis of Northrop's idea that phage replication was like the conversion of trypsinogen to trypsin (see Section 6.2), Northrop and Krueger hypothesized that the bacterial cell possessed a precursor that was converted into phage. According to Northrop, Delbrück was not studying virus *replication* but some simple conversion of precursor to active phage. Krueger published numerous papers claiming to have shown the existence inside the bacterial cell of the putative precursor. According to Delbrück, the quantitative assay method that Krueger used was flawed, rendering the conclusion of a precursor meaningless (Delbrück, 1942). "For several years...my papers were largely directed at destroying Northrop and Krueger" (Fischer and Lipson, 1988). Actually, Northrop's work was on a lysogenic system, but, as discussed in Chapter 7, at this time Delbrück did not believe in lysogeny. Northrop was the editor of the *Journal of General Physiology*, where Delbrück published his early phage papers. Delbrück experienced difficulty having one of his papers accepted because it did not agree with Northrop's theories. (This correspondence is in the Northrop file in the Delbrück archives.)

Delbrück completed his Rockefeller Foundation fellowship in the fall of 1939. Due to the worsening political situation in Germany, he remained in the United States. With further help from the Rockefeller Foundation, which was impressed with his research, a position was found for him in the Physics Department at Vanderbilt University, where he remained from 1940 throughout World War II. Although he taught physics, his research continued on phage, and it was during this period that he initiated the important collaborations with S.E. Luria and A.D. Hershey (see below).

Phage replication, although remaining a major Delbrück interest, did not really become understood until after detailed biochemical work was carried out, especially using radioactive tracers, in the post-World-War-II period (see Section 6.12). The areas of phage research that were to first come to fruition under Delbrück's influence were those related to genetics and radiobiology, and involved the first important collaborations between Luria and Delbrück.

6.4 S.E. LURIA

Salvador E. Luria was trained in medicine in Italy and began to work on phage in Enrico Fermi's laboratory in Rome (Luria, 1984). He became acquainted with Delbrück's paper on target theory but initiated research on phage independently of Delbrück, with the idea that it might provide a model system for understanding the gene. He left Italy at the height of the racial terrors brought about by Mussolini's collaboration with Hitler, finding his way to the Institute of Radium in Paris. Continuing his phage research, he worked with Eugène Wollman (an early student of lysogeny and the father of Elié Wollman, see Section 7.5) on the kinetics of the inactivation of phage by X rays. This work showed that phage inactivation resembled an elementary quantum process and was in agreement with the view that phages are monomolecular structures: "...the inactivation should be conceived as a quantic transition of such a molecule. In this respect, our results are closely similar to those obtained by Timoféef-Ressovsky on gene radio-mutations in Drosophila" (Wollman, Holweck, and Luria, 1940). Completely independently of Delbrück, Luria developed the Poisson relationship for phage infection and showed: "These results prove that bacterial lysis can perhaps be determined by the action of a single bacteriophage particle" (Luria, 1939). The statistical analysis of the infection process made possible the development of a precise quantitative assay that permitted a careful kinetic analysis of the radiation inactivation process.

Forced to leave France when the Germans entered in 1940, Luria found his way to the United States, where, helped financially by the Rockefeller Foundation, he received a fellowship at the College of Physicians and Surgeons of Columbia University, in the laboratory of the physicist Frank M. Exner. Luria and Exner (1941) showed that the kinetics of X-ray inactivation of various phages depended on the suspending medium and could be separated into two events, direct and indirect. When phage was suspended in distilled water, there was an "indirect" effect as the result of the production of a short-lived inactivating agent. If the phage was suspended in solutions of proteins, this indirect effect was nullified and only the "direct" effect of radiation on the phage particles was observed. To use radiation to probe the "target" of inactivation, it was obviously critical to eliminate the indirect effect. In reviewing these results, Delbrück (1942) stated that Luria's results were "...of the utmost importance, both for the virus problem and for radiation biology in general." (It was later shown that ionizing radiation induced free radicals that inactivated the phage. Constituents of broth scavenged such reactive molecules, so that in broth inactivation was a "pure" ionizing radiation effect.)

Luria and Delbrück

Luria (1984) has described in detail how his collaborations with Delbrück began, first at Cold Spring Harbor Laboratory and subsequently at Indiana University and Vanderbilt University. Familiar with each other's work from the literature, they arranged to meet at the American Association for the Advancement of Sciences meeting in Philadelphia in December 1940. They became immediate friends and began their collaboration in Luria's laboratory at Columbia.

> From the start Delbrück struck me as a dominant personality. Tall, and looking even taller because of his extreme thinness, moving and speaking sparingly and softly but with great precision, he conveyed the impression that whatever he said had been carefully thought out. His seriousness was occasionally broken by sparks of amusement, often produced by unexpected contrasts, and especially at the expense of someone's pretentiousness. His humor was usually gentle but could be deflating, although never cruel...He was certainly not lavish with approval. When I

published my first book, *General Virology*, he never mentioned it to me except to point out a misprint in a footnote (Luria, 1984).

(Delbrück always marched to his own drummer. I heard him lecture several times, but the most memorable was in the late 1960s at Indiana University when he emptied a packed lecture hall by talking not about molecular genetics but Copernicus. The title of his lecture, "A Renaissance Philosopher," was sufficiently vague so that most expected a topic of interest to biologists. As the talk wore on and it became clear that the talk was indeed to be about a renaissance philosopher, students started to leave. Soon the departure trickle turned to a flood. Despite the clatter of upturned writing boards, Delbrück pressed onward and finished on a satisfying note. By then, the room was only one-third full!)

6.5 BACTERIOPHAGE HAVE TAILS

One of the more exciting developments of the war years was the use of the electron microscope to study bacteriophage (Anderson, 1966). The first U.S. electron microscope was installed by the Radio Corporation of America (RCA) at its Camden, New Jersey research laboratories; Anderson received a National Research Council fellowship to use this new instrument to study biological problems. Luria came to the RCA laboratory to discuss the possibility of estimating the sizes of some bacteriophages to check the cross-sections estimated by X-ray killing, and Anderson and Luria immediately entered into a collaboration. To everyone's surprise, phage particles proved to be complex structures, consisting of a round "head" and a much thinner "tail," giving them a "peculiar sperm-like appearance" (Luria and Anderson, 1942). Other important observations: (1) Each phage had a characteristic morphology, and the size corresponded with the size predicted by the X-ray inactivation studies; (2) the characteristic structures were absent in uninfected cultures; (3) the particles were adsorbed to sensitive host strains but not to insensitive strains; (4) adsorption of the phage particle to the host cell appeared to be by the tail; (5) the same kind of particle could be observed during the lysis of infected cells. The discovery of the complex morphology of bacteriophages, of course, opened more questions than it answered, although it was clear that no one could any longer think of a phage as a soluble enzyme or low-molecular-weight material (see also the footnote on J.J. Bronfenbrenner in Section 6.8).

In the summer of 1942, RCA installed an electron microscope for Anderson at the Marine Biological Laboratory (MBL) at Woods Hole, and Delbrück and Luria both came to do further studies on phage structure (Luria, Delbrück, and Anderson, 1943). An important conclusion of this work was that replication must occur *inside* the cell because until the lysis process occurred, no evidence for phage particles on the surface of the cell could be seen. Because the number of phage particles seen attached to the outside was very small, it was evident that only very few particles "entered" the cell, a fact of "greatest consequence." The complexity of phage particles was at variance with Stanley's idea that viruses were "simple" nucleoprotein structures (Anderson, 1966).

6.6 LURIA AND DELBRÜCK AT COLD SPRING HARBOR

The Cold Spring Harbor Laboratory (Fig. 6.3) has had a long and interesting history, having served since early in the 20th century as one of the major sites

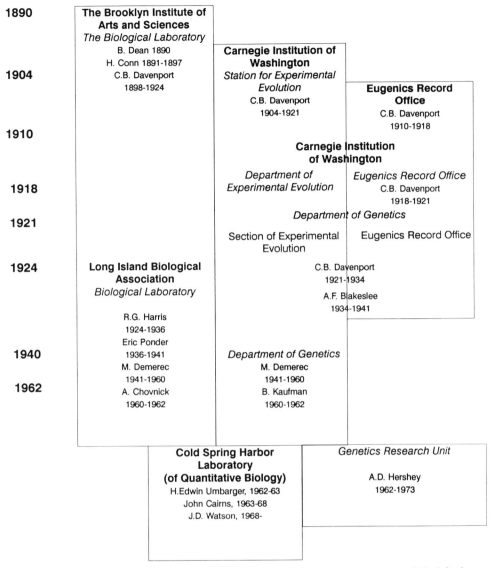

Figure 6.3 Overview of the history of the Cold Spring Harbor Laboratory. (Modified from Micklos, 1988.)

of U.S. genetics research (Micklos, 1988). Founded originally in 1890 when the Brooklyn Institute of Arts and Sciences set up a summer field laboratory, it was expanded in 1904 when the Carnegie Institution of Washington established a Department of Experimental Evolution under the direction of Charles B. Davenport. Although the Carnegie laboratory at Cold Spring Harbor had several departments, eventually only its Department of Genetics remained. For many years, the two institutions, the Cold Spring Harbor Laboratory and the Carnegie laboratory, were administratively separate. In 1924, the Long Island Biological Association (LIBA) was established as an administrative structure to take over the laboratory from the Brooklyn Institute. For 38 years, the LIBA operated the laboratory together with the Carnegie Institution, but in 1962 the Carnegie connection was abolished and the laboratory was reorganized as an independent unit, operated for the LIBA by a board of directors. During the main period of the Luria/Delbrück involvement with Cold Spring Harbor, although the two laboratories were separate, the director of both LIBA and the Department of Genetics of the Carnegie laboratory was Milislav Demerec. Demerec was a distinguished *Drosophila* geneticist (see Section 2.9), interested in mutation and the nature of the gene, who recognized the fundamental value of the Luria/Delbrück work. During World War II, the Biological Laboratory turned to "defense research," carrying out extensive studies on mutation in *Penicillium chrysogenum* to obtain strains capable of producing higher yields of penicillin. Other antibiotics research work, especially on antibiotic resistance, followed, and by the time the war was over, Demerec had turned the Cold Spring Harbor laboratory almost completely toward microbial (primarily bacterial) genetics. Demerec provided some early financial support for Luria and from time to time tried to hire either Luria or Delbrück as a permanent staff member.

In defense of the reorientation of the research program to microbial genetics, Demerec wrote the following in his report to the LIBA directors in 1946:

> You may properly ask what is the purpose of this research, and why we use microorganisms in our investigations. The answer to this is very simple. By using as experimental material these microorganisms, which are easy to handle, biologists and biochemists are trying to unravel two of the most intricate puzzles of all living matter; namely, the mechanism of heredity, and the reproduction of living substances. Since the fundamental laws of nature are general, discoveries made by working with these minute organisms help us to understand the life processes of higher living beings.[1]

Demerec invited Luria and Delbrück to spend the summer of 1941 at Cold Spring Harbor, where they used mixed infection to study the growth of bacteriophage (see below). Although this work would later lead to important studies on phage genetics, at that time the possibility that genetic recombination between viruses might occur was not considered, and the interest was on the use of mixed infection to reveal aspects of virus growth. In the fall of 1941, Delbrück returned to Vanderbilt and Luria to Columbia, but in March 1942, Luria was awarded a Guggenheim Fellowship to work in Delbrück's laboratory at Vanderbilt.

After the spring in Nashville with Delbrück, both Luria and Delbrück returned to Cold Spring Harbor for the summer of 1942, where they continued their work on mixed infection. That summer Luria received an offer of a faculty position at Indiana University (Bloomington), "...a place I had never heard of," and he moved to Indiana in January 1943 (Luria, 1984). The work on phage-resistant mutations discussed in Section 4.8 was carried out by Luria primarily at Bloomington, with the theoretical work done by Delbrück at Vanderbilt. Luria and Delbrück continued to collaborate at Cold Spring Harbor in the summers during the war years, and by 1944, Demerec had also begun to work on phage.

---

[1]Minutes of the Long Island Biological Association, July 30, 1946. It should be noted that Demerec wrote these words soon after the completion of a historic meeting, the 11th Symposium on Heredity and Variation in Microorganisms, where Lederberg and Tatum gave their first paper (see Section 5.3) and Hershey and Delbrück announced the discovery of genetic recombination in phage.

Although the conditions for research at Cold Spring Harbor were primitive (Luria, 1984), Luria and Delbrück favored research there and continued to spend summers there throughout the 1950s. First, sophisticated facilities were not required for the work they were doing, but simply many petri plates, some culture media, and incubator and sterilizing equipment. Second, Cold Spring Harbor was a small place, so that interaction between scientists was easy. Third, although the living quarters were primitive, they could live with their families right on the grounds, and it was thus a short walk to the laboratory. A good beach nearby, excellent sailing, tennis courts, a reasonable library, and a nearby town for living supplies meant that everything one needed was at hand. Finally, it was an easy train ride into Manhattan, so that numerous guests passed through the laboratory in the summer.

The first "phage course" was started by Delbrück in the summer of 1945. The intent of this course was to familiarize workers with phage research in order to encourage research on all kinds of phage problems. The course was taught that first summer primarily by Delbrück, but lectures were also given by Luria.

Fischer and Lipson (1988) and Cairns, Stent, and Watson (1966) give some detailed descriptions of what the Cold Spring Harbor Laboratory was like in the early days of the phage group. People who took the course or were summer visitors and later went on to make significant contributions on phage or other aspects of molecular genetics included:[2] A.H. Doermann (1945), Rollin D. Hotchkiss (1945), Stuart Mudd (1945), Werner Maas (1945), Herman Kalckar (1945), Mark H. Adams (1946), Seymour S. Cohen (1946), E.A. Evans, Jr. (1946), B. Vennesland (1946), V. Bryson (1946), H.B. Newcombe (1946), H. Gaffron (1946), Harriet Taylor (later Ephrussi-Taylor) (1946), George Streisinger (1947), Leo Szilard (1947), Aaron Novick (1947), Albert Kelner (1947), G.H. Beale (1947), Philip Morrison (the physicist) (1947), Richard B. Roberts (1947), Wolf Vishniac (1947), B.D. Davis (1948), J.S. Gots (1948), Gunther Stent (1948), Margaret Lieb (1948), Seymour Benzer (1948), Sol H. Goodgal (1949), G. Bertani (1949), W.F. Goebel (1949), G. Lark (1949), and Norton Zinder (1949). Several years later, a bacterial genetics course was established by Demerec and taught by a distinguished group of visitors.

The "Phage Treaty"

In the summer of 1944, under Delbrück's influence, the group of phage workers at Cold Spring Harbor decided to concentrate research on a restricted group of bacteriophages that attacked *Escherichia coli*. This agreement was the so-called "phage treaty of 1944." Up until then, every investigator had a private collection of phages and host bacteria, making comparisons between laboratories difficult. Delbrück insisted that researchers concentrate on a set of seven phages active against the same host, *Escherichia coli* strain B (Table 6.1)(Fig. 6.4). The set of phages, called T1, T2, T3, T4, T5, T6, and T7 (T for "type"), were distinguished serologically and by the use of

[2]Data from the Annual Report of the Biological Laboratory of Cold Spring Harbor. List of participants in the phage course and in some early phage symposia can be found in a brochure published by the Cold Spring Harbor Laboratory for the dedication of the Max Delbrück laboratory, August 29, 1981. This brochure also has numerous photographs of Cold Spring Harbor phage workers.

Table 6.1 The Phages of the T System

Name	Plaque size	Morphology head (nm)	tail (nm)	Latent period (min)	Burst size
T1	medium	50	150 × 15	13	180
T2	small	65 × 80	120 × 20	21	120
T3	large	45	invisible	13	300
T4	small	65 × 80	120 × 20	23.5	300
T5	small	100	tiny	40	300
T6	small	65 × 80	120 × 20	25.5	200–300
T7	large	45	invisible	13	300

All phages grow on *E. coli* strain B. Phage T1 was called α in early work by Delbrück and Luria and P28 by Bronfenbrenner. Phage T2 was first called γ by Delbrück and Luria and PC by Bronfenbrenner. Phage T7 was first called δ by Delbrück and Luria. The other phages were first isolated by Demerec and Fano (1945). (Modified from Demerec and Fano, 1945; Delbrück, 1946.)

host strains resistant to each phage (Demerec and Fano, 1945; Delbrück, 1946). It later developed that T2, T4, and T6 (often called, for convenience, the T-even phages) were related morphologically and serologically, as were T3 and T7. The T-even phages also turned out to be related biochemically and genetically. Although the "phage treaty" made sense in the early days, concentrating on only these virulent phages led to the neglect of the whole problem of lysogeny (see Chapter 7). Delbrück's intransigence in this matter became a legend: "Once this decision was made, Max held to it firmly. He would not look at results of experiments done with strains other than those included in the treaty. Though Max never found this easy, he felt a good researcher must tame his curiosity" (Fischer and Lipson 1988). According to Doermann (1983a), Delbrück called the T system "Snow White and the Seven Dwarfs!"

6.7 DELBRÜCK AT CALTECH

By 1945, the war was over and new opportunities were arising. Delbrück had welcomed his Vanderbilt position in 1940, but he became increasingly dis-

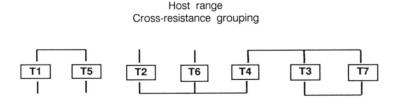

Figure 6.4 Classification of the T system of phages by cross resistance and serology. *Escherichia coli* strain B resistant to T1 is also resistant to T5 (B/1,5), and a strain resistant to T4 is also resistant to T3 and T7 (B/3,4,7). Note also that although T2, T4, and T6 are related serologically, they are not related by cross resistance. T3 and T7 are related serologically as well as by cross resistance. T1 and T5 are related by cross resistance but not by serology. (Modified from Delbrück, 1946.)

satisfied there. Among other things, his access to good students was limited, and his departmental affiliation was with physics rather than biology. He was now 40 years old and at a turning point in his life. Over the next year and a half, Delbrück received a number of inquiries from other institutions offering him staff positions, including the University of Manchester (England) and Cold Spring Harbor. In response to the latter offer from Demerec, Delbrück outlined the kind of research program he envisioned.

> In reply to your letter...I would like to submit at some length my views about the ways in which in my opinion a healthy program of research in this field might be organized.
>
> At present, phage research is branching out in two directions. The first might be called the study of model viruses and results in this field are of import for virus research in general. The second line, which uses phages merely as an experimental tool in the isolation and characterization of bacterial mutants leads to bacterial genetics, and, as you have shown, is related to the problems of drug resistance of bacteria. It also leads, as E.H. Anderson has shown, to problems of bacterial physiology, and eventually, it may be hoped, to physiological genetics similar to Beadle's studies.
>
> Although these two directions seem to be divergent they can, in fact, not be pursued separately, for reasons which are familiar to anybody who is working in the field. The person who works on phage growth is necessarily interested in the bacterial mutants, as he uses them as indicator strains, while the person who works on bacterial mutants is interested in the physiological basis of resistance to phage.
>
> From the point of view of fundamental research and from the point of view of applied science, both of these lines of research with phages are in my opinion worthy of a broad development, in fact of a broader development than could be undertaken by any single institution. It calls for the gradual training of highly qualified research workers who can introduce this research at other institutions. I am anxious to foster such a development, because, unless the field of phage research is really opened up, the promising results obtained up until now will remain isolated curiosity pieces and will once more sink back into the oblivion of small print addenda in text books.
>
> In a small way the phage course this summer was of some help in this direction...To sum up, I consider it of equal importance to bring new men into phage research as to do the research ourselves....[3]

Although the offer of a position at Cold Spring Harbor did materialize, Delbrück was reluctant to accept it without the possibility of a vital connection to a biology department at a major university. The Manchester position was to set up a new program in biophysics, and Delbrück was attracted to Manchester at least in part because it had been the university where Niels Bohr did his pioneering work with Ernest Rutherford on the quantum theory of the atom. After a visit to Manchester in the summer of 1946, Delbrück was offered the position and decided to accept it. However, in late 1946 while he was preparing to leave Vanderbilt, a much better offer came, a professorship in biology at CalTech. Delbrück quickly modified his plans and by January 1947 he was in Pasadena.

The CalTech environment was ideal for Delbrück's intellectual development, and for the flourishing "phage group." George Beadle was now head of biology, and Linus Pauling was head of chemistry. In the post-World-War-II era, a vigorous new program in genetics research was under development.

---

[3]Letter from Max Delbrück to M. Demerec dated November 6, 1945. (Reproduced, with permission, from the CalTech archives.)

Continuing to spend the summers at Cold Spring Harbor and the academic year at CalTech, Delbrück had found exactly what he was looking for. Over the next 10 years, Delbrück's laboratory became the "nerve center" of the burgeoning field of molecular biology. "Everyone" in the field came to visit, either for a brief stay or a year or two. Among the many prominent scientists who spent time at CalTech were A.D. Hershey, S.E. Luria, A.H. Doermann, André Lwoff, Elie Wollman, William Hayes, J.D. Watson, Seymour Benzer, G. Bertani, Renato Dulbecco, Sidney Brenner, Jacques Monod, Francois Jacob, Robert L. Sinsheimer, Peter Starlinger, Gunther Stent, George Streisinger, and Jean Weigle. Although Delbrück himself carried out little experimental work, his "presence" served to galvanize the field. "He cared little about who was first to bring off a particular experiment. The essential thing was that it had been done" (Jacob, 1988).

6.8 A.D. HERSHEY

In addition to various scientists mentioned above who became involved with Luria and Delbrück on phage research, an important early worker was Alfred Day Hershey. Trained as a chemist (Ph.D., University of Chicago, 1934), Hershey started as an assistant of J.J. Bronfenbrenner[4] in the Department of Bacteriology at Washington University School of Medicine in St. Louis, becoming an assistant professor in 1938 and an associate professor in 1942. Hershey worked first on problems of interest to Bronfenbrenner, such as the kinetics of inactivation of phage by antibody, but he gradually became an independent investigator pursuing his own interests on phage research. After some preliminary correspondence and exchange of strains, in early 1943, Delbrück invited Hershey to visit Vanderbilt and give a seminar. This initiated a lengthy correspondence and friendship that lasted until Delbrück's death in 1981.[5]

In April 1943, Delbrück and Luria both traveled to St. Louis to visit Hershey and give seminars. This was presumably the first meeting of the three future Nobel laureates and can be considered the starting point of what came to be called the phage group (Fischer and Lipson, 1988). Hershey moved to the Department of Genetics of the Carnegie Institution of Washington at Cold Spring Harbor in 1950 and remained there for the rest of his career, retiring in 1974. His work on phage genetics and the role of phage DNA in infection are discussed below, and his work on phage *lambda* is discussed in Chapter 7.

[4]Brief biographical information on J.J. Bronfenbrenner can be found in Varley (1986). Bronfenbrenner believed that phage were really soluble but behaved as particles when adsorbed to various colloidal substances present in lysates. Measuring the diffusion constant of phage, Bronfenbrenner calculated the average particle size as 6 μm, but stated that this was the size of the carrier particle rather than the "soluble" phage. After Luria and Anderson obtained electron micrographs that demonstrated phage particles of definite morphology, Bronfenbrenner gradually abandoned his colloidal attachment hypothesis. As reported by Anderson (1966), when Bronfenbrenner was first shown the electron micrographs, he exclaimed: "Mein Gott! They've got tails." And later he inquired: "...if these tails should represent organs of locomotion, it is quite possible that the diffusion of the particles might be speeded up to such an extent as to nullify the results of our diffusion experiments..."

[5]In his initial letters, Delbrück used the salutation "Dr. Hershey," later simply "Hershey," and finally "Al." Luria has written of Delbrück's initial impressions of Hershey: "Drinks whiskey but not tea. Simple and to the point. Likes living in a sailboat for three months, likes independence" (Luria, 1966).

6.9 PHAGE GENETICS: THE FIRST STEPS

When Luria, Delbrück, and Hershey began their work, they did not consider that it would be possible to do "traditional" genetics research with phage, and it was certainly not thought that hereditary mechanisms in viruses might bear any relationships to those of higher organisms. In the first work of the phage group, the focus was on the mode of replication of the phage particle; these studies involved the first mixed infection experiments. The isolation of phage mutants by Hershey and by Luria opened up the possibility of doing genetics with phage. Although Burnet and Lush (1936) had isolated phage mutants some years before, their work had not been followed up, and, of course, had not been done with one of the phages designated by the "phage treaty."

Phage Mutants

Variability in phage had been first considered in a review by Burnet (1930). He noted that phages differed in strain specificity and virulence, and that in any study of variation, it was necessary to begin with a "pure" phage, obtained by single-plaque isolation. Host range was extensively used to classify phages. However, it was not possible to distinguish resistance to virulent phage from acquisition of resistance as a result of lysogenization. It was clear, though, that individual phage particles were capable of a slight modification in character. "Many of these modifications are transient, the descendants showing only the average activity, but if conditions are such as to allow the greater activity of a particular modification, the average character of the phage will tend toward this modified type. The whole aspect of these reactions is so fundamentally similar to that of bacterial variation that they offer extremely impressive evidence for the living nature of the units concerned" (Burnet, 1930).

Host strains resistant to bacteriophage had been frequently isolated during the extensive phage work in the 1930s and 1940s (see Section 4.7; also Yang and White, 1934; Luria and Delbrück, 1943). These phage-resistant bacteria were used by Luria (1945a,b) to isolate *host range* or *h* mutants. As discussed in Section 4.8, in 1943 Luria and Delbrück had developed the fluctuation test to analyze the process by which *Escherichia coli* became resistant to phage infection and had shown that resistance arose as a result of spontaneous mutation. However, the phage-resistant host could itself be infected by mutants of the phage (Demerec and Fano, 1945). In the terminology used, the T2-resistant mutant of *Escherichia coli* strain B was called B/2 (pronounced "bee bar two") and the phage mutant able to infect B/2 was called T2*h*. (The wild type is thus called T2h^+.)

By plating large populations of phage on resistant hosts, Luria (1945b) was able to isolate in pure form a series of phage mutants that were indistinguishable from parents either morphologically or serologically. Their particle size (X-ray sensitivity) also remained the same and they all infected the original host, *Escherichia coli* strain B. Using these phage mutants, a new series of phage-resistant bacterial mutant strains was again isolated. Luria showed with the fluctuation test that the phage variants arose by mutation, producing clonal groups with very large variations in their numbers. Thus, the resistant bacteria acted as selective agents for phage mutants in the same way that the phage in turn acted as selective agents for bacterial mutants. In addition to revealing the origin of phage variants, the results of the fluctuation tests suggested that the virus was multiplying autocatalytically, in the same way that bacterial cells do.

Another consequence of these studies on phage variants was that they focused on the manner in which phage and host interact. Strain B/2 did not adsorb T2h^+ but did adsorb T2h. Thus, a change of the bacterial surface must have taken place, making the cell resistant to the virus, and this change was compensated for by a change in the virus. Although studies of cell surface/phage particle interaction had little permanent impact on genetics, they did focus on an interesting aspect of the virus infection process, and would eventually lead to important studies on how animal and human viruses penetrate their host cells.

Another class of phage mutants isolated were called *rapid lysis* or *r* mutants. The *r* mutants were first noticed by Hershey (1946a), who saw a small fraction of T2 plaques that appeared different from normal plaques. On *Escherichia coli* strain B, normal phage gave small plaques, with a clear center surrounded by a large diffuse halo. The *r* mutants, on the other hand, produced plaques with a large, clear center and a sharp edge. These mutants were designated *r* and the wild type *r*$^+$. A number of other plaque mutants were also detected and were to prove useful in genetic studies (Doermann, 1953; Stent, 1963).

The physiological basis of the *r* phenotype originates from the phenomenon of *lysis inhibition* found in the wild type. If a cell infected with wild-type phage (r$^+$) is secondarily infected with another r$^+$ phage particle after intracellular growth of the primary phage is well under way, lysis of the infected cell is greatly delayed. Rapid lysis (*r*) mutants do not exhibit lysis inhibition. The diffuse outer halo of plaques of the wild type arises because on the agar plate after a few rounds of replication, phage particles released from the initial infections will reinfect other infected cells and inhibit lysis. Since secondary infection does not result in lysis inhibition in the *r* mutant, each infected cell produces phage in a normal time period. (The phenomenon of lysis inhibition had first been noted by Delbrück in connection with his one-step growth experiments, and the mechanism had been deduced by A.H. Doermann in Delbrück's laboratory.) Lysis inhibition was only found with the T-even phages.

Hershey showed that the *r* mutation occurred without a change in the host range or antigenic specificity of the phage and that it occurred at a frequency of about one per thousand duplications of the virus. The reverse mutation, from *r* to *r*$^+$, also occurred, with a frequency of about 10^{-8}. Hershey also showed that there were many distinct *r* mutants, with similar phenotypes but different genotypes. The *r* mutation was to play a major role in Benzer's important studies on the fine structure of the gene (see Section 6.10).

A plating "trick" that Delbrück developed permitted the detection of *h*, *r*, and *hr* mutants on the same plate. With T2, the plating indicator was a mixture of *Escherichia coli* B and B/2. The *hr*$^+$ plaques were small and clear because both B and B/2 were affected by phage with the *h* gene. *h*$^+$*r* gave large turbid plaques because B/2, being unaffected, grew. *hr* gave large clear plaques and *h*$^+$*r*$^+$ gave small turbid plaques.

Mixed Infection

The discovery that phages could be *crossed* arose out of studies on mixed infection of a single cell with more than one phage particle. The initial interest in mixed infection was to study the mechanism of virus growth, using superinfection with a second virus after replication of the first virus was under way (Delbrück and Luria, 1942). In the initial work, two unrelated phages were used (T1, T2), and it was found that an infected cell would produce one kind of virus or the other, but never both. This phenomenon, called *mutual exclusion*, was then thought to be part of a general mechanism of virus infection in which only *one* virus particle entered the cell. The excluded virus could, however, greatly reduce the yield of the successful virus, a phenomenon called the *depressor effect* (Delbrück, 1945). However, Hershey showed in both mass cultures and in single cells that in a mixed infection with the *same* phage strain, for example, $T2r^+$ and $T2r$, *both* types of phage particles were produced. Therefore, mutual exclusion did not occur when the two infecting phages were closely related. Furthermore, Delbrück showed that mixed infections with pairs of *related* viruses, such as two members of the T-even group, gave only partial mutual exclusion.

The first experiment that showed evidence of breakdown of mutual exclusion was that of Hershey, reported to Delbrück in a letter in the spring of 1945 (see Delbrück and Bailey, 1946). Still considering the work as a study of mutual exclusion, Hershey plated phage of r^+ and r genotypes. The plaques that developed could be classified morphologically into *three* classes: wild type ($^+$), mutant (r), and mottled (mixed, r and $^+$). When mottled plaques were picked and cultured, they gave about 50% of the two types.

Crossing Phage

The work on mixed infection just described led to the discovery of genetic recombination of phage. The studies were performed simultaneously in Delbrück's laboratory at Vanderbilt and Hershey's laboratory in St. Louis, and Delbrück and Hershey remained in close contact by frequent visits and extensive correspondence.[6]

In Hershey's work, reported at the 1946 Cold Spring Harbor symposium, mixed infections with both *host range* and *rapid lysis* mutations were studied (Hershey, 1946b). Since these were two independent phenotypes, their use in genetic analysis was relatively easy. Forward and reverse mutation of these two phenotypes occurred independently of each other, thus showing that each phenotype was governed by a *separate* genetic site. Using mixed infection with h^+r and hr^+, Hershey could show that *new* phenotypes arose that were genetically hr and h^+r^+ (Table 6.2). Although both Hershey and Delbrück initially interpreted their results in terms of "induced mutation" (see below), Hershey was less cautious than Delbrück about considering it to be evidence of genetic crossing. "A new and promising approach...is by means of mixed-infection experiments in which one observes an apparent segregation of genetic factors. Whether this phenomenon is related to crossing over, and indeed whether it implies actual exchange of genetic material at all, are

---

[6]Copies of much of this correspondence are preserved under Hershey's name in the Delbrück archives at CalTech.

Table 6.2 Incidence of Various Phenotypes in Progeny of Mixed Clones from Infection of h^+r and hr^+

Mixed clone	Number of isolates of phenotype			
	h^+r^+	hr^+	h^+r	hr
1	1	10	10	1
2	0	11	6	5
3	0	11	6	5
4	2	9	8	3
5	0	11	10	1
6	0	11	6	5
7	0	11	4	9
8	0	11	3	8
9	0	11	11	0
10	2	9	10	1
Total	5	105	71	39
Percent	2	48	32	18

A mixed infection of *E. coli* strain B by T2h^+r and T2hr^+ was made. About 65% of the plaques were mottled (mixed plaques). These plaques arose from bacterial cells in which both phage grew. Of these mottled plaques, 10 were examined for other phenotypes. The results are shown above. h^+r^+ and hr represent new phenotypes. *Controls*: 20 clones of each of the original stocks and 5 secondary clones from each of 10 plaques of hr^+ yielded no variant types. The stock of hr^+ contained about 1% r mutants. (Simplified from Hershey, 1946.)

pressing and important questions. But whatever the mechanism responsible for the impressed mutations, it is probable that the purely genetic study of their patterns will yield valuable clues to the genetic structure of viruses" (Hershey, 1946b).

Delbrück's work was less elegant than Hershey's since instead of mutants of the same phage, mixed infections were made with two related phages, T2r^+ and T4r (Delbrück and Bailey, 1946). These phages can be distinguished serologically and also by patterns of host resistance (see Fig. 6.4). An analysis of the progeny showed not only the parental types, but also a new type, T4r^+. "The wild-type particles, therefore, represented a new type, created during the mixed infection" (Delbrück and Bailey, 1946).

Initially, these new types were interpreted as the result not of genetic recombination but of induced mutation. The term *induced mutation* was commonly used at this time also to explain the results of Avery's transformation experiments in pneumococcus (see Section 9.6). The first interpretation of the phage results as crossing-over was actually made by Hermann Muller. Delbrück had reported the above results at a mutation conference in New York City on January 26, 1946, six months before the Cold Spring Harbor symposium. As usual, Muller was willing to place a firm genetic interpretation on the results: "In my opinion, the most probable interpretation of these virus...results then becomes that of actual entrance of the foreign genetic material already there, by a process essentially of the type of crossing-over, though on a more minute scale...In view of the transfer of only a part of the genetic material at a time...a method appears to be provided whereby the gene constitution of these forms can be analysed, much as in the cross-breeding tests on higher organisms" (Muller, 1947). However, Muller's important paper was not published until well after the papers for the Cold Spring Harbor meeting were written, and the title of the Delbrück/Bailey paper uses the term "induced" mutation.

Delbrück was clearly uncomfortable with the idea of "induced mutation," as he wrote: "Perhaps one might dispute the propriety of calling the observed changes 'induced mutations.' In some respects they look more like transfers, or even exchanges, of genetic materials. We do not pretend to be able to put forward convincing arguments for either point of view" (Delbrück and Bailey, 1946).

On September 2, 1946, after the Cold Spring Harbor symposium where the work was first reported, Delbrück wrote Hershey: "We have not done much on induced mutations...The main reason...is the fact that I do not like any more our present set-up of working with two independent wild types and only one genetic marker. That is really as primitive as the first Mendel experiments...I have therefore been thinking of getting other mutants. We have tried an analogue of the neurospora technique to obtain 'biochemical' mutants (since Tom Anderson found that T4 and T6 are 'tryptophane-deficient' I thought other deficiencies might occur). We have tested, to date, 108 UV survivors of T2, by a very simple technique, so far no luck."[7] On September 11, 1946 he wrote again: "The tests for biochemical mutants were negative, as far as we have gone, which is far enough to discourage us from going further, for the time being. We are therefore thrown back into the arms of your r,w,s mutations. On the other hand I would much prefer to work with pairs of independent wild types, where there are clear host range differentiations which are not confused by mutations, and where one can check the origin of new types serologically. On the other hand, as I wrote in the last letter, I am afraid of the many other factors in the genetic factors that might transfer in the crosses, and confuse the picture...I hope you will do something for this problem. I have a bad conscience for not having done much, or at least not much constructive."[8]

Soon, Hershey's work would show definitively that the new genetic types arising from mixed infections were the result of genetic recombination. The first detailed studies on phage genetics were carried out by Hershey and Rotman (1948, 1949) on crosses between *r* and *h* mutants of T2. By this time, Hershey had accepted the idea that a mixed infection experiment was equivalent to a genetic analysis:

> The experimental procedure used is the one-step growth experiment of Delbrück and Luria, with mixed multiple infection...For simplicity of language, we shall speak of this experiment as a cross, described as heterologous when new types appear among the yield of virus, and homologous when only the parent types are recovered (Hershey and Rotman, 1948).

By careful analysis using essentially standard genetic techniques, Hershey and Rotman were able to demonstrate linkage and crossing-over and developed the first genetic map (Fig. 6.5). Care was taken to ensure that all *r* mutants were of independent origin. When crosses between two *r* mutants

[7]Letter in CalTech archives, quoted by permission.
[8]Letter in CalTech archives, quoted by permission.

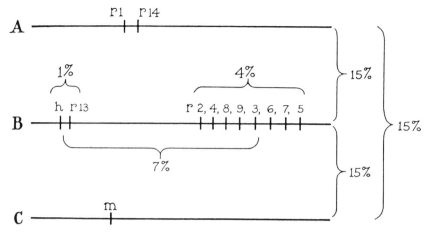

Figure 6.5 Linkage relations among T2 mutants. The percentages indicate yields of wild type in two-factor crosses. (Reprinted, with permission, from Hershey and Rotman, 1949.)

were made, it was found that they could be placed in two clearly defined classes, A and B. Crosses between any pair belonging to the separate classes yielded about 15% wild-type progeny, whereas crosses *within* a class yielded much lower wild-type progeny, between 0.5% and 8%. Crosses between any two *r* mutants revealed not only wild types, but also *new r* mutants distinguishable from the parental types by back cross (Table 6.3).

These were the days before genetic fine-structure analysis had shown that crossing-over *within* a gene could occur (see Section 6.10). Classical genetics would have concluded that, because recombination was occurring between *r* mutants, each *r* mutant corresponded to a *different* phage gene, since recombination was not supposed to occur *within* a gene. However, it seemed unlikely that each *r* mutant represented a separate gene, since a large number of *r* mutants had been isolated, and all affected the same phenotype, lysis

Table 6.3   Recombination Frequencies in T2 Crosses with Different *r* Mutants

Cross		Percent of genotype			
		h^+r^+	hr^+	h^+r	hr
$h \times r1$	input	0	53	47	0
	yield	12	42	34	12
$h \times r7$	input	0	49	51	0
	yield	5.9	56	32	6.4
$h \times r13$	input	0	49	51	0
	yield	0.74	59	39	0.94

Mixed infections with multiplicities of about five phage of each type per bacterium were used. After 5 min adsorption and 1 hr incubation, the samples were plated on mixed indicator plates and the frequency of all four genotypes was scored the next day. In the crosses, each *r* represents an independently isolated mutant. (Simplified from Hershey and Rotman, 1948.)

inhibition. This made it difficult to interpret the results in terms of conventional genetics.

> This remarkable circumstance might be interpreted as an indication that the different mutations result from a single type of chemical change occurring at different points in the genetic material of the viral particle. If this were so, the different *r* mutants might represent alterations of the same biochemical property (Hershey and Rotman, 1948).

One can see from his letters to Delbrück the evolution of Hershey's thinking over the period from spring 1945, when the first mixed infection experiments were done, until fall 1948, when the Hershey/Rotman paper appeared. At first, the experiments were described as "mutual exclusion," then as "induced mutation," later as dealing with the "segregation of factors," and finally with "genes," "loci," and "linkage."[9] Hershey had no formal genetics background. He was also developing a genetic system in completely new material. Caution gradually left as the results were generated and geneticists such as Hermann Muller introduced traditional genetics thinking.

Within several years, when he reviewed work on the genetics of bacteriophage, Delbrück was interpreting the data as the result of genetic recombination between "haploid organisms capable of exchanging or transferring their genetic material between each other during their simultaneous multiplication within the same bacterial cell...In consequence, a mixed infection corresponds to a genetic cross" (translated from Delbrück, 1949b).

The mechanism of genetic recombination in phage remained of considerable interest. The eclipse phenomenon and the development of biochemical knowledge of phage replication (see below) led Visconti and Delbrück (1953) to postulate that phage replicated as a pool of genomes that could undergo successive pairwise matings, each of which could lead to an exchange of genetic material. As virus maturation continued, genomes were removed from this pool into mature phage particles, where they could no longer participate in genetic recombination. Because of the multiple copies of the genome that develop within the infected cell, the phage cross could not be visualized as exactly analogous to a cross between two "haploid genomes." The outcome would depend not only on the linkage between the two loci being studied, but also on the number of mating events that occur before lysis takes place. This analysis required a different approach to the calculation of linkage relationships, which is described in detail by Stent (1963).

6.10 FINE STRUCTURE OF THE GENE: BENZER'S WORK

The study by Seymour Benzer on the fine structure of the *r*II locus of bacteriophage T4 can be considered the archetype of modern genetic analysis. It was also a major advance in understanding the relationship between the gene and the DNA molecule.

Benzer received his Ph.D. in physics from Purdue University in 1947, working on solid state physics. Although hired as an assistant professor of physics at Purdue, he was soon granted a leave of absence to pursue an

---

[9]In a personal communication, Joshua Lederberg, who attended the 1946 Cold Spring Harbor symposium, has written me as follows: "I remember that very vividly! Delbrück was insistent on mutual exclusion, so one phage had to be acting from *without*. By the time of the CSH, Hershey had pretty well decided it was recombination. He didn't want to confront Max too strongly."

interest in biology, with the goal of establishing a biophysics program at Purdue. Benzer's interest in biology and in Max Delbrück was awakened by a reading of Schrödinger's book, *What is Life*? Through a chance contact with Luria (Benzer, 1966), he was directed to the phage course at Cold Spring Harbor, which he took in 1948. After a year with Alexander Hollaender at the Oak Ridge National Laboratory, he spent two years with Delbrück at CalTech (1949–1951) and a year in Lwoff's laboratory at the Pasteur Institute working with Francois Jacob, before returning to Purdue to start a new biophysics laboratory. During his CalTech and Pasteur years, Benzer worked on various research problems using bacteriophage (see his work on induced enzyme synthesis in Section 10.7). In Paris, he made the chance observation that was to lead him into the fine-structure studies with T4 that are discussed below. After a short, highly productive 10 years of phage research, he switched to molecular neurobiology, and in 1975, he returned to CalTech as Boswell Professor of Neurosciences.

Classical Genetics and the Nature of the Gene

By the 1950s, genetic analysis in higher organisms had reached such a level of sophistication that doubt was being cast on the conventional corpuscular interpretation of the gene (Stadler, 1954). Although workers of the Morgan school had been content to think of genes as particulate entities, it was becoming clear that if the nature of gene action was to be understood, a finer resolution was needed. One of the strongest proponents of the use of genetic analysis to resolve the structure of the gene was G. Pontecorvo of the University of Glasgow, who applied his classical genetic background to research on the fungus *Aspergillus nidulans*. From this foundation, Pontecorvo wrote important theoretical reviews and books on gene structure and analysis (Pontecorvo, 1952, 1958).

One of Pontecorvo's key insights involved the use of genetic crosses for estimating the *size* of the gene, an approach that Hermann Muller had pioneered in the early 1920s. By the 1950s, the *resolution* of genetic analysis had vastly improved, but as Pontecorvo noted, this resolution depended on the number of meiotic products analyzed. There was no *a priori* limitation to this type of genetic analysis, since all one needed to do to push the resolution further was to analyze more progeny of a cross. Practically, however, this had its limits, and in 1952, recombination frequencies of the order of 10^{-5} were the lowest measurable in *Drosophila*. In microorganisms, however, the selective techniques introduced by Lederberg (see Chapter 5) had made it possible to extend the resolution at least two orders of magnitude further (Pontecorvo, 1952).

Frequency of crossing-over is expressed as the frequency of its occurrence *between* two genes at meiosis; that is, the proportion of the products of meiosis that show recombination. The usual unit of crossing-over is the *centi*Morgan, equal to 1% crossing-over, but subdivisions of the *milli*Morgan and *micro*Morgan are possible. For such an analysis to succeed, one needed not only lots of progeny, but also many genetic markers.

In *Drosophila*, crude calculations with assumptions about the length of the chromosome had indicated that a gene corresponded to a chromosome length of about 100 Angstroms (Pontecorvo, 1952). Another method based on

X-ray inactivation of a gene and assumptions about how radiation interacted with matter had led to estimates of the size of a gene of 20–60 Angstroms.

An important discussion that had arisen from the *Drosophila* and *Aspergillus* work was that with certain genes, when the resolution was pushed far enough, evidence existed for recombination *within* a single allele. This result seemed difficult to resolve if a genetic locus was defined as a unit that could undergo recombination with another locus, but not with itself. In some cases, loci were found that definitely affected the same function, yet could recombine. The simplest interpretation was that the loci were *not* alleles, but that recombination *within* a single allele was taking place. Such phenomena went under the name *pseudoallelism* (closely linked genes having similar effects; Lewis, 1951). A related phenomenon was the so-called *position effect*, also first observed in *Drosophila*, in which a gene was seen to have a different effect depending on its position in relation to other genes.

The rII Mutation

The rapid lysis phenotype was first discovered by Hershey and exploited in genetic mapping by Hershey and Rotman. *r* mutants are very easy to detect and isolate, due to their distinctive plaque morphology on *Escherichia coli* strain B and their frequency of mutation, which is high enough that no special selection procedure is necessary in order to find them. As discussed in Section 6.9, Hershey had isolated a large number of *r* mutants and had shown that recombination between some of them occurred. What led Benzer into phage genetics was the discovery that one class of *r* mutants, which he called *r*II, were *conditional* lethals, able to grow well on strain B but unable to grow on strain K-12 when it was lysogenic for bacteriophage *lambda*. If K-12 was cured of *lambda*, the resulting bacterial strain no longer restricted *r*II and, in fact, *r*II grew similar to wild-type T4. Benzer called the restricting strain K and the permissive strain K-12S or simply S.[10] The differentiation of the three classes of *r* mutants by the use of three different *Escherichia coli* hosts is given in Table 6.4 and the genetic map is given in Figure 6.6 (Benzer, 1957).

Benzer realized that the restriction phenomenon involving strain K could be used as a tool for analyzing in detail the genetic structure of the *r*II region. Two *r*II mutants could be crossed on strain B and wild-type recombinants could be detected in the lysate, even if they were present in very small numbers, by plating on strain K, where neither parent could grow. Because of the extreme sensitivity of the system (high resolution), even if two *r*II mutants were very close together on the genome, recombinants could be detected. Previously, large numbers of plates would have been required to measure recombination frequencies, but they could now be easily measured on single plates. The first description of the *r*II system was published in Benzer (1955), and the details were published in Benzer (1957).

In 1955, when Benzer first realized the possible significance of the *r*II locus, the Watson and Crick structure for DNA was well established and interest was high in relating DNA structure to protein structure. At that time,

---

[10]The details of how *lambda* prophage prevents growth of *r*II mutants are still not completely understood, but it is known that a *lambda* gene called *rex*, near the immunity region, is involved. *rex*⁻ lysogens of *lambda* do not restrict *r*II. The product defined by the *r*II gene seems to play some role in membrane function (Singer, Shinedling, and Gold, 1983).

Table 6.4 Differentiation of the Three Classes of *r* Mutants of Phage T4

Phage strain	Bacterial host strain		
	B	S	K
Wild	wild	wild	wild
*r*I	*r*	*r*	*r*
*r*II	*r*	wild	no growth
*r*III	*r*	wild	wild

The plaque morphologies on the various host strains are given. (B) *E. coli* strain B; (S) *E. coli* K-12 nonlysogenic for *lambda*; (K) *E. coli* K-12 lysogenic for *lambda*. (Modified from Benzer, 1957.)

sequencing of DNA was impossible, but the Sanger method for sequencing proteins had been developed; X-ray crystallography of proteins had been perfected; and a number of amino acid sequences of proteins were being determined (for review, see Anfinsen, 1959). An approach that presented itself, therefore, was to isolate the *r*II protein from various mutants, determine amino acid sequences, and then establish the colinearity of alterations in amino acid sequence with the locations of the mutations in the genetic map. If the DNA bases could also be identified, then the genetic code could even be solved (Benzer, 1966).

From his analysis of the *r*II system, Benzer was able to determine that the sensitivity of the system was such that genetics *could* be related to DNA.

An outline of Benzer's proposed research is given in a letter he wrote to Delbrück on February 3, 1955, now in the Delbrück archives at CalTech. In

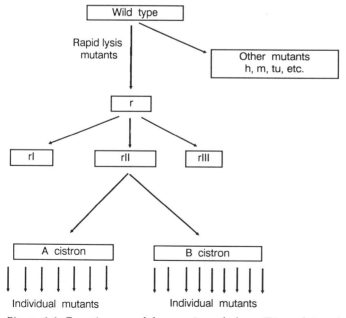

Figure 6.6 Genetic map of the *r* region of phage T4, as determined on *Escherichia coli* strain B. (Based on Benzer, 1957.)

this outline, he discussed the Watson/Crick model for DNA and noted that there was no evidence in DNA for any structural units larger than those of the single nucleotide link. He noted that the resolving power of his method would be sufficient to detect recombination between mutants that are only one nucleotide link apart and indicated that a precision mapping study would permit analysis of genetic recombination to the nucleotide level.

Unfortunately, the *r*II protein proved extremely difficult to find,[11] but the mapping progressed famously. Benzer recognized that it was important to relate the various *r*II mutations to a *functional* unit. In *Drosophila* genetics, the functional unit was defined genetically by means of the *cis-trans* (complementation) test (Pontecorvo, 1958). In a complementation test, two mutants affecting the same function are brought together in the same cell in separate genetic units, as a heterozygote or (in haploid fungi) a heterocaryon (see Section 5.1 for a discussion of complementation in *Neurospora*), and the phenotype is determined. If the wild-type phenotype is restored, the mutants are said to be complementary and hence are located in different functional units. If the mutants do not complement, they are in the same unit. Genetic analysis shows that noncomplementing mutations are invariably closely linked and affect the same function (enzyme). (Complementing mutants can affect related biochemical steps but can also, of course, affect completely different functions.) In traditional genetics, noncomplementing mutants would have been considered alternative states in the same gene, but a difficulty of interpretation would arise if the genetic resolution was sufficient that recombination between them took place. It was gradually being realized that the gene as a functional unit (as defined by complementation tests) and the gene as a unit of recombination (as defined by crossing-over) were two different things.

In a complementation test, since the mutations are in different genetic elements, they are said to be in a *trans* position. As a control, the two mutants to be compared are also placed on the same genetic element, in the *cis* configuration. In the *cis* configuration, a genome nondefective in the *rII* region is present in one genetic element, and the two defective mutations are together on the other genetic element. In the absence of polarity effects, the *cis* configuration produces a nondefective (or almost nondefective) phenotype.

When Benzer carried out *cis-trans* tests with *r*II mutants, he found that they could be grouped in two separate classes, which he called A and B. When the mutational sites of the two groups were mapped, it was found that all A mutations mapped together in one region and all B mutants in another, adjacent region. Mixed infection of strain K with any two A-type mutants or any two B-type mutants failed to yield progeny (except for rare recombinants), but mixed infection with one A and one B yielded A and B progeny at essentially wild-type levels. The complementation test thus made it possible to redefine a gene in terms of function, and the idea that parts of genes were inseparable by recombination was dropped.

To make clear the distinction between the gene as a unit of crossing-over and the gene as an element of complementation, Benzer used the term *cistron*

---

[11]Benzer (1966) has presented the list of scientists who passed through his laboratory at Purdue and sought, in vain, the *r*II protein.

for the latter (Benzer, 1957). The term *cistron* was of considerable utility when the fine-structure studies on the *r*II locus were being carried out, but as knowledge of how genes and proteins advanced, it became clear that *cistron* and *gene* are equivalent. For a while, the two terms were used interchangeably, but the term *cistron* is now no longer used extensively.

The fine structure of the *r*II region was revealed to consist of several hundred independently mutable and recombinable sites that could be mapped on a strictly linear structure. Benzer introduced an important simplified mapping procedure: the use of overlapping deletions (Benzer, 1957). Deletion mutants, detected initially because they did not revert, were mapped by recombination, and then other point mutations were mapped by crossing with the deletions. If wild-type recombinants are produced when a point mutant and a deletion are crossed, then the point mutant must have a map position *outside* the region of the deletion. By use of this procedure, mapping of large numbers of mutants was greatly simplified. Once a mutant had been localized to a particular region by deletion mapping, it could then be mapped in relation to other mutants of this same region, using two-factor and (preferably) three-factor crosses. "With a sufficient number and appropriate distribution of deletions, one could hope to order a large length of map. The reader will note an analogy (not altogether without significance!) to the technique used by Sanger to order the amino acids in a polypeptide chain by means of overlapping peptide segments" (Benzer, 1957).

Over the next five years, Benzer mapped the *r*II region in exquisite detail. "It was an example of what we called 'Hershey Heaven.' This expression comes from a reply that Alfred Hershey gave when Garen once asked him for his idea of scientific happiness: 'To have one experiment that works, and keep doing it all the time'" (Benzer, 1966).

The mutation and mapping studies of the *r*II region showed several interesting aspects. First, mutability was nonrandom among the various sites. Mutations occurred at certain sites over and over again, whereas other sites had fewer mutations, and some, by inference from map distances, had none. Second, certain sites showed large numbers of mutants. These sites were called "hotspots," presumably reflecting some inherent instability of the DNA at those locations.

This work, using forward mutations, was also seen as the first step toward a determination of the genetic code. By study of induction of reverse mutations with different chemical mutagens, the specific bases on each site might be deduced (Benzer, 1961). Fortunately, biochemical approaches to the genetic code intervened, so that this formidable undertaking was not necessary. However, the *r*II studies did provide strong support for the idea of nonsense codons (Benzer and Champe, 1962).

The acceptance and importance of Benzer's work can be indicated by the following quotation from Pontecorvo (1958):

> The most obvious wrong idea...is that of the particulate gene, i.e., of the genetic material as beads on a string in which each bead is an ultimate unit of crossing over, of mutation and of specific activity [function]. This picture was not merely crude: it was wrong because it implied an unnecessary, and almost certainly nonexistent, structural differentiation between the beads and the string...What has replaced it is the picture of a nonrepetitive linear structure of building blocks of only a few different kinds, the unique groupings of which determine functions.

Each of these functions we now call a "cistron." The analogy of the genetic material with a written message is a useful commonplace. The important change is that we now think of the message as being in handwritten English rather than in Chinese. The words are no longer units of structure, of function, and of copying, like the ideographic Chinese characters, but only units of function emerging from characteristic groupings of linearly arranged letters. Miscopying has now become misspelling: a mistake in letters or in their order, not usually a mistake in words. In this analogy, letters correspond to mutational sites exchangeable by crossing over, words correspond to cistrons, and misspellings to mutations (Pontecorvo, 1958).

This clarification, which took place between Pontecorvo's two reviews of 1952 and 1958, was due almost solely to Benzer's work on the genetic fine structure of the *r*II locus of bacteriophage T4. The concepts that Benzer developed were also applied almost immediately to bacterial genes, using the high resolution analysis that became available by the discovery of transduction (see Section 8.5). Soon this work would make study of colinearity of gene and protein possible (see Section 10.14).

6.11 THE PHAGE GENETIC MAP

Benzer's work showed the significance of conditional lethal mutations for carrying out genetic analysis. Up until the early 1960s, only a few phage genes had been mapped, and those that had been studied were of little value in understanding the mechanism of phage replication. It should be noted that much of the biochemistry of phage replication (see Section 6.12) was worked out before the phage genes were identified.

Two groups, one at CalTech under R.S. Edgar, and the other at the University of Geneva, Switzerland, under R.H. Epstein and E. Kellenberger, adapted Benzer's approach and sought other types of conditional lethal mutants. One type, called *amber*, formed plaques on *Escherichia coli* strain CR63 but not on strain B, although wild-type T4 forms plaques on both strains (Epstein, Bolle, Steinberg, Kellenberger, Boy de la Tour, Chevalley, Edgard, Susman, Denhardt, and Lielausis, 1963). It is now known that the *amber* codon is a nonsense codon that signals a *stop* site during translation. These phage *amber* mutations could replicate in bacterial strain CR63 because this strain has a suppressor mutation that permits the *amber* codon to be translated. Because such nonsense mutations are always point mutations, and can occur virtually anywhere in the genome, the use of the host containing an *amber* suppressor permitted isolation of phage mutants that mapped virtually anywhere in the genome.

The other type of mutation, generating a temperature sensitive (*ts*) phenotype, was found in a more direct way. It had been known in *Neurospora* (see Section 5.1) as well as in *Escherichia coli* that mutations could occur that would make proteins temperature sensitive. Such mutants would be recognized as phenotypes that were normal at the lower (permissive) temperature and mutant at the higher (restrictive) temperature. In the T4-*Escherichia coli* system, the permissive temperature used was 30°C and the restrictive temperature was 42°C. By using these two procedures, mutants in virtually any gene in the T4 genome could be obtained and mapped. Furthermore, protein deficiencies could be determined by biochemical studies. Within 15 years after these studies, over 130 genes had been identified in T4, and more genes were still anticipated (Doermann, 1983b; Mosig, 1983).

Phage mutants blocked in the expression of early, middle, and late proteins (see Section 6.12), and in various control functions, were identified. A large number of mutants in the protein coat of the virus were isolated, and the whole machinery of virus assembly was worked out. Certainly, the legacy from Benzer's pioneering work is impressive!

Further genetic analysis, using three-factor crosses, showed that the genome of the T-even phages could be characterized as formally circular (Streisinger, Edgar, and Harrar-Denhardt, 1964; Streisinger, 1966), although the duplex DNA of T4 is actually linear. This apparent circularity arises because different DNA molecules in a phage population contain redundant regions at their ends and the terminal nucleotide sequences of these ends differ from one particle to another. Thus, if the overall gene sequence is considered to be A...Z, one phage molecule might start with AB... and end with ...ZAB, another might start with GH... and end with ...FGH, etc. Genetic recombination between molecules exhibiting such terminal redundancies gives the *appearance* of circularity even though the molecules are linear. When genetic analysis first led to evidence of circularity, it was initially assumed, following the example of the *Escherichia coli* chromosome (see Section 5.11), that the T4 chromosome was circular. The idea of circularity of the genetic map would perhaps have been rejected, despite the data, if it had not been for the *Escherichia coli* data.

Phenotypic Mixing

In addition to mutual exclusion and genetic recombination, another phenomenon discovered in mixed infection experiments was *phenotypic mixing*. This phenomenon, which occurs only when the two infecting phages are related, involves the packaging of the virus genome of one virus type within the coat of the other type. Thus, particles can have different phenotypes but the same genotype. This phenomenon, first discovered by Novick and Szilard (1951), complicated the studies on virus genetics but is of general interest in virology. Knowledge of phenotypic mixing aided in the interpretation of the transduction phenomenon discovered the next year (see Chapter 8).

6.12 BIOCHEMISTRY OF PHAGE REPLICATION

Phage genetics and phage biochemistry were twin disciplines that advanced together, leading ultimately to a detailed understanding of the molecular basis of phage replication. The 1953 symposium on viruses at Cold Spring Harbor, organized by Delbrück, can be viewed as the major turning point in the conversion of phage research from essentially the "black box" approach to the primarily biochemical and molecular approach of later workers (Delbrück, 1953). What had happened to change ideas about phage multiplication? First, Doermann had discovered that phage entered an "eclipse" period, during which it was fundamentally changed. Second, Hershey and Chase had shown that only the DNA of the phage had to enter the cell for virus replication to occur. (Both of these developments are discussed in detail below.) Third, the structure of DNA was enunciated by Watson and Crick. The first full exposition of the Watson/Crick model was given at this same

1953 symposium. The impact of the Watson/Crick model can be seen by the following comments of Delbrück in the introduction to the symposium:

> Special mention should be made of a last minute addition to the program...The discovery of a structure for DNA proposed by Watson and Crick a few months ago, and the obvious suggestions arising from this structure concerning replication seemed of such relevance to many questions to be discussed at this meeting that we thought it worth while to circulate copies of three letters to Nature concerning this structure among the participants before the meeting, and to ask Dr. Watson to be present at the meeting (Delbrück, 1953).

The Prejudice against Biochemistry

Much has been made of the avoidance by the Delbrück school of "real" biochemistry (Cohen, 1975, 1984), for it was only through biochemistry that some key aspects of phage replication could be elucidated. But in the early days of phage research, most of the necessary biochemical techniques (especially the widespread availability of radioisotopes) were not available, so it would have been difficult to do "meaningful" biochemistry on phage. Without biochemistry, Luria, Delbrück, and Hershey studied phage in the way a geneticist would, by means of mixed infections and comparisons with related but distinct phage strains. This work was to lead to some of the most important discoveries in the phage field (Benzer's *r*II work is one of the best examples) and was also to influence developments in bacterial genetics itself (see Section 4.8). In the days before 1953, "real" biochemistry was not only shunned, it was actively discouraged. This point has been well made by Seymour Cohen (1968):

> All members of the small group of [phage] biochemists, including Earl Evans, Frank Putnam, and Lloyd Kozloff...were convinced of the need to use...[biochemistry] in attaining our goals. For many years, however, it was not possible to establish either the validity or the pressing importance of the biochemical methodology among most of the committed biologists. Indeed, it was not until the relatively late period of study described in this volume that many of the same biologists chose to adopt chemical approaches to the problems whose outlines they had drawn; by that time these approaches had become patently inescapable if the biological phenomena were to be dissected to their molecular bases.
>
> The rejection of "biochemistry" and the adoption of the term "molecular biology" by the genetically oriented microbiologists in the last decade is one of the most striking and interesting phenomena of modern biology. This event is of no less interest to historians of biochemistry than to those of cellular biology. After 1952, the phage biologists relinquished the idea that cellular biology had a mystical core impenetrable to biochemical concepts. Nevertheless, in formulating more and more precisely the problems of inheritance that interested them, they felt that they could afford to bypass innumerable apparently irrelevant facts and techniques of biochemistry (Cohen, 1968).

That this is not "sour grapes" can be shown by the following remark of Delbrück, written in 1950 as an introduction to a "syllabus" on phage research:

> This syllabus has been read by most of those whose work has been cited and their comments have been taken to heart in preparing the revised version here printed. No general criticism was expressed in any of these comments with one exception. Dr. S.S. Cohen felt that the syllabus did not give proper weight to the biochemical approach, to which he himself has so pre-eminently contributed, and that this

failure of a proper appreciation of the biochemical results had led to a false appraisal of the present situation. While the authors of the syllabus do not accept this criticism they would like to draw particular attention to the recent review article by Dr. Cohen...and which should be considered as a complementary report, giving particular emphasis to the biochemical approach to phage (Delbrück, 1950).

Those involved in the preparation of the 1950 syllabus were listed as S. Benzer, M. Delbrück, R. Dulbecco, W. Hudson, G.S. Stent, J.D. Watson, W. Weidel, J.J. Weigle, and E.L. Wollman. S.E. Luria was also a participant of the general meeting. The principal phenomena presented in the syllabus concerned phage morphology, serology of phage, virus mutants, radiobiology, adsorption, effect of environmental conditions on lysis and burst size, mixed infection, photoreactivation, multiplicity reactivation, lysis inhibition, and lysis from without, but no biochemistry. The review Delbrück refers to is Cohen (1949).

As noted earlier in this chapter, Delbrück had been motivated to enter biology by Niels Bohr's interest in the application of the complementarity principle to biological systems. Since complementarity set limits to how far a system can be broken down, this strong philosophical view was certainly an underpinning to Delbrück's reluctance to "do" biochemistry. Whether this influence slowed the growth of the field, as Cohen (1968) insists, is uncertain.

Chemistry of Phage

One of the key advantages in the use of phage as a molecular system was that particles could be purified rather easily from bacterial lysates. This was especially the case with the T-even phages because of their large size. Purification from high-titer lysates could be achieved by simple centrifugation, alternating low-speed centrifugation to remove bacterial debris with high-speed centrifugation to sediment the phage particles. The availability of high-speed centrifuges in the 1930s and 1940s made it possible for a number of laboratories to purify phage and study its biochemistry. By the 1930s, some understanding of the chemistry of DNA was also available, from the pioneering work of Phoebus A.T. Levene and others (Portugal and Cohen, 1977; see Section 9.8 for a brief discussion of DNA chemistry).

The first chemical analyses of purified phage particles were carried out in the early 1930s in Germany by Martin Schlesinger. He obtained weighable quantities of the phage WLL and found by direct chemical analysis about equal amounts of protein and DNA (Schlesinger, 1934). Schlesinger showed that his purified phage contained relatively large amounts of phosphorus and gave a strong positive reaction with the Feulgen test (specific for DNA). He concluded that phage is a deoxyribonucleoprotein consisting of about half protein and half DNA. However, Levene's *tetranucleotide hypothesis*, which stated that DNA was composed of equal amounts of the four nucleotides, was still the accepted explanation for the structure of DNA (see Section 9.8), so that Schlesinger's conclusion led to little further understanding of the biochemical nature of phage. Because of Schlesinger's departure from Germany, followed by his untimely death in 1936, further work on the chemistry of phage ceased (see Stent, 1963, for a brief biographical sketch of Schlesinger). As noted above, the impact of Delbrück's ideas probably also inhibited a general interest in the "chemistry," as opposed to the "biophysics," of phage.

The advent of chromatography and radioisotope methodology in the aftermath of World War II greatly changed research on phage chemistry. Chromatography made possible the analysis of the amino acids and purine and pyrimidine bases of phage, and radioactive phosphorus, ^{32}P, made it possible to trace the DNA through the infected cell. Although it had been shown that phage could be treated as a genetic system, and hence was relevant to an understanding of the genetics of higher organisms, without phage biochemistry the research would have remained abstract. In this regard, one key figure was Hershey, who provided a connection between the Delbrück phage group and the biochemists. The other key figure was the biochemist Seymour S. Cohen.

Seymour S. Cohen

Some of Seymour Cohen's comments about the importance of biochemistry in phage research have been given above. A number of autobiographical passages in his book (Cohen, 1968) and in Cohen (1987) present some of his background. Cohen received his Ph.D. in 1941 in biological chemistry at Columbia University, working under Erwin Chargaff on phospholipids and nucleic acids in the microsomal fraction of lung tissue. He received an early fellowship from the National Foundation for Infantile Paralysis and spent a year working on virus chemistry with Wendell Stanley at the Rockefeller Institute (see Section 6.3 and Cohen, 1979). With World War II intervening, Cohen worked on vaccines for rickettsia. He became a faculty member at the University of Pennsylvania in 1943 and turned to the study of the biochemistry of viruses. At that time, Thomas F. Anderson was there, and Cohen learned of the T-even phages from him. He took Delbrück's phage course in 1946, the second year it was offered, and after this concentrated his research on the biochemistry of the T-even phages. Shunning the strictly black box approach of Delbrück, Cohen did early work on phage replication using radioisotopes (Cohen, 1947).

Biochemistry of T-even Phage Replication

Cohen chose the T-even bacteriophages to study because of the extensive data available on the physiology and genetics of these phages. Two major types of experiments were done: (1) tracing nucleic acid through the infection cycle with ^{32}P, and (2) use of inhibitors such as 5-methyl tryptophan to inhibit protein synthesis and examine how this inhibition affected DNA synthesis.

In the ^{32}P experiments, uninfected cells were fully labeled and then infected with nonradioactive virus particles in a nonradioactive medium. The virus progeny were then analyzed for ^{32}P. The results showed that the newly released virus particles contained little or no ^{32}P, whereas if unlabeled cells were infected with unlabeled virus in a radioactive medium, the virus particles released were highly radioactive. These results showed that the DNA of the virus is synthesized *de novo* from inorganic phosphate present in the culture medium (Cohen, 1947). Cohen also showed by careful balance experiments that RNA did not participate "directly" in the synthesis of phage DNA. Using inhibitors and careful kinetic analyses, Cohen also showed that phage DNA synthesis could be divided into two phases: (1) a latent period after infection when protein synthesis was necessary for DNA synthesis, and (2) a subsequent period during which DNA synthesis occurred in the absence of further protein synthesis. These results provided the first evidence for the requirement of new proteins, presumably new enzymes, for phage DNA synthesis.

Another important conclusion, which ran counter to a number of ideas at that time, was that the energy as well as the substance for virus synthesis was supplied in an "entirely normal fashion by the parasitized cell." These experiments, which were extensively confirmed in later years, focused attention in a biochemical way on the role of the virus particle as a genetic unit. "Thus the virus appears to be synthesized by the cell according to the models (templates) which it provides for the host's enzymes." Although it was subsequently to be shown that, especially in the T-even phages, numerous virus-specific enzymes were involved, the conclusion that the main role of the virus was instructive was a critical one in focusing research in this critical period up until the time of the Hershey/Chase experiment. Cohen (1949) summarized this early important work, focusing especially on processes that might provide clues for how virus growth could be specifically inhibited.

Following up on Cohen's work, Putnam and Kozloff (1950) and Kozloff (1953) (see also Evans, 1953), using ^{32}P-labeled phage, showed that a fraction of the infecting DNA ended up in the progeny particles. Since most of the DNA of the infecting phage did not appear in progeny, this experiment was first considered to demonstrate that the parental phage DNA molecule was completely degraded during the latent period, and its degradation products built up into new particles (see also Kozloff, 1966). Other experiments on nucleic acid transfer were done by Maaløe and Watson (1951) and Watson and Maaløe (1953). However, subsequent experiments with more sophisticated techniques did show that there was some molecular continuity between infecting phage and progeny (Levinthal and Thomas, 1957).

Unusual Bases in T-even DNA

One of the surprising findings from Cohen's group was that the T-even phages contained a previously unknown pyrimidine base, 5-hydroxymethyl cytosine (HMC) (Wyatt and Cohen, 1953a,b). Cohen has presented a good description of the discovery of HMC (Cohen, 1968). Among other consequences, this important discovery showed that because of this chemical difference between host DNA and phage DNA, some important aspects of the biochemistry of phage replication could be analyzed directly (by assaying for HMC in infected cells). Hershey, among others, quickly exploited this approach, estimating the time of initiation of synthesis of phage DNA and the size of the pool of precursor phage DNA (Hershey, 1953a).

Further chemical analysis showed that the T-even phages also contained the hexose sugar *glucose*, which was attached to the hydroxymethyl group of HMC (Sinsheimer, 1954). Although neither the HMC nor the glucose affected the double-stranded structure, their presence emphasized the chemical uniqueness of phage.

The Eclipse Period

The one-step growth experiment of Ellis and Delbrück had focused attention on the events that were occurring during the latent period. In 1948, A.H. Doermann[12] developed a method by which infected cells could be gently lysed and the infectivity of these artificial lysates could be analyzed (Fig. 6.7). Surprisingly, Doermann found that the infectivity of the original infecting phage was lost at the beginning of the infection period, since no infecting phage particles were found within the first ten minutes after infection. After this ten-minute period, called the *eclipse*, there was a dramatic increase in the number of infective phage particles until the final crop of progeny had been produced. "This result is in agreement with the gradually emerging concept that a profound alteration of the infecting phage particle takes place before reproduction ensues" (Doermann, 1952).

Another type of phenomenon, called *multiplicity reactivation*, focused on the uniqueness of the intracellular growth phase (Luria, 1947). In this phenomenon, two phage particles that have been inactivated by ultraviolet radiation (UV) are able to reproduce when both infect the *same* cell, although they are unable to reproduce when infecting separate cells. Luria interpreted multiplicity reactivation as a sort of genetic exchange, assuming that a number of different viral subunits each multiplies independently, and the progeny particles subsequently become assembled into a complete virus particle. In this model, UV would "hit" each virus in a separate part of the genome, and by "pooling" two damaged virus genomes, an undamaged virus would be produced. Although the idea that the phage genome comes apart into subunits is incorrect, the discovery of multiplicity reactivation encouraged Hershey to consider that the appearance of new phage types during mixed infection (see Section 6.9) was really due to genetic exchange.

6.13 THE HERSHEY/CHASE EXPERIMENT

By 1952, genetic and biochemical research had clearly shown that phage replication was something quite apart from cellular replication. The stage was set for the Hershey/Chase experiment, a key development that was to galvanize the phage community.

The Hershey/Chase experiment showed that only the DNA of the phage had to enter the cell for virus replication to occur. Although in retrospect this experiment appears technically flawed, it was to have profound influence on the subsequent development of ideas about molecular genetics. Hershey (1966) has described the antecedents to this important experiment.

> The admissible view of the time was...[that] the phage particle is a dual structure (mainly protein and DNA) in which differentiation of nuclear and cytoplasmic functions is vaguely discernible. After phage infection, the bacterial metabolic system falls under the control of the nuclear apparatus of the phage, producing mainly phage-specific materials. The infecting phage particle does not itself survive infection, losing at least its ability to infect another bacterium. Thus intracel-

---

[12]A.H. (Gus) Doermann had been a student of George Beadle at Stanford, working on *Neurospora* genetics. Coming under the influence of Delbrück, he worked at Vanderbilt, Cold Spring Harbor, and CalTech, before settling at the Oak Ridge National Laboratory in the post-World-War-II period. He was later a faculty member at the University of Rochester and the University of Washington (Seattle). The work on the eclipse was done when he was at Cold Spring Harbor.

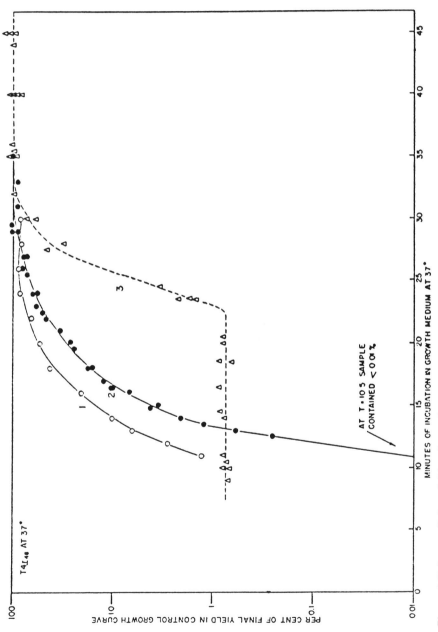

Figure 6.7 Intracellular viable phage during the latent period, as determined by premature lysis with cyanide (solid line). The dashed line is a control one-step growth curve. (Reprinted, with permission, from Doermann, 1952.)

lular phage growth is characterized by a phase of eclipse, lasting about half the life of the infected bacterium, during which the cell does not contain demonstrable phage particles. During the phase of eclipse, genetic recombination takes place, apparently preceded by replication of genetic determinants of the phage. Phage growth therefore consists of two stages: replication of some noninfective form of virus, followed by conversion of the products of replication back into finished phage particles which do not themselves participate in reproduction.

In truth, the facts available in 1950 did not demand any fundamental distinction between growth of phage and...growth of malaria parasites, which also multiply in eclipse. The conviction of phage workers about the uniqueness of phages was, for the time being, an article of faith, the origin of which seems obscure... (Hershey, 1966).

Before the Hershey/Chase experiment is discussed, some further background information must be given.

Structure and Osmotic Properties of Phage Particles

Electron microscopy had shown that most phage had tails, but even in the T system there was a wide range in morphologies (see Table 6.1). Thomas F. Anderson in particular was using electron microscopy and biophysical techniques to study the nature of phage. Among other things, he made the important observation that when T-even phage particles were suspended in a high concentration of sodium chloride and then rapidly diluted, they were converted into inactive "ghosts" (Anderson, 1949). Other solutes gave similar effects, and since no inactivation occurred if the dilution was allowed to take place slowly, Anderson concluded that the particles possessed an osmotic membrane (Anderson, 1953). Roger Herriott, the Johns Hopkins biochemist, had shown that the ghosts formed by osmotic shock consisted mainly of protein (Herriott, 1951). In addition, Herriott found that the phage DNA was rendered sensitive to DNase by osmotic shock. These facts led to the conclusion that the DNA of the phage particle is on the inside (Anderson, 1953). Because phage particles had no lipid, the outer semipermeable membrane was clearly protein in nature.

An additional technical fact that was to be important was the observation by Anderson (1949) that phage particles could be "stripped" from their host cell surfaces by violent agitation in a high-speed blender.

Phage as Transforming Principle

By the early 1950s, the significance of Avery's transformation experiments was widely discussed (see Section 9.10), and the possible analogy to phage replication had not been missed. Hershey (1966) quotes from a letter he received in 1951 from Roger Herriott: "I've been thinking—and perhaps you have, too—that the virus may act like a little hypodermic needle full of transforming principles; that the virus as such never enters the cell; that only the tail contacts the host and perhaps enzymatically cuts a small hole through the outer membrane and then the nucleic acid of the virus head flows into the cell" (Hershey, 1966).

According to Hershey (1966), another motivation for initiating the experiments was the work of Putnam and Kozloff and of Watson and Maaløe (discussed above) on the transfer of phosphorus atoms from parental to progeny DNA. "Their discovery seemed to offer a very direct method for analyzing the material basis of heredity..." In addition to the availability of ^{32}P as a label for DNA, radioactive sulfur, ^{35}S, was available as a label for protein.

The Experiment

Hershey has recalled the experiment in detail:

> The blender experiment has been described incorrectly in print several times, and in the hope of preserving its essential simplicity I shall recall it here. A chilled suspension of bacterial cells recently infected with phage T2 is spun for a few minutes in a Waring Blender and afterwards centrifuged briefly at a speed sufficient to throw the bacterial cells to the bottom of the tube. One thus obtains two fractions: a pellet containing the infected bacteria, and a supernatant fluid containing any particles smaller than bacteria. Each of these fractions is analyzed for the radiophosphorus in DNA or radiosulfur in protein with which (in different experiments) the original phage particles have been labeled. The results are:
>
> 1. Most of the phage DNA remains with the bacterial cells.
> 2. Most of the phage protein is found in the supernatant fluid.
> 3. Most of the initially infected bacteria remain competent to produce phage.
> 4. If the mechanical stirring is omitted, both protein and DNA sediment with the bacteria.
> 5. The phage protein removed from the cells by stirring consists of more-or-less intact, empty phage coats, which may therefore be thought of as passive vehicles for the transport of DNA from cell to cell, and which, having performed that task, play no further role in phage growth (Hershey, 1966).

The key experiment of Hershey and Chase (1952) is shown in Figure 6.8. It can be seen that after infection very little of the phage DNA phosphorus is removed by the blender, whereas most of the phage protein sulfur is released. Although 20% of the protein sulfur is not removed by blending, none of this remaining sulfur was found in the phage progeny, so it was concluded that this sulfur represented a residue that had no function in phage multiplication.

The genetic significance of these observations was clearly expressed in the original paper:

> Our experiments show clearly that a physical separation of phage T2 into genetic and non-genetic parts is possible...The chemical identification of the genetic part must wait, however, until some [further] questions...have been answered (Hershey and Chase, 1952).

However, at the 1953 Cold Spring Harbor symposium, more than a year after the above experiments were published, Hershey (1953b) was still not certain about the significance of the experiments:

> There are now three types of evidence suggesting a genetic role for DNA. (1) The average DNA content of chromosomes correlates with species and with ploidy, not with the tissue of origin. (2) Specific heritable effects can be produced in certain bacteria by exposing them to DNA from specific sources. (3) DNA plays some dominant though unidentified role in T2 infection. None of these, nor all together, forms a sufficient basis for scientific judgement concerning the genetic function of DNA. The evidence for this statement is that biologists (all of whom, being human, have an opinion) are about equally divided pro and con. *My own guess is that DNA*

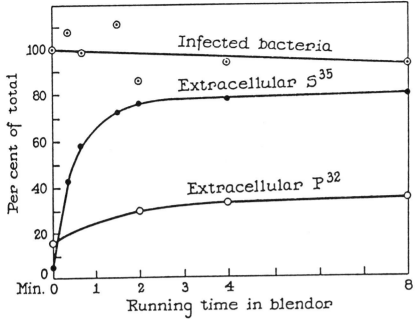

Figure 6.8 The Hershey/Chase experiment. Removal of ^{35}S and ^{32}P from bacteria infected with radioactive phage, during agitation in a high-speed blender. The upper line shows that the blending does not inactivate the infected bacteria. (Reprinted, with permission, from Hershey and Chase, 1952.)

> *will not prove to be a unique determiner of genetic specificity*, but that contributions to the question will be made in the near future only by persons willing to entertain the contrary view [emphasis added] (Hershey, 1953b).

Although Hershey was cautious in his interpretation, others were less inclined to doubt. At this same symposium, James Watson discussed the significance of DNA with the following words:

> It would be superfluous at a Symposium on Viruses to introduce a paper on the structure of DNA with a discussion on its importance to the problem of virus reproduction. Instead we shall not only assume that DNA is important, but in addition that it is the carrier of the genetic specificity of the virus (for argument, see Hershey, this volume) and thus must possess in some sense the capacity for exact self-duplication (Watson and Crick, 1953).

Watson had the benefit of having thought about the structure of DNA and especially about how mutations might arise from base changes. Coming at the same time as the elucidation of the structure of DNA, the Hershey/Chase experiment not only pointed the direction for research on phage replication, it also focused on DNA as the genetic material in a way that Avery's transformation experiments, despite their much more convincing chemistry, had never quite done (see Chapter 9). Although both the Hershey/Chase experiment and the Avery work on transformation provided support for the genetic role of DNA, it was the Watson/Crick model that forced the issue. This model had profound influence not only on concepts of the nature of virus, but also on the role of DNA as the genetic material. A few years later, the doctrine that nucleic acid is the carrier of viral heredity was confirmed by studies on the RNA of tobacco mosaic virus. Certainly molecular genetics would have developed without the impetus of the Hershey/Chase experiment, but probably somewhat more slowly.

We see in this experiment a direct line of descent from the initial work of Ellis and Delbrück (1939), the work of Delbrück and Luria on phage mutations, the work of Delbrück and Hershey on genetic recombination in phage, Doermann's work on the eclipse, Cohen and Kozloff's work on transfer of atoms from infecting phage to progeny, and Anderson's work on electron microscopy and osmotic shockability of phage. Knowledge of Avery's work on transformation also informed the research. Despite Delbrück's prejudice against chemistry, and despite Delbrück's influence on all of these workers, the chemical studies ruled the day; but only with the genetic work behind it could the interpretation of the chemical work be profound.

The Hershey/Chase experiment and the Watson/Crick model came to be seen as major turning points in the development of the field of molecular biology. Phage research would never be the same: The genie had been let out of the "black box." Biochemistry would assume equal partnership with genetics.

6.14 HOST AND VIRUS PROTEIN SYNTHESIS

The role of RNA in protein synthesis is discussed in Section 10.12. One of the important early experiments was that of Volkin and Astrachan (1956) on the synthesis of RNA after T2 infection. Although it had been known since the work of Kozloff, Evans, and Cohen (see Section 6.12) that *net* RNA synthesis ceased after T-even infection, Volkin and Astrachan showed, using ^{32}P, that some RNA synthesis did occur after infection and that the RNA formed had the same base composition as the phage DNA. (Fortuitously, the base composition of the T-even phages differed from that of the *Escherichia coli* host.) The Volkin-Astrachan experiment, although not then understood, was one of the important antecedents of the messenger RNA concept (see Section 10.12).

From knowledge of the formation immediately after infection of phage-specific RNA and the presence of a new base in phage DNA that was absent in the host, it seemed reasonable that phage-specific enzymes might play a role in the multiplication process. Although a few studies were done in the 1950s, it was not until the formation of the messenger RNA concept that a rationale for studying the regulation of phage protein synthesis existed. As discussed in Section 10.12, T-even infection was used as a model in the early 1960s to study mRNA synthesis. Numerous studies through the rest of the 1960s focused on phage protein synthesis. Such studies became more intelligible after Edgar, Epstein, and their colleagues (see Section 6.11) discovered numerous conditional mutants affecting head and tail proteins. By the late 1960s, a large number of virus-specific proteins were known, and the timing of synthesis had been worked out (Cohen, 1968).

6.15 THE RESTRICTION/MODIFICATION PHENOMENON

Among the numerous discoveries of general biological significance that came out of phage research is the one called *host-controlled modification* or *phenotypic modification*. Discovered almost simultaneously in the early 1950s in four separate laboratories (Stent, 1963), for many years the phenomenon was no more than a curiosity, but it has led in the 1970s and 1980s directly to *genetic engineering* and the techniques of *recombinant DNA* research (see Chapter 11).

Host-controlled modification of phage was first studied in detail by Luria and Human (1952; see also Luria, 1984). Luria and Human were comparing the growth of a single phage on two different hosts when they found that infectivity varied depending on the host the phage had previously been grown on. The key characteristic of host-controlled modification is that it is strictly phenotypic (nonhereditary). The phage grows much better on one host than the other, and the second host is said the *restrict* the growth of the phage. The small number of phage particles produced on the restricting host are *modified* so that they are now able to infect anew this same host at high efficiency. Thus, passing the phage through the restricting host results in a phage population that is no longer restricted. That the modification induced by the restricting host is nonhereditary is shown by the fact that if the phage is passed through the permissive host, it again grows only at low efficiency on the restricting host.

Soon after its discovery, host-controlled modification had been detected in numerous phage/host systems (Luria, 1953). It was shown initially that the inability of modified phage to infect was not due to lack of adsorption. Rather, the effect occurred after the injection of the DNA. Second, the presence of rare plaques on the restricting host was due to a variation in physiology of host cells rather than variation in phage. Presumably, such rare host cells lacked the ability to restrict the phage, and once a restricted phage has gone through a round of infection, it is normal and is able to continue the infection cycle (this early work is reviewed in detail by Hayes, 1964).

The first clue to the mechanism of host-controlled modification came from studies on T-even phage (Hattman and Fukasawa, 1963). The DNA of these phages is glucosylated via the hydroxymethylcytosine residues (see Section 6.12), and this glucosylation renders the DNA resistant to attack by nucleases in the *Escherichia coli* strain B host. It happens that mutants of *Escherichia coli* B selected for combined resistance to phages T3, T4, and T7 (B/3,4,7) lack the enzyme UDPG pyrophosphorylase, necessary for the glucosylation reaction. Thus, when B/3,4,7 is infected with T2 or T6 grown on B, normal phage growth occurs (because the infecting phage DNA is glucosylated), but the progeny particles are nonglucosylated and are unable to undergo another cycle in B/3,4,7. However, phage with nonglucosylated DNA can be grown on *Shigella*, which does not degrade the unglucosylated DNA.

The most extensive studies on restriction were done with the *lambda* system. *Escherichia coli* K-12 lysogenized with phage P1 restricts the growth of *lambda*. When *lambda* DNA is injected into K-12 that has been lysogenized with P1, it is rapidly broken down. By use of a methionine-requiring strain of *Escherichia coli* as host, it was shown that if methionine is withheld during infection, some of the released phage lack the modification even though they were grown on a modifying host. Because methionine serves as a methyl donor for bacterial DNA, it was inferred that modification was due to DNA methylation (Arber, 1965). Subsequent work showed that the host had an enzyme called a *restriction enzyme* that degraded DNA, and that this restriction enzyme did not degrade the DNA if it had been modified by methylation. Thus, restriction and modification exist as a paired system, the function of which is to protect host DNA but destroy foreign DNA.

The discovery of host-controlled modification in phage not only led to modern knowledge of restriction enzymes (see Chapter 11), but it also provided strong evidence for the role of secondary modification in DNA.

6.16 CONCLUSION

Phage research has come a long way since the first one-step growth experiments of Ellis and Delbrück. Perhaps the greatest indication of Delbrück's legacy is the fact that in 1983 it took a whole book to summarize the ongoing research on a single phage, bacteriophage T4 (Mathews, Kutter, Mosig, and Berget, 1983). Not all phage work has been of direct relevance to bacterial genetics, but concepts arising from phage pervade modern molecular biology. In this brief summary, some of the key ideas that came from phage will be listed. The items are divided into two separate categories, those that derive directly from the Delbrück "black box" approach, and those that make use of biochemical techniques. Some of these ideas will be presented in more detail in subsequent chapters.

Nonbiochemical Studies

Studies that dealt primarily with an analysis of the infection cycle and with "pure" genetics fall into this category.

1. The infection cycle of adsorption, eclipse, maturation, and lysis was discovered by nonbiochemical means. An understanding of the infection cycle was essential for the quantitative biochemical work. The ideas that arose from this work found broad application in virology, especially as applied to animal and human viruses. It is likely that concepts arising from phage research considerably hastened the development of animal virology.
2. Phage genetics began as "pure" biological research. Mixed infection, multiplicity reactivation, genetic recombination, conditionally lethal mutants, and mapping all made use of classical biological procedures and could take place without any knowledge of the chemical nature of the entities involved. Even the restriction/modification system was initially a nonbiochemical study.
3. Genetic fine structure (Benzer's work) took place without any knowledge of the biochemistry involved. Even many years later, the proteins controlled by the *r*II locus are not known, but the approach to fine-structure mapping was revolutionary.
4. The discovery of *amber* (nonsense) mutants in phage T4 had important implications for deciphering the code.

The above categories provided the essential base for the biochemical work, some of which occurred at the same time, but most of which occurred later. Although it is clear that Delbrück in particular "discouraged" biochemical work, he appreciated its results, and many members of the phage group even carried on biochemical work. Delbrück himself ceased working solely on phage after 1954 when he turned to *Phycomyces*, but his laboratory remained a center for phage research for many years. Perhaps the greatest contribution of the phage group was the demonstration that the study of viruses would provide important insights for genetics. This was clearly emphasized very early by both Luria and Delbrück in discussion of Hermann Muller's paper at the 1941 Cold Spring Harbor symposium (Muller, 1941).

The Delbrück *Festschrift* (Cairns, Stent, and Watson, 1966) celebrates the successes of the "black box" approach. It also reveals how narrow the outlook of the phage group actually was. Operating as mostly a closed body, the phage group had the strength of coherence to a defined set of goals, but it must have represented a formidable obstacle to any "outs," such as Seymour Cohen. The Delbrück *Festschrift*, now often viewed as a fairly definitive history of the origins of molecular biology, is an unfair treatment of the biochemical approach to molecular biology.

Biochemical Work

Once the phage infection cycle was understood, it was possible to use chemical techniques to study phage multiplication and its regulation.

1. The early biochemical work of Cohen provided strong evidence for the relative importance of host and virus in the infection cycle. The fact that the host provided the essential machinery for virus replication, such as energy generation, made it seem likely that the main role of the virus itself was instructive (that is, genetic).
2. The Hershey/Chase experiment provided strong support for the idea of DNA as the genetic material, an idea that first arose from studies on pneumococcus transformation (see Chapter 9).
3. Physiological and biochemical studies, especially the Kozloff/Evans/Cohen work and the Volkin/Astrachan experiment, provided some of the strongest evidence for the messenger RNA concept (see Section 10.12).
4. Benzer's work on the fine structure of the gene led to biochemical work that proved the colinearity of gene and protein (see Section 10.14).
5. Numerous enzymes involved in DNA replication were first discovered from studies on phage replication.
6. The discovery of the mechanism of host-controlled modification, although initially only a "black box" phenomenon, led eventually to the discovery of restriction enzymes and the modern age of biotechnology (see Chapter 11).

Virulent and Temperate Phage

Finally, a few words should be added in anticipation of the next chapter. All of the work described in the present chapter has dealt with what are now called "virulent" phages, although at the time the early work was carried out this distinction was not clear. Even though Burnet and McKie (1929) had pointed out that there were two kinds of phage, Delbrück and other members of the phage group did not accept this distinction. As discussed in the next chapter, it was only after André Lwoff demonstrated unequivocally the nature of the lysogenic system that Delbrück and the rest of the phage group accepted lysogeny as a "real" phenomenon.

REFERENCES

Anderson, T.F. 1949. The reactions of bacterial viruses with their host cells. *Botanical Reviews* 15: 464–505.

Anderson, T.F. 1953. The morphology and osmotic properties of bacteriophage systems. *Cold Spring Harbor Symposia on Quantitative Biology* 18: 197–203.

Anderson, T.F. 1966. Electron microscopy of phages. pp. 63–78 in Cairns, J., Stent, G.S., and Watson, J.D. (ed.), *Phage and the Origins of Molecular Biology*. Cold Spring Harbor Laboratory of Quantitative Biology, Cold Spring Harbor, New York.

Anfinsen, C.B. 1959. *The Molecular Basis of Evolution*. John Wiley, New York. 228 pp.

Arber, W. 1965. Host-controlled modification of bacteriophage. *Annual Review of Microbiology* 19: 365–378.

Benzer, S. 1955. Fine structure of a genetic region in bacteriophage. *Proceedings of the National Academy of Sciences* 41: 344–354.

Benzer, S. 1957. The elementary units of heredity. pp. 70–93 in McElroy, W.D. and Glass, B. (ed.), *A Symposium on the Chemical Basis of Heredity* Johns Hopkins Press, Baltimore.

Benzer, S. 1961. On the topography of the genetic fine structure. *Proceedings of the National Academy of Sciences* 47: 403–415.

Benzer, S. 1966. Adventures in the *r*II region. pp. 157–165 in Cairns, J., Stent, G.S., and Watson, J.D. (ed.), *Phage and the Origins of Molecular Biology*. Cold Spring Harbor Laboratory of Quantitative Biology, Cold Spring Harbor, New York.

Benzer, S. and Champe, S.P. 1962. A change from nonsense to sense in the genetic code. *Proceedings of the National Academy of Sciences* 48: 1114–1121.

Blaedel, N. 1988. *Harmony and Unity. The Life of Niels Bohr*. Science Tech Publishers, Madison, Wisconsin. 323 pp.

Bordet, J. 1922. Concerning the theories of the so-called "bacteriophage." *British Medical Journal*, August 19, 1922, p. 296.

Burian, R.M., Gayon, J., and Zallen, D. 1988. The singular fate of genetics in the history of French biology, 1900–1940. *Journal of the History of Biology* 21: 357–402.

Burnet, F.M. 1930. Bacteriophage and cognate phenomena. pp. 463–509 in *A System of Bacteriology in Relation to Medicine*, Volume 7. His Majesty's Stationery Office, London.

Burnet, F.M. and Lush, D. 1936. Induced lysogenicity and mutation of bacteriophage within lysogenic bacteria. *Australian Journal of Experimental Biology and Medical Sciences* 14: 27–38.

Burnet, F.M. and McKie, M. 1929. Observations on a permanently lysogenic strain of *B. enteritidis Gaertner*. *Australian Journal of Experimental Biology and Medical Science* 6: 277–284.

Cairns, J., Stent, G.S., and Watson, J.D., eds. 1966. *Phage and the Origins of Molecular Biology*. Cold Spring Harbor Laboratory of Quantitative Biology, Cold Spring Harbor, New York. 340 pp.

Carlson, E.A. 1971. An unacknowledged founding of molecular biology: H.J. Muller's contributions to gene theory, 1910–1936. *Journal of the History of Biology* 4: 149–170.

Cohen, S.S. 1947. The synthesis of bacterial viruses in infected cells. *Cold Spring Harbor Symposia on Quantitative Biology* 12: 35–49.

Cohen, S.S. 1949. Growth requirements of bacterial viruses. *Bacteriological Reviews* 13: 1–24.

Cohen, S.S. 1968. *Virus-Induced Enzymes*. Columbia University Press, New York. 315 pp.

Cohen, S.S. 1975. The origins of molecular biology. *Science* 187: 827–830.

Cohen, S.S. 1979. Some contributions of the Princeton Laboratory of the Rockefeller Institute on proteins, viruses, enzymes, and nucleic acids. *Annals of the New York Academy of Sciences* 325: 303–306.

Cohen, S.S. 1984. The biochemical origins of molecular biology. *Trends in Biochemical Sciences* 9: 334–336.

Cohen, S.S. 1986. Finally, the beginnings of molecular biology. *Trends in Biochemical Sciences* 11: 92–93.

Cohen, S.S. 1987. Approaching the biochemistry of virus multiplication. *BioEssay* 7: 88–91.

Delbrück, M. 1942. Bacterial viruses (bacteriophages). *Advances in Enzymology and Related Subjects.* 2: 1–32.

Delbrück, M. 1945. Interference between bacterial viruses. III. The mutual exclusion effect and the depressor effect. *Journal of Bacteriology* 50: 151–170.

Delbrück, M. 1946. Bacterial viruses or bacteriophages. *Biological Reviews* 21: 30–40.

Delbrück, M. 1949a. A physicist looks at biology. *Transactions of the Connecticut Academy of Arts and Sciences* 38: 173–190. (Reprinted in Cairns, Stent, and Watson, 1966.)

Delbrück, M. 1949b. Génétique du bactériophage. *Colloques Internationaux du Centre National de la Recherche Scientifique.* 8: 91–103.

Delbrück, M., ed. 1950. *Viruses 1950.* California Institute of Technology, Pasadena. 147 pp.

Delbrück, M. 1953. Introductory remarks about the program. *Cold Spring Harbor Symposia on Quantitative Biology* 18: 1–20.

Delbrück, M. 1959. Preface. pp. v–vii in Adams, M.H. (ed.), *Bacteriophages.* Interscience Publishers, New York.

Delbrück, M. and Bailey, W.T. 1946. Induced mutations in bacterial viruses. *Cold Spring Harbor Symposia on Quantitative Biology* 11: 33–37.

Delbrück, M. and Luria, S.E. 1942. Interference between bacterial viruses. I. Interference between two bacterial viruses acting upon the same host, and the mechanism of virus growth. *Archives of Biochemistry* 1: 111–141.

Demerec, M. and Fano, U. 1945. Bacteriophage-resistant mutants in *Escherichia coli.* *Genetics* 30: 119–136.

d'Herelle, F. 1917. Sur un microbe invisible antagoniste des bacilles dysentériques. *Comptes rendus Académie Sciences* 165: 373–375.

d'Herelle, F. 1921. *Le bactériophage. Son role dans l'immunité.* Masson, Paris. (English translation, with some new material: *The Bacteriophage; its Role in Immunity.* Williams and Wilkins, Baltimore, 1926.)

d'Herelle, F. 1949. The bacteriophage. *Science News* No. 14, pp. 44–59.

Doermann, A.H. 1952. The intracellular growth of bacteriophages. I. Liberation of intracellular bacteriophage T4 by premature lysis with another phage or with cyanide. *Journal of General Physiology* 35: 645–656.

Doermann, A.H. 1953. The vegetative state in the life cycle of bacteriophage: evidence for its occurrence and its genetic characterization. *Cold Spring Harbor Symposia on Quantitative Biology* 18: 3–11.

Doermann, A.H. 1983a. Introduction to the early years of bacteriophage T4. pp. 1–7 in Mathews, C.K., Kutter, E.M., Mosig, G., and Berget, P.B. (ed.), *Bacteriophage T4.* American Society for Microbiology, Washington, D.C.

Doermann, A.H. 1983b. Mapping of mutations and construction of multi-mutant genomes. pp. 302–311 in Mathews, C.K., Kutter, E.M., Mosig, G., and Berget, P.B. (ed.), *Bacteriophage T4.* American Society for Microbiology, Washington, D.C.

Duckworth, D.H. 1976. Who discovered bacteriophage?. *Bacteriological Reviews* 40: 793–802.

Ellis, E.L. 1966. Bacteriophage: one-step growth. pp. 53–62 in Cairns, J., Stent, G.S., and Watson, J.D. (ed.), *Phage and the Origins of Molecular Biology.* Cold Spring Harbor Laboratory of Quantitative Biology, Cold Spring Harbor, New York.

Ellis, E.L. and Delbrück, M. 1939. The growth of bacteriophage. *Journal of General Physiology* 22: 365–384.

Epstein, R.H., Bolle, A., Steinberg, C.M., Kellenberger, E., Boy de la Tour, E., Chevalley, R., Edgard, R.S., Susman, M., Denhardt, G.H., and Lielausis, A. 1963. Physiological studies of conditional lethal mutants of bacteriophage T4D. *Cold Spring Harbor Symposia on Quantitative Biology* 28: 375–392.

Evans, E.A., Jr. 1953. The origin of the components of the bacteriophage particle. *Annales de l'Institut Pasteur* 84: 129–142.

Fischer, E.P. and Lipson, C. 1988. *Thinking About Science. Max Delbrück and the Origins of Molecular Biology.* W.W. Norton, New York. 334 pp.

Fischer, P. 1985. *Licht und Leben. Ein Bericht über Max Delbrück, den Wegbereiter der*

Molekularbiologie. Universitätsverlag Konstanz, Konstanz, Federal Republic of Germany. 282 pp.

Fleming, A. 1922. On a remarkable bacteriolytic element found in tissues and secretions. *Proceedings of the Royal Society of London, Series B* 93: 306–317.

Hattman, S. and Fukasawa, T. 1963. Host-induced modification of T-even phages due to defective glucosylation of their DNA. *Proceedings of the National Academy of Sciences* 50: 297–300.

Hayes, W. 1964. *The Genetics of Bacteria and their Viruses.* John Wiley, New York. 740 pp.

Hayes, W. 1982. Max Ludwig Henning Delbrück, 1906–1981. *Biographical Memoirs of Fellows of the Royal Society* 28: 58–90.

Herriott, R.M. 1951. Nucleic-acid free T2 virus "ghosts" with specific biological action. *Journal of Bacteriology* 61: 752–754.

Hershey, A.D. 1946a. Mutation of bacteriophage with respect to type of plaque. *Genetics* 31: 620–640.

Hershey, A.D. 1946b. Spontaneous mutations in bacterial viruses. *Cold Spring Harbor Symposia on Quantitative Biology* 11: 67–77.

Hershey, A.D. 1953a. Nucleic acid economy in bacteria infected with bacteriophage T2: II. Phage precursor nucleic acid. *Journal of General Physiology* 37: 1.

Hershey, A.D. 1953b. Functional differentiation within particles of bacteriophage T2. *Cold Spring Harbor Symposia on Quantitative Biology* 18: 135–139.

Hershey, A.D. 1966. The infection of DNA into cells by phage. pp. 100–108 in Cairns, J., Stent, G.S., and Watson, J.D. (ed.), *Phage and the Origins of Molecular Biology*, Cold Spring Harbor Laboratory of Quantitative Biology, Cold Spring Harbor, New York.

Hershey, A.D. and Chase, M. 1952. Independent functions of viral protein and nucleic acid in growth of bacteriophage. *Journal of General Physiology* 36: 39–56.

Hershey, A.D. and Rotman, R. 1948. Linkage among genes controlling inhibition of lysis in a bacterial virus. *Proceedings of the National Academy of Sciences* 34: 89–96.

Hershey, A.D. and Rotman, R. 1949. Genetic recombination between host-range and plaque-type mutants of bacteriophage in single bacterial cells. *Genetics* 34: 44–71.

Jacob, F. 1988. *The Statue Within.* Basic Books, New York.

Kay, L. 1985. Conceptual models and analytical tools: the biology of physicist Max Delbrück. *Journal of the History of Biology* 18: 207–246.

Kay, L.E. 1986a. *Cooperative individualism and the growth of molecular biology at the California Institute of Technology, 1928–1953.* Ph.D. thesis, Johns Hopkins University, Baltimore. 308 pp.

Kay, L.E. 1986b. W.M. Stanley's crystallization of the tobacco mosaic virus, 1930–1940. *Isis* 77: 450–472.

Kozloff, L.M. 1953. Origin and fate of bacteriophage material. *Cold Spring Harbor Symposia on Quantitative Biology* 18: 209–220.

Kozloff, L.M. 1966. Transfer of parental material to progeny. pp. 109–115 in Cairns, J., Stent, G.S., and Watson, J.D. (ed.), *Phage and the Origins of Molecular Biology.* Cold Spring Harbor Laboratory of Quantitative Biology, Cold Spring Harbor, New York.

Levinthal, C. and Thomas, C.A. 1957. The molecular basis of genetic recombination in phage. pp. 737–743 in McElroy, W.D. and Glass, B. (ed.), *The Chemical Basis of Heredity.* Johns Hopkins Press, Baltimore.

Lewis, E.B. 1951. Pseudoallelism and gene evolution. *Cold Spring Harbor Symposia on Quantitative Biology* 16: 159–174.

Luria, S. 1939. Sur l'unité lytique du bactériophage. *Comptes rendus des Séances de la Société de biologie.* 130: 904–905.

Luria, S.E. 1945a. Mutations of bacterial viruses affecting their host ranges. *Genetics* 30: 84–99.

Luria, S.E. 1945b. Genetics of bacterium-bacterial virus relationship. *Annals of the Missouri Botanical Garden* 32: 235–242.

Luria, S.E. 1947. Reactivation of irradiated bacteriophage by transfer of self-reproducing units. *Proceedings of the National Academy of Sciences* 33: 253–264.

Luria, S.E. 1953. Host-induced modifications of viruses. *Cold Spring Harbor Symposia on Quantitative Biology* 18: 237–244.

Luria, S.E. 1966. Mutations of bacteria and of bacteriophage. pp. 173–174 in Cairns, J.,

Stent, G.S., and Watson, J.D. (ed.), *Phage and the Origins of Molecular Biology*. Cold Spring Harbor Laboratory of Quantitative Biology, Cold Spring Harbor, New York.

Luria, S.E. 1984. *A Slot Machine, A Broken Test Tube: An Autobiography*. Harper and Row, New York. 228 pp.

Luria, S.E. and Anderson, T.F. 1942. The identification and characterization of bacteriophages with the electron microscope. *Proceedings of the National Academy of Sciences* 28: 127–130.

Luria, S.E. and Delbrück, M. 1943. Mutations of bacteria from virus sensitivity to virus resistance. *Genetics* 28: 491–511.

Luria, S.E. and Exner, F.M. 1941. The inactivation of bacteriophages by X-rays—influence of the medium. *Proceedings of the National Academy of Sciences* 27: 370–375.

Luria, S.E. and Human, M.L. 1952. A non-hereditary host-induced variation of bacterial viruses. *Journal of Bacteriology* 64: 557–569.

Luria, S.E., Delbrück, M., and Anderson, T.F. 1943. Electron microscope studies of bacterial viruses. *Journal of Bacteriology* 46: 57–77.

Maaløe, O. and Watson, J.D. 1951. The transfer of radioactive phosphorus from parental to progeny phage. *Proceedings of the National Academy of Sciences* 37: 507–513.

Mathews, C.K. 1971. *Bacteriophage Biochemistry*. Van Nostrand Reinhold, New York, 373 pp.

Mathews, C.K., Kutter, E.M., Mosig, G., and Berget, P.B. 1983. *Bacteriophage T4*. American Society for Microbiology, Washington, D.C.

Micklos, D. 1988. *The First Hundred Years*. Cold Spring Harbor Laboratory, Cold Spring Harbor, New York. 35 pp.

Mosig, G. 1983. Appendix: T4 genes and gene products. pp. 362–374 in Mathews, C.K., Kutter, E.M., Mosig, G., and Berget, P.B. (ed.), *Bacteriophage T4*. American Society for Microbiology, Washington, D.C.

Muller, H.J. 1922. Variation due to change in the individual gene. *American Naturalist* 56: 32–50.

Muller, H.J. 1941. Induced mutations in *Drosophila*. *Cold Spring Harbor Symposia on Quantitative Biology* 9: 151–167.

Muller, H.J. 1947. The gene. *Proceedings of the Royal Society* B134: 1–37.

Northrop, J.H. 1937. Chemical nature and mode of formation of pepsin, trypsin and bacteriophage. *Science* 86: 479–483.

Novick, A. and Szilard, L. 1951. Virus strains of identical phenotype but different genotype. *Science* 113: 34–35.

Pontecorvo, G. 1952. Genetic formulation of gene structure and gene action. *Advances in Enzymology* 13: 121–149.

Pontecorvo, G. 1958. *Trends in Genetic Analysis*. Columbia University Press. New York, 145 pp.

Portugal, F.H. and Cohen, J.S. 1977. *A Century of DNA*. MIT Press, Cambridge. 384 pp.

Putnam, F.W. and Kozloff, L.M. 1950. Biochemical studies of virus reproduction: IV. The fate of the infecting virus particle. *Journal of Biological Chemistry* 179: 243–250.

Schlesinger, M. 1934. Zur Frage der chemischen Zusammensetzung des Bakteriophagen. *Biochemische Zeitschrift* 273: 306.

Schrödinger, E. 1944. *What is Life*? Cambridge University Press, United Kingdom.

Singer, B.S., Shinedling, S.T., and Gold, L. 1983. Some complexities of T4 genes, gene products, and gene product interactions. pp. 327–333 in Mathews, C.K., Kutter, E.M., Mosig, G., and Berget, P.B. (ed.), *Bacteriophage T4*. American Society for Microbiology, Washington, D.C.

Sinsheimer, R.L. 1954. Nucleotides from T2r^+ bacteriophage. *Science* 120: 551–553.

Stadler, L.J. 1954. The gene. *Science* 120: 811–819.

Stanley, W.M. 1935. Isolation of a crystalline protein possessing the properties of tobacco-mosaic virus. *Science* 81: 644–645.

Stanley, W.M. 1938. Biochemistry and biophysics of viruses. pp. 446–546. in Doerr, R. and Hallauer, C. (ed.), *Handbuch der Virusforschung*. Julius Springer Verlag, Vienna.

Stanley, W.M. and Knight, C.A. 1941. The chemical composition of strains of tobacco mosaic virus. *Cold Spring Harbor Symposia on Quantitative Biology* 9: 255–262.

Stent, Gunther S. 1963. *Molecular Biology of Bacterial Viruses*. W.H. Freeman, San Francisco. 474 pp.

Streisinger, G. 1966. Terminal redundancy, or all's well that ends well. pp. 335–340 in Cairns, J., Stent, G.S., and Watson, J.D. (ed.), *Phage and the Origins of Molecular Biology*. Cold Spring Harbor Laboratory of Quantitative Biology, Cold Spring Harbor, New York.

Streisinger, G., Edgar, R.H., and Harrar-Denhardt, G. 1964. Chromosome structure in phage T4, I. The circularity of the linkage map. *Proceedings of the National Academy of Sciences* 51: 775–779.

Timoféeff-Ressovsky, N.W., Zimmer, K.G. and Delbrück, M. 1935. Ueber die Natur der Genmutation und der Genstruktur. *Nachrichten der Gesellschaft der Wissenschaft, Göttingen Math.-phys. Kl.* Fachgruppe 6: 189–245.

Twort, F.W. 1915. An investigation on the nature of the ultramicroscopic viruses. *Lancet* 189: 1241–1243.

Varley, A. W. 1986. *Living Molecules or Autocatalytic Enzymes: The Controversy over the Nature of Bacteriophage, 1915-1925*. Ph.D. thesis, University of Kansas, Lawrence. 602 pp.

Visconti, N. and Delbrück, M. 1953. The mechanism of genetic recombination in phage. *Genetics* 38: 5–33.

Volkin, E. and Astrachan, L. 1956. Phosphorus incorporation in *Escherichia coli* ribonucleic acid after infection with bacteriophage T2. *Virology* 2: 149–161.

Watson, J.D. and Crick, F.H.C. 1953. The structure of DNA. *Cold Spring Harbor Symposia on Quantitative Biology* 18: 123–131.

Watson, J.D. and Maaløe, O. 1953. Nucleic acid transfer from parental to progeny bacteriophage. *Biochimica et Biophysica Acta* 10: 432–442.

Wollman, E., Holweck, F., and Luria, S. 1940. Effect of radiations on bacteriophage C_{16}. *Nature* 145: 935.

Wyatt, G.R. and Cohen, S.S. 1953a. The bases of the deoxyribonucleic acids of T2, T3, and T6 bacteriophages. *Annales de l'Institut Pasteur* 84: 143–146.

Wyatt, G.R. and Cohen, S.S. 1953b. The bases of the nucleic acids of some bacterial and animal viruses: The occurrence of 5-hydroxymethylcytosine. *Biochemical Journal* 55: 774–782.

Yang, Y.N. and White, P.B. 1934. Rough variation in *V. cholerae* and its relation to resistance to cholera-phage (type A). *Journal of Pathology and Bacteriology* 38: 187–200.

Zimmer, K.G. 1966. The target theory. pp. 33–42 in Cairns, J., Stent, G.S., and Watson, J.D. (ed.), *Phage and the Origins of Molecular Biology*. Cold Spring Harbor Laboratory of Quantitative Biology, Cold Spring Harbor, New York.

7
LYSOGENY

Lysogeny, the hereditary ability to produce phage, occupies a unique position at the junction of genetics and virology. Lysogeny has played an important role in the formulation of ideas about phage, as well as about bacterial genetics. For many years lysogeny remained a mystery, even a controversy, and an objective and dispassionate view was not possible. The discoverer of phage, Felix d'Herelle, refused to believe in the existence of a phage carrier state, whereas his key opponent, Jules Bordet, insisted that at least some viruses lacked virulent characteristics. Later, Max Delbrück, concentrating his research on virulent phage, rejected the idea of lysogeny, despite the convincing evidence provided by F.M. Burnet and Eugène and Elisabeth Wollman in the 1930s. It was only gradually, over a 30-year period, that the nature of the lysogenic state and the relation of virulent to temperate phage became understood, mainly through work after World War II by André Lwoff, Elie Wollman,[1] and Francois Jacob (Galperin, 1987; Jacob, 1988). The clarification of the nature of lysogeny led ultimately to the development of a strong link between the two major camps, the virological/biophysical, under Delbrück, and the genetic/biochemical, under Lwoff and Monod. The implications of research on lysogeny for gene regulation, cancer, and animal virology are widespread and extremely important.

7.1 THE NATURE OF LYSOGENY

Bacteriophage able to bring about the lysogenic state are called *temperate*, in contrast to the *virulent* phages discussed in the last chapter. Lysogeny is a *genetic* property of a bacterial cell, but it is a peculiar genetic property, because if the lysogenic character is expressed, it may be a lethal event for the cell. When a lysogenic *cell* expresses its genetic character, it produces, in the normal way, 50 to 200 phage particles and lyses. Because under normal growth conditions the lytic phenomenon is expressed in only a relatively small fraction of the cells in a bacterial population, the lysogenic *culture* appears normal. However, within such a culture, small numbers of phage

---

[1]The contributions of Elie Wollman to the understanding of lysogeny have been understated by most reviewers. In addition to providing an important intellectual link between his parents' work and the post-World-War-II studies on lysogeny, Elie Wollman made numerous early discoveries, including the critical one that linked the phage genome with the bacterial genome. Elie Wollman spent two years at CalTech in Delbrück's laboratory, thus also providing an important connection between the phage group and the Pasteur Institute. Wollman's personality and style of work are discussed in some detail by Jacob (1988).

particles are present, arising from the rare cells that have lysed spontaneously.

If the lysogenic culture appears normal, how does one recognize the lysogenic state? The phage particles produced by the lysogen are able to infect related but nonlysogenic strains, called *indicator strains*, and cause productive infections. In a standard plaque assay, an indicator strain is used to detect the presence of phage particles in the culture, the plaques arising from the productive infections of cells of the indicator strain. One of the peculiarities of the lysogenic system is that such a plaque assay with a temperate phage is exactly equivalent to the phage assay procedures used in the study of virulent phages.

It should be emphasized that a lysogenic culture can be maintained indefinitely without any indication of its peculiar properties because the principal criterion of the lysogenic state is lysis of a related sensitive strain. Thus, critical to a detection of lysogeny is the availability of a culture sensitive to the temperate phage that can serve as an indicator strain.

7.2 EARLY STUDIES ON LYSOGENY

Once understood, the nature of lysogeny seems relatively simple, but because of its "subtle" nature, it is not surprising that in the early days of phage research, the whole idea seemed so strange, even, to Max Delbrück, ludicrous.

Bordet's Work

As discussed in Section 6.1, d'Herelle's original idea that bacteriophage was a virus was strongly rejected by the Belgian Jules Bordet, one of the leading bacteriologists and immunologists of his day. Although Bordet initially objected to d'Herelle's interpretation because it implied a role for phage in immunity, strong philosophical differences pushed Bordet to adopt an extreme view of phage. Bordet insisted that certain bacterial strains could produce phage active against other strains, whereas d'Herelle insisted that Bordet's phage-producing cultures were *contaminated* with phage, and this contamination could be eliminated by careful purification. The contrast is thus between an *extrinsic* origin of phage (d'Herelle) and an *intrinsic* origin (Bordet). The idea that there were actually two kinds of phages, virulent and temperate phages, was not conceived until the 1950s, by Lwoff and colleagues. A brief analysis of the connection between the early phage work and Mendelian genetics is given by Burian, Gayon, and Zallen (1988), and Galperin (1987) has discussed the history of French research on lysogeny in some detail.

The process of *autolysis* or "self-lysis" was well known to medical bacteriologists. Certain bacteria, such as the pneumococcus and meningococcus, autolyzed so readily that one needed to take special precautions to maintain viable cultures (see Section 9.1). In addition, many Gram-negative bacteria could be induced to autolyze by various chemical treatments. Such autolysates were often used by medical bacteriologists as sources of antigens for immunological studies, and as an immunologist Bordet was, of course, quite familiar with such procedures. (The phenomenon of antibody-induced lysis already had been described by Richard Pfeiffer for *Vibrio cholerae* in the 19th century.)

Bordet discovered lysogeny in the following way: Approaching d'Herelle's phenomenon in an immunological context, Bordet injected a culture of *Escherichia coli* into the peritoneal cavity of a guinea pig and withdrew the peritoneal exudate that resulted. This exudate, rich in leukocytes, was then added to an *Escherichia coli* culture, and lysis was observed. The ability of the exudate to cause lysis was transmissible with bacterial filtrates. Under some conditions, the lysis caused by the exudate was not complete and a "secondary culture" developed which could itself induce lysis of the original culture. Thus, Bordet, in this "backhanded way," discovered lysogeny. d'Herelle, working with virulent phage, had also obtained secondary cultures that resisted the action of the phage, but he interpreted these, probably correctly, as survivors that had been able to resist the attack of the phage (see Sections 4.8 and 6.9 for a discussion of host mutants).

On the basis of his own research and his preconceived notions, Bordet interpreted the "d'Herelle phenomenon" as a type of autolysis, with the difference that the factor involved was *transmissible* (Bordet and Ciuca, 1920a,b). He thought lysis was due to a "hereditary nutritive vitiation consisting in the production of a kind of lytic ferment." Thus, in his very first work, Bordet viewed the phenomenon in a hereditary manner:

> The most outstanding manifestations of the living state are reproduction, which assures the formation of a new being, and heredity, which ensures that this being resembles its parents. Heredity consists, in reality, in the transmission of the variations which confer on each species its special characteristics, and which permit it to be distinguished from other species, races, or families...All variation depends, clearly, on an immediate factor, which operates directly within the cell, but which is also influenced by the external environment. In the case in which this external influence is transmissible when the cell divides, one understands that the variation is due to an intracellular factor which has been released by the external factor and which perpetuates itself readily, continuing to act (translated from Bordet and Ciuca, 1920a).

As were many bacteriologists of his day (see, for instance, Section 4.3), Bordet was a Lamarckian, believing that acquired characteristics could be readily transmitted. Since he believed that d'Herelle's bacteriophage was a lytic factor, Bordet found it no problem to conceive that this lytic factor, once introduced into a culture, could be readily transmitted to offspring.

> The variation becomes from that time not only hereditable, but contagious: the simple contact with a culture fluid that has been modified by the microbe is sufficient to transmit this modification to normal microbes of the same species, which, in their turn, bequeath this character to their offspring, which can themselves transmit the character to normal microbes, and so on, indefinitely. It is clear, finally, that the variation consists in a disturbance of the equilibrium between the construction of living material and its destruction, leading, therefore, to a transmissible variation which manifests itself in a pronounced aptitude on the part of the microbe to undergo autolysis (translated from Bordet and Ciuca, 1920a).

Thus, adopting a clear *intracellular* origin of the d'Herelle phenomenon, Bordet set the stage for the development of the concept of lysogeny. Even the term itself, which he coined in 1925, suggests nothing about a viral analogy.

Of great interest was an idea of Eugène Wollman that was presented in response to the Bordet and d'Herelle papers. Wollman, who was to become the principal French researcher on lysogeny, was a Russian working in Metchnikoff's laboratory at the Pasteur Institute. Wollman was to provide a direct link between the early work on phage and André Lwoff (see Sections 7.3 and 7.4). Wollman was struck by the parallel between Bordet's idea of a transmissible hereditary factor and Darwin's concept of pangenesis. Darwin had postulated (see Section 2.2) the existence of minute granules, which he called *gemmules*, that were capable of transmission from one generation to the next. Wollman then made the following prophetic statement:

> One can see...that if Darwin's hypothesis of pangenesis is applied to the d'Herelle phenomenon...the *gemmules* of Darwin become the *intracellular factors* of Bordet and Ciuca. The microbial culture then becomes the equivalent of the evolving "life system" (translated from Wollman, 1920).

Although Wollman was also a Lamarckian (as were most French scientists of this period, see Burian, Gayon, and Zallen, 1988), his idea could not but encourage research on phage as a genetic system. However, it was not until a few years later that Wollman and his wife began their active research on phage.

The Discovery of the Indicator System

Although its true significance was not to be realized until later, an interesting discovery in these early years by Lisbonne and Carrère (1922) was to open up the experimental study of lysogeny. After repeating Bordet's experiments on peritoneal exudates, Lisbonne and Carrère found that it was also possible to obtain a transmissible lysis completely *in vitro*, using simply *two* bacterial strains, one a *Shigella*, the other *Escherichia coli*. The *Shigella* was sensitive to a phage produced by the *Escherichia coli*. What Lisbonne and Carrère had discovered was an indicator system, whereby one organism, the *Shigella*, could be used to assay the phage produced by the other organism, the *Escherichia coli*. However, it was only some years later that this idea became understood sufficiently so that it could be used in the study of lysogeny. d'Herelle, blinded by his belief in the unicity of phage, dismissed the work of Lisbonne and Carrère as due to a contamination of the *Escherichia coli* strain with phage. Without studying their strains, d'Herelle argued that *Escherichia coli* isolated from nature was frequently mixed, carrying phage particles along, and that if Lisbonne and Carrère had purified their strain more carefully, the phage would be removed. d'Herelle's idea of *carrier strains* remained one of the accepted ideas of lysogeny even up until 1950 (see discussion of Delbrück's interpretation of lysogeny later in Section 7.4).

Within a few years, Bordet himself would adopt the indicator system for the study of lysogeny (Bordet, 1925), although he still insisted that phage was an enzyme rather than a particle. (It was in this 1925 paper that Bordet first used the term *lysogenic*.) An important contribution of Bordet was the recognition that phage was present in the *Escherichia coli* culture even *before* it came in contact with the *Shigella*. Furthermore, a neutralizing antiserum prepared against the phage did not eliminate the ability of the culture to produce phage. Thus, d'Herelle's explanation of a carrier state seemed untenable. Finally, a sensitive culture treated with phage could become itself lysogenic, but there was a strong element of specificity. *Escherichia coli* and the *Shigella* were closely related, but the coli phage could not be transmitted to unrelated organisms. The faculty to produce phage, Bordet wrote, is "inscribed in the heredity of the bacterium" (Bordet, 1925). Although this idea sounds surprisingly modern, this is fortuitous since Bordet's understanding of heredity was essentially Lamarckian. What Bordet called heredity was only the continuation of a purely individual physiology, a characteristic state that prolongs itself during the course of cell division. The main purpose of Bordet's statement was to destroy d'Herelle's idea that phage was particulate. Obviously, if the ability to initiate lysis was a hereditary property of the organism, then the idea of phage as virus was untenable.

Bordet did not realize that lysis was not the important fact in bacteriophagy. There are numerous reasons why bacteria can lyse. The significant thing about phage is its reproduction within the cell and the ability to produce a *specific* structure, the phage particle. These were the days before electron microscopy, so that the phage "particle" had not yet been observed. Bordet did not actually consider how the ability to produce phage might be maintained in a culture. Although the discussion here has dealt just with the French work, there was also extensive German work on lysogeny and the indicator system, from the laboratory of Gildemeister in Berlin (Gildemeister and Herzberg, 1924).

Burnet's Work

The manner in which the ability to produce phage was carried in a culture was first studied by Burnet and McKie (1929), who studied a strain of *Salmonella enteritidis* that produced a phage active against a strain of *Salmonella sanguinarium*. A number of *Salmonella* strains of these species were isolated and compared for phage production and sensitivity. The strains were also characterized serologically and in terms of their rough and smooth (S/R) properties. The lysogenic strain was streaked, and a number of colonies were picked and tested for ability to produce phage. Every isolate obtained was lysogenic. Phage production occurred only during active growth and did not occur in resting cultures. Attempts were made to determine if phage particles were preexistent within the lysogen by inducing lysis with an unrelated phage. Lysis by the unrelated phage actually reduced the amount of homologous phage liberated, thus showing that the "intrinsic phage" existed within the lysogenic bacteria as an independent unit. Comparisons of the amount of phage produced by rough and smooth lysogens showed that the rough strains produced much higher yields than the smooth strains. Some colonies produced so much phage that frank lysis occurred, and colonies on plates appeared mottled.

> Such results are compatible with the view that there is an unstably balanced struggle between phage and bacterium, a very small proportion of the bacteria always undergoing lysis of the ordinary type. But the relatively high lysogenic capacity of rough strains cannot be explained on this view, since the phage produced is incapable of lysing any rough cultures. The liberation of phage here must be determined entirely by internal factors uninfluenced by the presence of phage in the medium...The permanence of the lysogenic character makes it necessary to assume the presence of bacteriophage or its *anlage* in every cell of the culture, i.e., it is part of the hereditary constitution of the strain...The conclusion

we arrive at, therefore, is somewhat as follows: All bacteria in this permanently lysogenic type contain in their hereditary constitution a unit potentially capable of liberating phage...Phage may be...liberated...during the normal processes of growth. The evidence further suggests that unless the activation of the hereditable *anlage* takes place spontaneously, disruption of the cell by any means will not liberate phage. The essentials of the process seem to take place entirely within the cell (Burnet and McKie, 1929).

Burnet and McKie also went on to discuss the essential difference between the lysogenic state they studied and d'Herelle's classic bacteriophage lysis, clearly recognizing the distinction between what would later come to be called virulent and temperate phages:

Liberation of phage from a permanently lysogenic strain depends entirely on intracellular changes...Classical bacteriophage lysis, on the other hand, is primarily dependent on the nature of the phage used...In the first case we have the bacteriophage unit physiologically co-ordinated in the hereditary constitution of the bacterial strain, but liberated into the medium under certain obscure conditions. Toward a sensitive bacterium this same unit behaves as a predatory entity, which becomes specifically adsorbed to the surface, and multiplies genetically within the bacterial cell, which it eventually destroys.

   The difficulty of reconciling these two aspects of bacteriophage phenomena has been responsible for all the current controversy on the intimate nature of phage, whether it is an independent parasite or a pathologically altered constituent of normal bacteria. In our view, both these contentions have been completely proved, and the current attitude on both sides regarding them as irreconcilable alternatives is quite unjustified. According to the particular type of bacterium that is reacting with the phage concerned, it may be useful and convenient to regard the phage as an independent parasite or as a unit liberated from the hereditary constitution of some bacterium, the usage being determined wholly by its functional activity at the time (Burnet and McKie, 1929).

Despite such clear words, it would be 20 years before the controversy regarding lysogeny would be settled.

7.3 THE *BACILLUS MEGATERIUM* SYSTEM

In 1931, the Dutch bacteriologist L.E. den Dooren de Jong isolated a lysogenic strain of the spore-forming bacterium *Bacillus megaterium* (previously called *B. megatherium*) (den Dooren de Jong, 1931a,b). An asporogenous mutant of this strain, which he called *mutilate*, was nonlysogenic and sensitive to the phage produced by the original strain. Using the mutilate strain to assay phage, den Dooren de Jong was able to show that spores of the lysogenic strain gave rise to clones that were lysogenic. In an attempt to clarify the alternate interpretations of lysogeny, *carrier state* and *hereditary state*, den Dooren de Jong tried to inactivate any presumed phage particles by heating a sporulated culture to a temperature that would inactivate free phage but would have no effect on spores. "To my great satisfaction," den Dooren de Jong found that pasteurized cultures continued to give rise to phage when cultured, despite the fact that any free phage had been inactivated.

How can these results be explained? Since the high thermolability of phages is generally assumed, and has actually been proven in the present experiments, no other conclusion can be drawn than that in my experiments the phage obtained is a product of the living bacterial cell. Therefore, d'Herelle's theory that bacteriophage is an ultramicrobe is untenable (translated from den Dooren de Jong, 1931a).

Unfortunately, even den Dooren de Jong's "convincing" evidence was soon to be discounted. Several workers showed that the heat resistance of phage was greatly increased when it was in the dry state, and since spores were supposed to be more or less dehydrated, den Dooren de Jong's conclusion was considered to be invalid (Cowles, 1931; Vedder, 1932). Writing as late as 1946, Max Delbrück, apparently ignoring Burnet's ideas, rejected den Dooren de Jong's work, basing his position on the criticisms of Cowles and Vedder (Delbrück, 1946). However, den Dooren de Jong (1936), in countering the work of Cowles and Vedder, had done a beautiful experiment in which he added two unrelated phages to a culture and then heated it. Only the homologous phage, which the strain produced, survived the heating procedure. Although Delbrück (1946) cited this paper, he misinterpreted it ("den Dooren de Jong found that the inclusion of the virus in the spores is a specific reaction. *B. undulatus* and other spore formers will not include the virus of *B. megatherium*, and vice versa"). By using the term "include" he signaled that he did not realize that the phage got into the spore via an infection of the vegetative cell. Delbrück's prejudice in favor of phage as virus is revealed in this statement. It is interesting that Delbrück's contamination idea was exactly the same argument that d'Herelle had used in his rejection of lysogeny in the 1920s (see Section 7.1).

The Wollmans' Studies on Bacillus megaterium

As discussed earlier, at the very beginning of phage research Eugène Wollman adopted an hereditary theory of bacteriophage. Following upon Darwin's concept of pangenesis, Wollman (1920) put forward the hypothesis that some genes could have a certain stability in the external medium and could be transmitted from cell to cell. He considered that phages might be analogous to lethal genes. Beginning in 1932, Wollman and his wife Elisabeth initiated a series of studies on lysogeny with den Dooren de Jong's strain of *Bacillus megaterium* (see Wollman and Wollman, 1936). They began with the theory that there were *two* stages in lysogeny, *intracellular* and *extracellular*. The phage particle constituted the extracellular stage, but what was the intracellular stage? They carried out quantitative titrations of phage and bacterial numbers and found they were about equal in lysogenic cultures, from which they concluded (erroneously) that each lysogenic bacterium produces one phage particle (Wollman and Wollman, 1938). "The numerical relationships between bacteria and bacteriophages are hardly compatible with the notion of the virus as parasite" (translated from Wollman and Wollman, 1938). In another study, they attempted to see if phage was present preformed in cells, using gentle lysis to liberate such putative phage. *Bacillus megaterium* cells were sensitive to the enzyme *lysozyme*, but phage particles were not, so they treated cell suspensions with this enzyme in an attempt to determine if phage would be liberated during lysis with lysozyme. The results were clear-cut: With washed cell suspensions of lysogenic cultures, or with infected sensitive cultures, *no* phage was released following lysis by lysozyme (Wollman and Wollman, 1939). This showed that infectious phage particles were not present preformed within the cell. Implicit in the Wollmans' work was the concept of *infectious heredity*, which Wollman sometimes called *paraheredity* (Wollman, 1928).

Northrop's Work

As discussed in Section 6.2, John Northrop did pioneering work on the crystallization of enzymes. On the basis of Stanley's (erroneous) conclusion that a virus was simply a protein, Northrop studied the kinetics of phage production in growing populations of bacteria. He used den Dooren de Jong's lysogenic culture of *Bacillus megaterium*, and followed the release of both phage and the enzyme gelatinase. From measurements of bacterial growth, phage titer, and gelatinase concentration, Northrop concluded that since all three increased in parallel, phage was produced during bacterial growth and not during lysis (Northrop, 1939). (*Bacillus megaterium*, like most spore-forming bacilli, exhibits a pronounced lysis when it enters the stationary phase and sporulates.) He concluded, therefore, that phage is *secreted* by bacteria. In subsequent work, Northrop was joined by a co-worker, A.P. Krueger, who then subsequently continued this line of research for many years (reviewed rather critically by Delbrück, 1942, 1946). But as Lwoff (1953a) stated, the end result would be the same whether each bacterium produces and secretes one phage at each division or if one bacterium out of 100 produces 100 phages. The study of large populations would not yield a solution, and it would be necessary to study events in a single lysogenic bacterium. The same criticism, of course, held for Wollman's work, discussed above.

7.4 THE WORK OF ANDRÉ LWOFF

Rarely can it be said that the studies of a single scientist have changed the complete course of subsequent work, but this can be said about lysogeny and André Lwoff (Lwoff, 1953a). Until Lwoff's work, fuzzy thinking and weak experiments had dominated the field of lysogeny. The distinction between temperate and virulent phages had been made but not accepted. Even the Delbrück phage group had misunderstood the essential nature of lysogeny. This confusion was all to end as a result of the brilliant work of André Lwoff. It is hard now to appreciate the revolutionary impact of Lwoff's simple experiments.

Lwoff's Background

French but of Russian extraction, André Lwoff began his scientific career in 1921 at the age of 19 working with two protozoologists, Edouard Chatton at the Marine Biological Station at Roscoff and Félix Mesnil at the Pasteur Institute (Lwoff, 1971). (A nice characterization of Lwoff's personality is given by Jacob [1988], and some fond remembrances by a number of colleagues are given by Monod and Borek [1971].) From that time until World War II, Lwoff studied the nutrition of microbes, primarily protozoa. He developed the first synthetic medium for a ciliate (*Tetrahymena pyriformis*) and discovered several new growth factors (see Section 3.8). His nutrition research culminated in a book that he wrote during the German occupation of France (Lwoff, 1944). In this book, the nutritional requirements of protists were considered in terms of the concept of physiological evolution. One of the principal theses of this book was that parasitic microorganisms are characterized by lower biosynthetic abilities than free-living ones, and that the parasites have been derived from free-living organisms through successive loss of function via mutation. Lwoff was out of touch with American publications due to the war: This idea had already received strong support from the *Neurospora* work of Beadle and Tatum (see Chapters 4 and 5). (To a certain extent, the work of Beadle and Tatum on *Neurospora* was based on Lwoff's mutation research.)

Lwoff's Switch to Lysogeny

During the 1930s, Lwoff became intimately acquainted with the work on lysogeny at the Pasteur Institute by his colleagues Eugene and Elisabeth Wollman (discussed above). In an appendix to their 1936 paper (Wollman and Wollman, 1936), Lwoff contributed an analysis of lysogeny that strikingly foreshadows his later work (Lwoff, 1936).[2]

Lwoff begins his theoretical discussion of the Wollmans' work with a review of earlier work (including the 1922 paper of Muller that was quoted in Section 1.1) which considered that viruses might be "modified genes." He mentions also that the agent of Rous sarcoma had been called a "transmissible mutagen," thus tying virology, genetics, and lysogeny to cancer. He then considered the phenomenon of cytoplasmic inheritance, especially as it was found in higher plants. He noted also that genes were often considered to be the "generators of cellular enzymes," thus introducing an early version of the one-gene/one-enzyme hypothesis. He next considered a phenomenon of a mosaic virus disease of plants in which the virus appeared to exist in two states, active and inactive. According to Lwoff, such viruses resembled the bacteriophages involved in lysogeny. He then turned to another important phenomenon, that of "adaptive and constitutive enzymes," as described carefully by Karström (see Section 10.2 for a discussion of Karström's work). Attempting to pull all of these diverse phenomena together, he concluded:

> In summary, the action of genes can be explained by the hypothesis that genes act in cells not directly as catalysts, but as inducers that are activated in a cyclic manner. The gene appears to exist in two states, active and inactive. The reactivation of the gene appears to be linked to division. This hypothesis agrees perfectly with the properties of the lysogenic principle and the mosaic virus, both of which are activated as a result of cell division. This common relationship of the lysogenic principle and the mosaic virus to cell division is not just due to convergence...It appears to us to be an expression of the fundamental similarity between the "virus-gene" and the normal gene... (translated from Lwoff, 1936).

Lwoff did not actually turn to a study of lysogeny until after World War II. Throughout the rest of the 1930s and through the war period, he continued his research on microbial nutrition and wrote his book on microbial evolution (see above).

In 1946, Lwoff and his colleague Jacques Monod (see Section 10.4) attended the Cold Spring Harbor symposium on *Heredity and Variation in Microorgan-*

---

[2]Surprisingly, Lwoff does not cite this appendix in his later papers on lysogeny.

isms. This was the seminal symposium at which Lederberg and Tatum first announced their successful mating experiments (see Chapter 5) and Delbrück and Hershey reported on genetic recombination in phage (see Chapter 6). The organizer of the symposium, Milislav Demerec (see Section 2.9), was interested in reopening contact with European workers who had been scientifically isolated during World War II. Monod was known because he had spent time in the 1930s at CalTech, where he became acquainted with the genetics community. Lwoff's work was known because his work on nutrition had led him to isolate nutritional mutants from the bacterium *Moraxella lwoffi*. (See Lwoff, 1966 for some recollections of this meeting.)

Remembering Max Delbrück and the 1946 Cold Spring Harbor symposium, Monod later wrote:

> Max thought the work of Burnet, Gratia, E. Wollman, and den Dooren de Jong on lysogenic bacteria was worthless. He refused to believe the phenomenon, which he attributed to unrecognized contaminations. Having read some of these works, and knowing well Eugène Wollman, a careful and stubborn experimenter, André and I were very disturbed by Max's arbitrary rejection of this work, which because of the strangeness of the phenomena had actually piqued our interest. Returning to Paris, we had long discussions about what experiments to do that would convince Max of his error and would demonstrate the reality of lysogeny.
>
> To convince Max: that was not a simple thing to do. André spent many months fishing cells of *Bacillus megaterium* with a micromanipulator, to follow under complete microscopic control their descendants, carefully placing cells in microdrops, watching bacilli which suddenly dissolved without apparent cause, to prove that they and they alone liberated virus which they had inherited directly in a latent form from their ancestors...Max was convinced. As a result, lysogeny received permission to enter the phage church. The study was legitimized (translated from Monod, 1971).

Because of the work of den Dooren de Jong and the Wollmans, Lwoff selected *Bacillus megaterium* as material. It had the advantage that it was a large bacterium, which made it possible to study events in single bacteria. According to Lwoff (1966), one reason he elected to study lysogeny in single bacterial cells was because he disliked statistical analysis, which would be required if he were to study virus production in populations of cells.

The Phage Group's Rejection of Lysogeny

Max Delbrück's rejection of the basic idea of lysogeny has been recalled by Elie Wollman, the son of Eugène and Elisabeth Wollman:[3]

> I recall that once, looking into a bibliographical index, at Caltech, I came across a reference to one of my parents' papers (Wollman and Wollman, 1937, 1938). This paper reported that no infective phage can be recovered when *Bacillus megatherium* is infected with bacteriophage and later lysed with lysozyme. My parents' conclusion was that phage entered a noninfectious intracellular phase after infection ...The comment on the Caltech index card...was "Nonsense" (E.L. Wollman, 1966).

Delbrück had considered lysogeny to be some sort of carrier state in which virus, carried along in a resistant culture, multiplies on the rare sensitive cells "thrown off by mutation. In such cultures, therefore, both virus and bacteria will grow *pari passu*...such cultures would give the impression that the virus

---

[3]The Wollmans themselves were forcibly removed by the Germans from the Pasteur Institute in 1943. They died in a concentration camp.

particles are liberated from growing cells without destruction of the cells, since no mass lysis would be observable" (Delbrück, 1946).

Delbrück was not the only member of the phage group to reject lysogeny. As late as 1948, Hershey also rejected the essential nature of lysogeny:

> More complex relationships between bacterium and virus, which are not well understood, give rise to so-called lysogenic cultures. This term is used loosely to describe any association between bacteria and virus which permits both to persist in serial transplants...It is easy to believe that lysogenic cultures...are simply unstable lines of bacteria mutating from resistance to sensitivity to a contaminating virus, which in turn mutates occasionally to a form capable of attacking all the bacteria in the culture. The known genetic properties of viruses and bacteria are sufficient to account for this type of lysogenesis. Less clear is the significance of stable virus-bacterium associations in which virus-free bacterial clones are not readily obtained...How virus is transmitted from cell to cell in lysogenic cultures seemingly refractory to lysis remains to be clarified. It must be concluded, however, that the phenomenon of lysogenesis, frequently cited as evidence for spontaneous intracellular origin of virus, can equally well be explained as one type or another of association between exogenous virus and incompletely susceptible bacterium (Hershey and Bronfenbrenner, 1948).

It seems likely that these strong negative reactions on lysogeny by two key members of the phage group were conditioned primarily by their attitude that phage were *viruses*, not *genes*. As Anderson (1966) has stated: "Many of us were later involved in a plot to interest the 'real plant and animal virologists' in bacteriophages by calling them 'bacterial viruses'."

In the syllabus published with the Viruses 1950 conference held at Cal-Tech, Delbrück (1950) had permitted only a brief discussion of lysogeny. The text was written by Elie Wollman, the son of Eugéne and Elisabeth Wollman. In his discussion, Elie Wollman made clear the distinction between *carrier strains*, in which a host population is in equilibrium with an extracellular phage (the phenomenon that both Delbrück and Hershey were considering in the material quoted above), and a "true" *lysogenic strain*, which could not be freed from phage. According to Wollman: "Each single bacterial cell appears to be capable of carrying the potentiality for reproduction of the virus *without further intervention of external phage*, while at the same time undergoing the processes of assimilation and multiplication. The fact that there may be lysis...in which no infective particles are released to the medium, suggests that as in the early stages of the latent period the phage is present in the cell in a non-infective state."

This statement, written before Lwoff's influential work became known, expressed essentially the ideas that had been developed by Elie Wollman's parents in the 1930s. Although the idea of a noninfective state in the cell sounds modern, it should be noted that the concept was quite vague, as it had been for Lwoff in his 1936 discussion.

But even as late as 1953, after Lwoff had convincingly demonstrated the *genetic* nature of lysogeny and had discovered induction, Delbrück was writing:

> The term provirus was coined by Lwoff and Gutmann three years ago to describe the state of the phage in lysogenic bacteria...It is clear that provirus is a very shadowy character. It is defined as a condition of the bacterium which confers on it a potentiality. In many cases this potentiality is easily and clearly recognized, but there is no way of ever being certain that a given strain does *not* possess such a potentiality. Clearly, we want to get better acquainted with this shadowy charac-

ter. Above all, we want to know where he resides in the cell, and how many of them there are within any one cell (Delbrück, 1953).

Delbrück may have eventually turned to favor lysogeny because of the discovery of the eclipse period and because of the Hershey/Chase experiment (see Section 6.12 and 6.13), both of which showed that intracellular phage was distinct from extracellular phage. Once intracellular phage was seen to be only DNA, the prophage concept made more sense.

In analyzing the attitudes of the American phage group, Francois Jacob made the following comments:

> It is curious to note that, once more, generalization from the conclusions drawn from a study of virulent bacteriophages denied at the same time the very existence of lysogenic bacteria...Thus, the American authors were led to the conclusion that lysogenic *bacteria* did not exist, but lysogenic *cultures* constituted a population equilibrium between resistant individuals and sensitive individuals contaminated by an exogenous phage...The confirmation of the theory of bacteriophage-virus led the American authors to reject the idea of lysogenic bacteria, using the same arguments that d'Herelle had used twenty-five years earlier (translated from Jacob, 1954).

The Experiments of Lwoff and Gutmann

Although Delbrück and colleagues did not agree, at the time Lwoff initiated his work the following points had been reasonably well established about lysogeny:

1. Lysogeny is a property of *all* cells of a lysogenic culture, and in a spore-forming organism, of all the spores. When spores are heated to a temperature sufficient to kill all free phage, they retain the lysogenic state.
2. The lysogenic characteristic persists after repeated passage of the culture through an antiserum specific for the phage. Such antibody experiments ruled out the possibility of free phage being carried along by infection and lysis of a small fraction of sensitive mutants.
3. Bacteria in a lysogenic culture are able to adsorb the phage which they produce, but are immune to it.
4. Lysis with lysozyme or an unrelated phage does not liberate phage from lysogenic cultures.
5. After infection of a sensitive host strain, one can isolate bacteria resistant to the phage which are now able themselves to produce a phage identical with the original phage (whose presence can be detected with an indicator strain).

Because of the large size of *Bacillus megaterium*, micromanipulation was relatively easy. A good micromanipulator, the de Fonbrune, had been perfected in France and was available for this work. Single washed cells were isolated in microdrops, and the bacteria were allowed to grow and divide. The progeny were removed and placed in their own microdrops. Subsamples of the medium in which the cells had grown were tested at intervals for presence of phage. In one experiment, 19 separate cultures were established, each derived from a single cell whose progeny had been isolated. In all cases the cultures were lysogenic, but no phage particles were ever found in the culture medium within the microdrops. Since no free phage ever appeared in the medium, it was concluded that lysogeny was maintained in the absence of free, extrinsic phage. This direct experiment confirmed the conclusion obtained more indirectly from the antibody experiments.

However, in other cases, one or another of the cells established in microcultures would lyse, and in certain cases this lysis was accompanied by the release of phage. Lwoff and Gutmann distinguished two kinds of lysis, which they called *slow* and *fast*. "With slow lysis, a ghost persists and remains recognizable...Later, we recognized another type of lysis, rapid, in which the bacterium disappears suddenly without any recognizable ghost remaining. It is this type of lysis in which bacteriophage is liberated" (translated from Lwoff and Gutmann, 1950).

Although free phage particles were not associated with single lysogenic cells, they were found in lysogenic *cultures*. What then was the source of this phage? Occasionally, a single cell being observed in a microdrop would undergo spontaneous lysis. "A bacterium is there, and suddenly it disappears. When this happens, phage is found in the droplet, around 100 phages per lysed bacterium" (Lwoff, 1953a). The important conclusion was that only a small fraction of the lysogenic bacteria produce phage, and phage production results in lysis of the producing cell. Thus, phage are *not* excreted by living cells. "The production of bacteriophage is a lethal process: lysogenic bacteria only survive if they do not produce phage" (Lwoff, 1953a). Kjeldgaard (1971) has described the patience and care with which Lwoff and Gutmann carried out their micromanipulator experiments.

As discussed earlier in this chapter, Burnet and McKie had postulated that when a cell was lysogenized, the phage was maintained in this cell as a noninfectious "anlage." For this noninfectious structure endowed with genetic continuity, Lwoff and Gutmann (1950) used the term *probacteriophage*, later shortened to *prophage*. The nature of the prophage would be soon determined, primarily by the genetic experiments of Wollman and Jacob and others, but at the time Lwoff was writing, it was important to clarify what prophage was and what it was not. What it definitely was not was a symbiotic virus. Thus, Lwoff criticized fairly severely the discussion in E.M. Lederberg (1951) on virus lysogenicity, which stated that lysogeny dealt with symbiotic, latent phage. Lwoff stated: "May prophage really be considered as a latent phage? An egg is not a latent organism, it is an organism *in posse*, a potential organism. A chromosome is not a latent cell; a gene is not a latent enzyme. *Prophage* is the specific hereditary structure necessary for the production of phage. It is by no means a hidden phage...Prophage is not a bacteriophage." And finally: "Is bacteriophage a virus? What is a virus?" (Lwoff, 1953a). The paper of Lederberg (1951) is an excellent presentation of the thinking in bacterial genetics just before the beginning of the molecular age. It attempts to relate bacterial genetic phenomena to those in eucaryotes. Lederberg states that lysogenic cultures are frequently "infected with cryptic, symbiotic viruses." In this same discussion, he also suggests, prophetically, a parallel between lysogeny and transformation. "...in many ways the transforming agent behaves like a symbiotic intracellular virus."

Lwoff's conception of the nature of lysogeny was well illustrated by the life cycle presented in Lwoff and Gutmann (1950) (Fig. 7.1). This figure has all the elements of lysogeny, and very little would have to be changed to use this figure in a modern textbook.

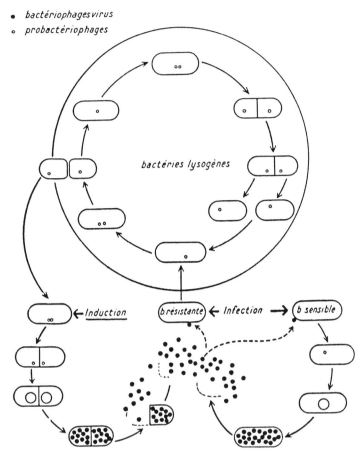

Figure 7.1 Diagram of the development of bacteriophage and probacteriophage (prophage) in sensitive and lysogenic *Bacillus megaterium*. (Reprinted, with permission, from Lwoff and Gutmann, 1950.)

According to a personal communication from Elie Wollman (1989), he was the first to propose the term "temperate." It was suggested at the Royaumont colloquium of 1952 (E.L. Wollman, 1953) because it could be used in both French and English. The derivation was based on Bach's *Well Tempered Clavier*. At the same meeting, an extensive definition of terms related to lysogeny was presented by Jacob, Lwoff, Siminovitch, and Wollman (1953).

Induction

Although only a small fraction of the lysogenic cells liberate phage under normal conditions, virtually the whole culture can be caused to liberate phage by treatment with an appropriate agent, a process called *induction* (Lwoff, Siminovitch, and Kjeldgaard, 1950). Induction changes lysogeny from a cellular to a population phenomenon.

Jacques Monod has recalled the background to Lwoff's discovery of the induction phenomenon in the following words:

But what to do then? André was upset by the randomness of the spontaneous lysis of a small fraction of the bacterial population. For others, this would have been a delicious excuse to carry out ponderous statistics. But André detested statistics ...While cultivating day after day his bacilli, André noticed that occasionally, and always without any apparent cause, a whole culture would suddenly lyse almost completely...All imaginable parameters that might have played a role in this lysis were studied: cations, anions, the source of the peptone, "the pH, the rH, and the SH" as one of our honorable Pasteur colleagues had said. André tirelessly tried something new each day, changing the conditions, exhausting his collaborators Louis Siminovitch and Niels Kjeldgaard, who produced a massive set of complicated results. When Max was shown all these data he could only declare himself "more bewildered than enlightened."

Until the day came when André exposed a culture for several seconds to a UV lamp which I was using for mutagenizing coli bacilli. The culture lysed a half-hour later. The experiment was repeated immediately. The same result. Then the next day, the following day, and the next week. No more erratic results. The phenomenon was reproducible with the regularity of a metronome. André had discovered induction, exorcised the spectre of statistics, and at the same time proved beyond doubt that *all* the cells of a clone carried the latent virus. The news caused a tremendous uproar. Induction made it possible to analyze indisputably the lysogenic system. Max himself immediately repeated the experiment and had to admit that even he was convinced (translated from Monod, 1971).

The discovery of induction was of the greatest importance for understanding the nature of lysogeny, since it made possible the study of the details of temperate phage production. Using induction, phage replication could be studied in a manner similar to that of virulent phage production. It was induction that was to convince Delbrück that lysogeny was a "real" phenomenon. Interestingly, not all lysogenic bacteria are inducible, so that induction by itself is not a critical aspect of the definition of lysogeny. However, with induction it was possible to convert lysogeny from a cellular to a population event, providing an experimental tool by which the phage production process could be studied *en masse*. Phage genetics became phage physiology. It was found that not only UV, but also ionizing radiations such as X-ray and gamma radiation could induce the lysogen.

An important point was that the ability of the culture to be induced depended on the physiological state, what Lwoff called the *aptitude*. Also, nonlysogenic strains of *Bacillus megaterium* that were sensitive to the phage did not lyse when irradiated, showing that lysis was *specific* for cells containing prophage. The demonstration that lysogeny made cells more sensitive to radiation served to focus attention, this early in the post-World-War-II "nuclear age," on lysogeny as a phenomenon of practical significance for understanding how radiation worked. Lysogeny was also of interest to those studying latent animal viruses and cancer. The relationship between mutagens, carcinogens, and inducers of lysogeny was described in detail by Lwoff (1953b).

Studies of bacteria in microdrops showed that after irradiation each bacterium would undergo one or two divisions and then lyse, liberating a large number (around 100) of phage particles. What was irradiation doing? These were the days before the Hershey/Chase experiment and the Watson/Crick structure for DNA, and Lwoff's general view was that radiation caused some kind of shock or stimulus to the cell. "The *B. megatherium* lysogen can be viewed as living in equilibrium with a specific particle endowed with genetic continuity: the probacteriophage. Under the influence of a modification of metabolism or under the action of an inducing shock in bacteria that have the appropriate aptitude, this equilibrium is disturbed" (translated from Lwoff, Siminovitch, and Kjeldgaard, 1950).

In this first paper on induction, an analogy was made between temperate phage and latent viruses of plants: "One can compare the lysogen to the latent plant virus found in the variety of potato known as *King Edward*, which, although innocuous for this variety, is pathogenic for other varieties of potato. This particle is transmissible from *King Edward* to other varieties by grafting. *Without a doubt one would obtain an analogous result if a B. megaterium lysogen was grafted onto a sensitive strain* [emphasis added] (translated from Lwoff, Siminovitch, and Kjeldgaard, 1950). This statement is a direct expansion of the ideas about the relationship between lysogeny and latent viruses of plants that Lwoff had presented in his addendum to the 1936 paper of the Wollmans (Lwoff, 1936). The suggestion of grafting is interesting because once high-frequency mating became available as a research tool, Jacob and Wollman were to do a type of "grafting" experiment via conjugation and discover the important phenomenon of *zygotic induction* (see Section 7.5).

Within several years, the technique of induction was to be used extensively in an analysis of the nature of lysogeny. Francois Jacob carried out extensive studies for his doctoral thesis with the bacterium *Pseudomonas pyocanea* (Jacob, 1954), and upon the discovery by the Lederbergs of lysogeny in *Escherichia coli* strain K-12, Elie Wollman and Jacob began their collaborative studies on this organism (see later). An important early discovery was that inducibility was controlled not by the host strain but by the phage (Ionesco, 1951), and that inducibility could be changed by mutation of the phage (Lwoff, 1953b). In the case of *lambda*, it is now known that the induction process is due to a complex series of events which occur as a result of DNA damage (Ptashne, 1986). Phage replication is held in check by the action of a phage repressor protein, and destruction of this repressor leads to a lytic response. A DNA repair system, the SOS system, is activated as a result of DNA damage, resulting in the activation of a recA protein and its conversion to a protease that cleaves the repressor. Induction thus can be seen as a result of an indirect action of a normal DNA repair system. Nothing is known about the mechanism of induction in Lwoff's *Bacillus megaterium* system.

Immunity

Another important study made at this time related to the phenomenon of *immunity*. As noted, lysogenic bacteria are immune to infection by the homologous phage, although lysogenic bacteria are able to adsorb such phage. Although it was initially thought that immunity to infection was an additional consequence of lysogenization, later work in *lambda* was to show that the same mechanism keeps both the prophage and the entering phage from initiating multiplication (Ptashne, 1986). By use of superinfection experiments with a mutant of a homologous *Pseudomonas* phage, Jacob (1954)[4]

---

[4]This was Jacob's doctoral thesis. In its preface, Lwoff relates how he was convinced by Jacob to accept him in his laboratory, despite Lwoff's initial reluctance (see also Jacob, 1988). Lwoff's decision to formally accept Jacob came because in the meantime he had discovered the phenomenon of induction and wanted the study extended to other systems.

showed that the superinfecting phage behaved as though its genetic material had persisted in the protoplasm for a period of time but was unable to undergo replication, eventually being diluted out and lost. Thus, the genome of the superinfecting phage entered the cell but did not replicate. Immunity to homologous phage was therefore not due to the destruction of the infecting phage, but to the fact that it was unable to develop. In this work, Jacob and Lwoff were using the mixed infection procedures that had been worked out a few years earlier for virulent phage by Luria, Hershey, and Delbrück.

"In lysogenic bacteria, both the prophage and the infecting phage are unable to develop. But when the prophage undergoes development, either spontaneously or as the result of an extrinsic induction, the infecting phage develops also: immunity is bound to the persistence of the prophage as such" (Lwoff, 1953a). Also, after induction, the immunity of the system was abolished (Jacob, 1954). These findings were to enter into the ultimate development of the *repression* model for phage immunity (see Section 10.10).

Unfortunately, Lwoff's early work was with *Bacillus megaterium* and Jacob's first work was with *Pseudomonas pyocanea*. In neither of these organisms was a genetic system available that could be used to analyze the nature of prophage. Further work would have to await the discovery of lysogeny in *Escherichia coli*.

7.5 LYSOGENICITY IN *ESCHERICHIA COLI*

The discovery of lysogenicity in *Escherichia coli* K-12 was made by Esther Lederberg (1951) and was described in some detail by Joshua Lederberg at the 1951 Cold Spring Harbor symposium (Lederberg, Lederberg, Zinder, and Lively, 1951).[5]

The Lederbergs had worked on the genetics of *Escherichia coli* K-12 for several years without realizing it was lysogenic. The discovery was accidental, occurring when during mutagenesis with UV a sensitive strain arose. During work with this sensitive strain, it was mixed with the parent strain and numerous plaques were seen. The Lederbergs soon showed that the phage was carried by all stocks except the unique sensitive strain.

Lederberg initially believed that lysogenicity was due to a kind of extranuclear heredity involving self-reproducing cytoplasmic entities called plasmagenes (see especially J. Lederberg, 1952). The relationship between viruses and plasmagenes had been discussed in detail by Beadle (1945) in his influential review on biochemical genetics. According to Beadle, the existence of lysogeny showed that viruses could arise from normal cell proteins by mutation. In the 1940s and early 1950s, interest in plasmagenes and cytoplasmic inheritance was high (Sapp, 1987), and Tracy Sonneborn's research on the *kappa* "particles" of *Paramecium* was closely followed.[6] One explanation for Sonneborn's *kappa* was that the particles were simply intracellular symbionts (an idea later shown to be correct). Thus, Lederberg, Lederberg, Zinder, and Lively (1951) considered that phage *lambda* might also be a cytoplasmic symbiont and named it λ "by analogy to a killer factor in *Paramecium*" (cytogenes had traditionally been designated by Greek letters).

[5]The bulk of this lengthy paper dealt with attempts to construct a genetic map in *Escherichia coli* (see Chapter 5). It also contained the first report of what would later be called *transduction*.
[6]Sonneborn gave the closing paper at the 1951 Cold Spring Harbor symposium where lysogenicity in *Escherichia coli* was first announced.

Lederberg, Lederberg, Zinder, and Lively (1951) also hypothesized that transforming agents might also be "akin to latent viruses whose lytic activity is no longer discernible" (see Section 9.10).

These were ideas that were clearly testable with genetic studies, although, immersed as the Lederbergs were in the middle of the exciting discoveries on the mechanism of mating (see Chapter 5), they did not complete their genetic studies on *lambda* for several years (Lederberg and Lederberg, 1953). In the meantime, *lambda* came under study at both the Pasteur Institute (E.L. Wollman, 1953) and at CalTech (Weigle and Delbrück, 1951). A useful brief overview of the history of *lambda* can be found in Hayes (1980).

Genetic Mapping of Lambda

An important experiment that could be done with *Escherichia coli* K-12 but not with any of the previously studied lysogenic bacteria was a genetic analysis of the location of the prophage. Although the French workers were convinced from the beginning of some kind of chromosomal location, Lederberg had assumed, following his extrachromosomal model of lysogeny, that there would be no chromosomal linkage. If so, its inheritance should appear to be independent of the inheritance of other genetic markers. However, no genetic evidence could be obtained that *lambda* was a cytoplasmic agent. Lysogenicity behaved precisely as if it were controlled by a nuclear factor, linked to other segregating factors. The results provided strong support for Lwoff's prophage concept (Lederberg and Lederberg, 1953).

Although the genetic analysis of *lambda* in these early experiments was complicated by the phenomenon of *zygotic induction* (see below) and by the fact that only low-frequency recombination was available, the studies of both the Lederbergs and Elie Wollman showed clearly that *lambda* behaved as if it were linked to another genetic factor, the gal_4 gene involved in galactose fermentation. By making crosses between *Escherichia coli* strains carrying different *lambda* mutants, it was shown that lysogenicity consisted of a single additional cellular component, the prophage, which was attached to the bacterial chromosome at a specific locus (Appleyard, 1953).

Zygotic Induction

The discovery of high-frequency mating (Hfr; see Chapter 5) made possible a more efficient analysis of the lysogenic character than was possible by the low-frequency system described above. Almost immediately, a new phenomenon was discovered by Jacob and Wollman (1954), called *zygotic induction*, which was found to occur when the prophage was inducible.

Zygotic induction was revealed as an asymmetry of the results during crosses between Hfr and F⁻. Whenever a chromosomal segment containing an inducible prophage was introduced into a recipient that was non-lysogenic, the prophage became induced, entered the vegetative phase, and produced mature phage, thus lysing the cell (Table 7.1). Because a cell that has undergone zygotic induction is killed, the genetic analysis is confused. Using various *lambda* mutants, it was possible to carry out crosses in which either or both of the mating types were lysogenic. As noted above, *lambda* is closely linked to the gal_4 gene. When an Hfr *lambda*⁺ *gal*⁺ was crossed with

Table 7.1   The Phenomenon of Zygotic Induction, as Shown by Asymmetry of Results Depending on Whether the *Lambda* Prophage is Contributed by the Hfr or F⁻

Experiment	Number of *lambda* plaques per 100 lysogenic bacteria
Controls	
Hfr *lambda*⁺	0.41
F⁻ *lambda*	2.6
Experiments	
Hfr *lambda*⁺ × F⁻ *lambda*⁻	52.1
Hfr *lambda*⁻ × F⁻ *lambda*⁺	3.3

In all crosses, the lysogenic parent was present at 1/20 of the nonlysogenic parent. After allowing 40 min for chromosome transfer, the mating mixture was diluted and plated on streptomycin agar containing a *lambda*-sensitive indicator that was resistant to streptomycin. (Data from Jacob and Wollman, 1961.)

an F⁻ *lambda*⁺, a normal frequency of gal⁺ recombinants was obtained, but when the F⁻ was sensitive to *lambda*, very few *gal*⁺ recombinants were obtained, and of those that were, very few were *lambda*⁺. Since *lambda* and *gal* are closely linked (Fig 7.2), this disparity is the consequence of the spontaneous induction of the prophage in a large fraction of the zygotes. Thus, when carried by the Hfr, the prophage behaved in crosses as a lethal nuclear unit (Jacob and Wollman, 1957). By using streptomycin to eliminate either the Hfr or F⁻ bacteria (with the alternate of the pair resistant to streptomycin) and assaying for infective cells by an overlay with an indicator strain, it was possible to show that the development of the prophage took place, not in the Hfr *lambda*⁺ cells, but in the F⁻ recipient bacteria. Inactivation of the F⁻ with streptomycin eliminated the infective centers, whereas inactivation of the Hfr had no such effect.

The discovery of zygotic induction provided several important insights. Concerning the conjugation process itself (see Chapter 5), zygotic induction provided a clear demonstration of the distinct polarities between Hfr and F⁻. Furthermore, it could be used to study the kinetics of the conjugation process itself. Also, because linkage relationships appeared skewed, an understanding of the phenomenon of zygotic induction permitted the use of proper crosses to construct genetic maps. From the standpoint of the phage itself, zygotic induction was to play a major role in the development of the *repression* model of enzyme regulation (see Sections 10.9 and 10.10), eventually leading to the development of the highly sophisticated and elaborate model of the *lambda* control circuit that has been elegantly described by Ptashne (1986).

Origin   T L Az T1 Lac   Gal Lambda

Figure 7.2 Genetic map of the Hayes Hfr strain of *Escherichia coli*, showing the order of transfer of selected markers and the close linkage of *lambda* to *gal*.

7.6 THE PROPHAGE AS A GENETIC ELEMENT

In considering the relation of the prophage to the bacterial cell, four alternative models were suggested by Elie Wollman (1953): (1) The prophage is an ensemble of cytoplasmic particles; (2) the prophage is an independent "nuclear" unit; (3) the prophage is a gene; (4) the prophage is an ensemble of cytoplasmic particles whose maintenance in the cell is conditioned by a gene (Fig. 7.3). These four models were subsequently simplified to two by Jacob (1954): the prophage as cytoplasmic particle or as a genetic element linked to a specific structure in the cell, probably the nucleus. Although the cytoplasmic model was first favored, mating experiments with lysogeny soon demonstrated the chromosomal location of the prophage. "[The chromosomal location of *lambda*] appeared somewhat surprising, since it would seem a priori that the noninfective structure which, in lysogenic bacteria, carries the genetic information of a virus, the bacteriophage, should be cytoplasmic rather than chromosomal. Although this problem was never seriously considered before 1950...the hypothesis of a cytoplasmic determination of lysogeny had been accepted implicitly..." (Jacob and Wollman, 1961).

The Genetics of Prophage Lambda

Once the linkage of lysogeny to chromosomal genes was shown, it was possible to approach the study of lysogeny in a totally new fashion. This was primarily the work of Jacob, Wollman, and collaborators at the Pasteur Institute and Allan Campbell at Stanford University (Campbell and Jacob also worked together). The *lambda* system provided the foundation of the research, but a number of new temperate phages were isolated from other *Escherichia coli* strains, as well as a variety of *lambda* mutants differing in the size or appearance of plaques, host range, or capacity for lysogenization. One particular class of *lambda* mutants, called *virulent*, were able to overcome the immunity system of K12(λ) and develop within lysogenized (and normally *lambda*-immune) cells. All of these phage mutants retained the antigenic specificity of *lambda*. Such virulent phage mutants constituted a conceptual link to the virulent phages of Delbrück and colleagues. Jacob (1955) used

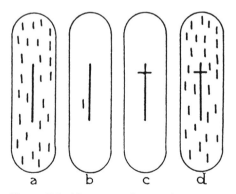

Figure 7.3 Alternate schemes for the state of the prophage in a lysogenic bacterium. (*a*) Ensemble of cytoplasmic particles; (*b*) independent nuclear unit; (*c*) a gene; (*d*) ensemble of cytoplasmic particles whose maintenance requires a gene. (Reprinted, with permission, from E.L. Wollman, 1953.)

transduction (see Chapter 8) with an unrelated temperate phage as carrier to show that the prophage could be handled as a genetic unit and transferred intact from a donor to a (presumably immune) recipient strain. By using the phage crossing techniques of Hershey and Delbrück (see Section 6.9), it was also possible to use these *lambda* mutants to develop a phage genetic map. All of the known characters of the phage could be arranged in a single linkage group (Jacob and Wollman, 1961). Genetic analysis by conjugation showed that the prophage had a chromosomal location, and unrelated prophages had separate locations. Thus, there was a site on the bacterial chromosome that was *complementary* to the genetic material of the prophage.

The Location of the Prophage

Was the prophage an addition to or a substitution in the bacterial chromosome? Two alternatives could be visualized: (1) Because of the obvious homology to a particular region of the chromosome, one might consider the prophage to be an allelic equivalent to a short segment of the chromosome. (2) However, because of the genetic complexity of the prophage (a large number of genes had been mapped), it seemed unlikely that the prophage would be simply allelic to a segment of the bacterial chromosome and hence integrated by a process analogous to crossing-over. In addition, if prophage was an allele of a bacterial gene, one might expect single mutations to occur from lysogeny to nonlysogeny and back. However, mutation from $lambda^-$ to $lambda^+$ had never been observed. Also, strains could be *cured* of prophage by heavy irradiation. Curing was a phenomenon that could not be explained by a simple mutational event, since in a cured cell *all* of the genes of the prophage were lost as a unit. Because of these considerations, the second (incorrect) hypothesis was adopted:

> One is led to conclude, therefore, that lysogenization corresponds to the *addition* of a new structure, the prophage, to the bacterial chromosome and that the curing of lysogeny corresponds to the *loss* of this structure. Between the prophage and the normal genetic constituents of the bacterial chromosome, there exists, therefore, a fundamental difference: the chromosome of a nonlysogenic bacterium does not contain a structure allelic to a given prophage. The characters $(ly)^+$ and $(ly)^-$ are not alleles in the usual sense of this term... (Jacob and Wollman, 1961).

Note that the term *addition* is used here as synonymous with insertion and does not imply a type of additive recombination. Jacob and Wollman (1961) attempted to distinguish between prophage *added* to the chromosome and prophage *inserted* into the chromosome by determining whether chromosomal markers *outside* the prophage site mapped farther apart in lysogenic than in nonlysogenic cells. Unfortunately, the genetic resolution was not sufficient to distinguish properly between a phage *added* to the chromosome, and a phage *inserted* into the chromosome. The insertion idea is correct, but in 1961, when this work was summarized, Jacob and Wollman concluded:

> [The prophage] is an added structure, located on the bacterial chromosome at a specific site, but not really part of the chromosome. Although specifically attached to the chromosome, it remains at least in its major part, extrinsic to it. It is gained as a whole unit by lysogenization, and it is also lost as a whole unit (Jacob and Wollman, 1961).

The Campbell Model

About this same time, an alternative model of *lambda* integration was proposed by Campbell (1962), deriving initially from work of Calef and Licciardello (1960). The Campbell model, which turned out to be correct, hypothesized that the *lambda* genome must be circular or cyclize, and it becomes integrated into the bacterial chromosome by crossing over. The result is that the genes of the ring become inserted linearly into the continuity of the host chromosome, but in a different order.

What were the genetic data that suggested to Campbell such a radical model? A special feature of *lambda* is that a map can be obtained in two entirely different ways: by phage genetics or by bacterial genetics (Campbell, 1971). A phage map can be obtained by carrying out mixed infections with genetically marked phage (see Section 6.9). A prophage map, on the other hand, can be derived from mating or transduction using lysogenic bacteria. These two maps turned out to be different, and different in an interesting way: The two gene orders were found to be *circular permutations* of one another. These differences provided the evidence that suggested to Campbell that the prophage was inserted *into* the bacterial chromosome rather than attached *to* the chromosome (Fig. 7.4). Jacob and Wollman (1957, 1961) had also done mapping of this sort, but they made the (reasonable) assumption that the gene order of the prophage was the same as that of the free phage and that prophage recombination followed its own rules rather than those of the bacterial chromosome. The pattern of recombination appeared to be too independent to be compatible with prophage insertion into the chromosome. However, their results are also compatible with the correct model, the gene order in the prophage being different than it was in the free phage. The data of Jacob and Wollman were first interpreted in this light by Calef and Licciardello (1960) and subsequently extended by Campbell and others (see Campbell, 1969, 1971).

It might seem strange that Campbell would postulate circularity for *lambda* DNA, but there were several precedents. First, "ring" chromosomes were well known in classical genetics, so the idea of a circle was not too farfetched. Second, Jacob and Wollman had provided strong evidence that the genetic map of *Escherichia coli* was circular (see Section 5.11) and had de-

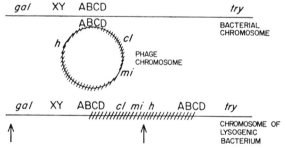

Figure 7.4 Campbell's model for the mechanism of integration of *lambda* into the bacterial chromosome. The circular phage DNA attaches to a specific site on the host and undergoes reciprocal crossing-over. Note that the order of the phage genes is changed in the lysogen from the order in the phage. (Reprinted, with permission, from Campbell, 1962.)

veloped a model for conversion of F^+ to Hfr that required the opening of the circular chromosome by insertion of the F factor. Third, genetic mapping suggested that virulent phage T4 was formally a circle (see Section 6.11). Thus, a circular model found considerable favor in the bacterial genetics "community."

According to Campbell: "At the time I proposed the model, I did not believe it was correct. It was simply the result of my best efforts to understand the genetic data then available while avoiding the distasteful notion that the episome [prophage] was stuck onto the outside of the chromosome" (Campbell, 1969).

Eventually, the Campbell model would be placed on a firm molecular basis from studies on *lambda* DNA. Hershey and Burgi (1965) showed that although *lambda* DNA is linear, it cyclizes because it has single-stranded cohesive ends that find each other inside the cell once released from the phage head. Studies on *lambda* mutants showed that integration into the host genome required two genes, the *att* site, which attaches to the host chromosome, and the *int* site, which codes for a site-specific enzyme called *integrase* and which carries out the specific cutting reaction. There is also an *att* site in the host chromosome to which the *lambda att* site is specifically complementary (Echols, Gingery, and Moore, 1968). Eventually, molecular analyses of *lambda* DNA, and correlation of physical structure with genetic structure, would lead to important insights into the biochemistry of genetic recombination (Hershey, 1971), but these matters take us too far afield. Note that in 1962 when Campbell was writing, these molecular details were not known, or even suspected.

Importance of Lambda

Research on bacteriophage *lambda* has had far-reaching effects on the development of molecular genetics. The *lambda* system provided one of the first detailed connections between genetics and DNA structure (Hershey, 1971). It provided the best model for studying the molecular basis of genetic recombination. Studies on the immunity system in *lambda* were a key element in the development of the *repressor* or *negative control* model of gene regulation by Jacob and Monod (see Section 10.10). Research on *restriction* and *modification* using *lambda* (see Section 6.15) led to the discovery of restriction enzymes and, eventually, to their use in *genetic engineering* (see Chapter 11). *Lambda* has become one of the principal vectors in *recombinant DNA research*. Phage *lambda* was the first rather complex virus whose behavior could be really understood at the molecular level. Numerous proteins of *lambda* have been isolated and purified, and their sequences have been determined. The complete DNA base sequence, 48,513 base pairs, has been determined. Many ideas from *lambda* research have found their way into studies on the molecular genetics of eucaryotes.

7.7 THE EPISOME

As discussed in Section 5.16, Jacob and Wollman (1961) recognized conceptual similarities between the prophage and the sex factor, and coined the term *episome* to describe such entities. An episome was defined as a genetic structure that is *added* to the genome and which, inside the cell, may exist in two distinct, perhaps mutually exclusive states, *autonomous* and *integrated*. Both temperate phages and sex factors have the ability to replicate independently of the chromosome, but also to become integrated in a specific manner into the chromosome.

It was recognized almost from the beginning that the study of lysogeny might have important implications for general virology and cancer. The existence of "latent viruses" in plants and animals had been known for some time (Lwoff, 1936), and the relevance of viruses to cancer had been often discussed. It was thus natural to consider the relevance of the prophage concept in the broader context of the *provirus*.

It had already been established that lysogenized cells differed antigenically and in perhaps other ways from nonlysogenized cells. The conversion of nontoxigenic to toxigenic *Corynebacterium diphtheriae* had also been related to lysogenization with a specific phage. This phenomenon, called *lysogenic* or *phage conversion*, will be discussed in Section 8.6. The key point here is that two cells, otherwise genetically identical, may exhibit functional differences as a consequence of acquisition of an extra piece of genetic material from a prophage. Although it would be some years before the details would be worked out, there was some reason to believe that in animals some latent viruses might become integrated into the host genetic material. Jacob and Wollman thus saw the episome concept in broad terms:

> The conclusion is thus reached that, by appropriate genetic events, all intermediary categories may be formed between the viruses (extrinsic, infectious, plasmids) and the normal genetic determinants of a cell (intrinsic, noninfectious, and integrated). Between heredity and infection, between cellular pathology and cellular physiology, between nuclear and cytoplasmic heredity, the episomes, as studied in bacteria, provide therefore the link which, although frequently postulated on theoretical grounds, had not hitherto received experimental support...These considerations...represent perhaps one of the main contributions of bacterial genetics (Jacob and Wollman, 1961).

REFERENCES

Anderson, T.F. 1966. Electron microscopy of phages. pp. 63–78 in Cairns, J., Stent, G.S., and Watson, J.D. (ed.), *Phage and the Origins of Molecular Biology*. Cold Spring Harbor Laboratory of Quantitative Biology, Cold Spring Harbor, New York.

Appleyard, R.K. 1953. Segregation of *lambda* lysogenicity during bacterial recombination in *E. coli* K-12. *Cold Spring Harbor Symposia on Quantitative Biology* 18: 95–97.

Beadle, G.W. 1945. Biochemical genetics. *Chemical Reviews* 37: 15–96.

Bordet, J. 1925. Le problème de l'autolyse microbienne transmissible ou du bactériophage. *Annales de l'Institut Pasteur* 39: 717–763.

Bordet, J. and Ciuca, M. 1920a. Exsudats leucocytaires et autolyse microbienne transmissible. *Comptes rendus Société de Biologie* 83: 1293–1295.

Bordet, J. and Ciuca, M. 1920b. Le bacteriophage de d'Herelle, sa production et son interprétation. *Comptes rendus Société de Biologie* 83: 1296–1298.

Burian, R.M., Gayon, J., and Zallen, D. 1988. The singular fate of genetics in the history of French biology, 1900–1940. *Journal of the History of Biology* 21: 357–402.

Burnet, F.M. and McKie, M. 1929. Observations on a permanently lysogenic strain of *B. enteritidis Gaertner*. *Australian Journal of Experimental Biology and Medical Science* 6: 277–284.

Calef, E. and Licciardello, G. 1960. Recombination experiments on prophage host relationships. *Virology* 12: 81–103.

Campbell, A. 1962. Episomes. *Advances in Genetics* 11: 101–145.

Campbell, A. 1969. *Episomes*. Harper and Row, New York. 193 pp.

Campbell, A. 1971. Genetic structure. pp. 13–44 in Hershey, A.D. (ed.), *The Bacteriophage Lambda*. Cold Spring Harbor Laboratory, Cold Spring Harbor, New York.

Cowles, P.B. 1931. The recovery of bacteriophage from filtrates derived from heated spore-suspensions. *Journal of Bacteriology* 22: 119–123.

Delbrück, M. 1942. Bacterial viruses (bacteriophages). *Advances in Enzymology and Related Subjects.* 2: 1–32.

Delbrück, M. 1946. Bacterial viruses or bacteriophages. *Biological Reviews* 21: 30–40.

Delbrück, M. (ed.). 1950. *Viruses 1950*. California Institute of Technology, Pasadena. 147 pp.

Delbrück. M. 1953. Introductory remarks about the program. *Cold Spring Harbor Symposia on Quantitative Biology* 18: 1–2.

den Dooren de Jong, L.E. 1931a. Studien über Bakteriophagie. I. Ueber Bac. megatherium und den darin anwesenden Bakteriophagen. *Zentrallblatt für Bakteriologie* I. Abteilung, Originale 120: 1–15.

den Dooren de Jong, L.E. 1931b. Studien über Bakteriophagie. II. Mitteilung: Fortsetzung der Untersuchungen über den Megatherium-Bakteriophagen. *Zentrallblatt für Bakteriologie* I. Abteilung, Originale 120: 15–23.

den Dooren de Jong, L.E. 1936. Studien über Bakteriophagie. VI. Beruht die Fähigkeit eines Bakteriophagen, Bakteriensporen zur Phagenbildung zu reizen, auf einer Infektion mit diesem Bakteriophagen? *Zentrallblatt für Bakteriologie* I. Abteilung, Originale 136: 404–409.

Echols, H., Gingery, R., and Moore, L. 1968. Integrative recombination function of phage λ: evidence for a site-specific recombination enzyme. *Journal of Molecular Biology* 34: 251–260.

Galperin, C. 1987. Le bactériophage, la lysogénie et son déterminisme génétique. *History and Philosophy of the Life Sciences* 9: 175–224.

Gildemeister, E. and Herzberg, K. 1924. Zur Theorie der Bakteriophagen (d'Herelle Lysine). 6. Mitteilung über das d'Herellesche Phänomen. *Zentralblatt für Bakteriologie*, I. Abteilung, Originale 93: 402–420.

Hayes, W. 1980. Portraits of viruses: bacteriophage lambda. *Intervirology* 13: 133–153.

Hershey, A.D., ed. 1971. *The Bacteriophage Lambda*. Cold Spring Harbor Laboratory, Cold Spring Harbor, New York. 792 pp.

Hershey, A.D. and Bronfenbrenner, J. 1948. Bacterial viruses: bacteriophages. pp. 147–162 in Rivers, T.M. (ed.), *Viral and Rickettsial Infections of Man*. Lippincott, Philadelphia.

Hershey, A.D. and Burgi, E. 1965. Complementary structure of interacting sites at the ends of λ DNA molecules. *Proceedings of the National Academy of Sciences* 53: 325–328.

Ionesco, H. 1951. Systemes inductibles et non inductibles chez *Bacillus megatherium* lysogene. *Comptes rendus Academie des Sciences* 233: 1702–1704.

Jacob, F. 1954. *Les bactéries lysogènes et la notion de provirus*. Masson, Paris. 176 pp.

Jacob, F. 1955. Transduction of lysogeny in *Escherichia coli*. *Virology* 1: 207–220.

Jacob, F. 1988. *The Statue Within*. Basic Books, New York.

Jacob, F. and Wollman, E. 1954. Induction spontanée du développement du bactériophage λ au cours de la recombinaison génétique chez *E. coli* K12. *Comptes rendus Academies des Sciences* 239: 455–456.

Jacob, F. and Wollman, E. 1957. Genetic aspects of lysogeny. pp. 468–500 in McElroy, W.D. and Glass, B. (ed.), *The Chemical Basis of Heredity*. Johns Hopkins Press, Baltimore.

Jacob, F. and Wollman, E. 1961. *Sexuality and the Genetics of Bacteria*. Academic Press, New York.

Jacob, F., Lwoff, A., Siminovitch, A. and Wollman, E. 1953. Définition de quelques termes relatifs a la lysogénie. *Annales de l'Institut Pasteur* 84: 222–224.

Kjeldgaard, N.O. 1971. The unmasking of the unseen. pp. 88–93 in Monod, J. and Borek, E. (ed.), *Of Microbes and Life*. Columbia University Press, New York.

Lederberg, E.M. 1951. Lysogenicity in *E. coli* k-12. *Genetics* 36: 560. (Abstract.)

Lederberg, E.M. and Lederberg, J. 1953. Genetic studies of lysogenicity in *Escherichia coli*. *Genetics* 38: 51–64.

Lederberg, J. 1951. Inheritance, variation, and adaptation. pp. 67–100 in Werkman, C.H. and Wilson, P.W. (ed.), *Bacterial Physiology*. Academic Press, New York.

Lederberg, J. 1952. Cell genetics and hereditary symbiosis. *Physiological Reviews* 32: 403–430.

Lederberg, J., Lederberg, E.M., Zinder, N.D., and Lively, E.R. 1951. Recombination analysis of bacterial heredity. *Cold Spring Harbor Symposia on Quantitative Biology* 16: 413–443.

Lisbonne, M. and Carrère, L. 1922. Antagonisme microbien et lyse transmissible du bacille de shiga. *Comptes rendus Société de Biologie* 86: 569–570.

Lwoff, A. 1936. Remarques sur un propriété commune aux genes, aux principes lysogènes, et aux virus des mosaiques. *Annales de l'Institut Pasteur* 56: 165–170.

Lwoff, A. 1944. *L'évolution physiologique. Étude des pertes de fonction chez les microorganismes.* Hermann Éditions, Paris. 308 pp.

Lwoff, A. 1953a. Lysogeny. *Bacteriological Reviews* 17: 269–337.

Lwoff, A. 1953b. L'Induction. *Annales de l'Institut Pasteur* 84: 225–241.

Lwoff, A. 1966. The prophage and I. pp. 88–99 in Cairns, J., Stent, G.S., and Watson, J.D. (ed.), *Phage and the Origins of Molecular Biology*. Cold Spring Harbor Laboratory of Quantitative Biology, Cold Spring Harbor, New York.

Lwoff, A. 1971. From protozoa to bacteria and viruses. Fifty years with microbes. *Annual Review of Microbiology* 25: 1–26.

Lwoff, A. and Gutmann, A. 1950. Recherches sur un *Bacillus megatherium* lysogène. *Annales de l'Institut Pasteur* 78: 711–739.

Lwoff, A., Siminovitch, L., and Kjeldgaard, N. 1950. Induction de la production de bacteriophages chez une bactérie lysogeène. *Annales de l'Institut Pasteur* 79: 815–859.

Monod, J. 1971. Du microbe à l'homme. pp. 1–9 in Monod, J. and Borek, E. (ed.), *Of Microbes and Life*. Columbia University Press, New York.

Monod, J. and Borek, E., eds. 1971. *Of Microbes and Life*. Columbia University Press, New York.

Northrop, J.H. 1939. Increase in bacteriophage and gelatinase concentration in cultures of *Bacillus megatherium*. *Journal of General Physiology* 23: 59–79.

Ptashne, M. 1986. *A Genetic Switch. Gene Control and Phage λ*. Cell Press/Blackwell Scientific Publications, Cambridge, Massachusetts. 128 pp.

Sapp, J. 1987. *Beyond the Gene. Cytoplasmic Inheritance and the Struggle for Authority in Genetics*. Oxford University Press, New York. 266 pp.

Vedder, A. 1932. Die Hitzeresistenz von getrockenen Bakteriophagen. *Zentralblatt für Bakteriologie* I. Abteilung, Originale 125: 111–114.

Weigle, J.J. and Delbrück, M. 1951. Mutual exclusion between an infecting phage and a carried phage. *Journal of Bacteriology* 62: 301–318.

Wollman, E. 1920. A propos de la note de MM. Bordet et Ciuca (phénomène de d'Herelle, autolyse microbienne transmissible de J. Bordet et M. Ciuca, et hypothèse de la pangénèse de Darwin). *Comptes rendus Société de Biologie* 83: 1478–1479.

Wollman, E. 1928. Bactériophage et processus similaires. Hérédité ou infection? *Bulletin de l'Institut Pasteur* 26: 1–14.

Wollman, E. and Wollman, E. 1936. Recherches sur le phénomène de Twort-d'Herelle (bactériophage out autolyse hérédo-contagieuse). *Annales de l'Institut Pasteur* 56: 137–164. (Plus an appendix by André Lwoff, pp. 165–170.)

Wollman, E. and Wollman, E. 1937. Les phases des bactériophages l'facteurs lysogènes. *Comptes rendus Société de Biologie* 124: 931–934.

Wollman, E. and Wollman, E. 1938. Recherches sur le phénomène de Twort-d'Hérelle (bactériophagie ou autolyse hérédo-contagieuse). V. *Annales de l'Institut Pasteur* 60: 13–57.

Wollman, E. and Wollman, E. 1939. Les "phases" de la fonction lysogène. Action successive du lysozyme et de la trypsine sur un germe lysogène. *Comptes rendus Société de Biologie* 131: 442–445.

Wollman, E.L. 1953. Sur le déterminisme génétique de la lysogénie. *Annales de l'Institut Pasteur* 84: 281–293.

Wollman, E.L. 1966. Bacterial conjugation. pp. 216–225 in Cairns, J., Stent, G.S., and Watson, J.D. (ed.), *Phage and the Origins of Molecular Biology*. Cold Spring Harbor Laboratory of Quantitative Biology, Cold Spring Harbor, New York.

8
TRANSDUCTION

By *genetic transduction* we now mean the transfer of genetic information from one cell to another by a virus particle. The discovery of transduction by Zinder and Lederberg was one of the major accomplishments of the "classical" period of bacterial genetics. Not only did its discovery markedly broaden ideas about the nature and significance of viruses, it also led to an understanding that the three major mechanisms of bacterial recombination—mating, transformation, and transduction—had in common the fact that they were all *fragmentary* unidirectional processes. The discovery of transduction thus helped to confirm the marked differences between the common natural recombination processes in procaryotes and those of eucaryotes. Transduction also became a major tool in fine-structure genetic mapping of bacteria.

We now distinguish two types of transduction, which are called generalized and specialized. In *generalized transduction*, the first kind discovered, bacterial DNA from anywhere in the genome is packaged accidentally into phage particles, by means of which the DNA can be transferred to sensitive cells (Masters, 1985). The phage is essentially a nonspecific carrier of host genes; virtually any gene of the host can be transduced, although the efficiency of gene transfer is low. The injected DNA can undergo homologous recombination with the recipient DNA, resulting in genetic recombination, or the DNA can remain unrecombined in the cytoplasm (probably in a circular form). In the latter state, if the gene is expressed, a linear mode of inheritance is seen, and the process is called *abortive transduction*.

Specialized transduction involves a more specific association between prophage and a set of host genes that are closely linked to the prophage attachment site. Only these closely linked genes are transduced, although often at high frequency.

8.1 TRANSDUCTION IN *SALMONELLA*

Transduction was discovered by Norton Zinder and Joshua Lederberg as a result of a systematic search for genetic recombination in the genus *Salmonella*. Lederberg began studying *Salmonella* genetics soon after assuming his first faculty position at the University of Wisconsin. Work with *Salmonella* genetics seemed to Lederberg a logical extension of his work with *Escherichia coli*, since the *Salmonella* group is a close taxonomic neighbor of *Escherichia* but has much greater medical significance. Thousands of serotypes of *Sal-*

monella have been recognized. "If the genetics of *Salmonella* could be given the same footing as that of *E. coli*, one could have powerful tools to study a broad range of problems: natural history and evolution, epidemiology, pathogenicity, serogenetics. In addition, the very proliferation of serotypes...had all the earmarks of a naturally occurring recombination process, and had been a prior encouragement...to the very possibility of genetic exchange in bacteria" (J. Lederberg, personal communication.)[1]

Norton Zinder

Soon after setting up his laboratory at Wisconsin, Lederberg received his first doctoral student, Norton Zinder.[2] Zinder came to Wisconsin in 1948 from Columbia University, where he had received his undergraduate degree in biology. Like Lederberg, Zinder had fallen under the influence of Francis Ryan, and it was Ryan who urged Zinder to go to Wisconsin to work for his Ph.D. Lederberg was only 23 years old, and Zinder was 19.

Zinder was supported primarily by a research grant to Lederberg. Records from the Research Adminstration-Financial of the University of Wisconsin-Madison indicate that Lederberg obtained his first research grant, on the genetics of *Salmonella*, from the U.S. Public Health Service (in those days part of the Federal Security Agency) in 1948 with an annual budget of $3780.00. This grant, with modifications and modest increases, continued for all the years that Lederberg was at Wisconsin. In 1952, the justification for continued support included the following statement: "Initial requests for research support...were at the rather modest level of about $4000 per annum. This grant, applied primarily to work on *Salmonella* transduction, was sufficient to enable one graduate student [Zinder] to assist in this research...For some years, little substantial progress could be reported from this project, and there might have been some question whether even the modest investment would be recovered. During the last year, however, the picture has changed completely to give experimental findings of considerable general interest."

Zinder received his Ph.D. in 1952 with a joint degree in Genetics and Medical Microbiology. He immediately joined the Rockefeller Institute (now Rockefeller University) where he has been ever since. After an early research career exploring transduction, Zinder discovered RNA bacteriophages and concentrated his research on this group of viruses (see Section 5.14).

Discovery of Transduction

The classification of *Salmonella* had been worked out in detail by medical bacteriologists, most especially by F. Kauffmann and P.B. White in Europe and P.R. Edwards in the United States. A large number of isolates were available, including various serologically characterized strains. In addition, K. Lilleengen in Scandinavia had used various bacteriophages to "type" numerous strains of the species *Salmonella typhimurium*.

---

[1]Quoted from an unpublished draft of a reminiscence on genetic exchange in *Salmonella*.

[2]I am grateful to Professor Norton Zinder for providing me a transcript of a lecture that provides some of the information on the personal history of his involvement in the discovery of transduction.

Zinder and Lederberg obtained representatives of all of Lilleengen's 20 "phage types," as well as additional strains from Kauffmann, Edwards, and others.[3] Using the penicillin-selection method (Lederberg and Zinder, 1948; see Section 4.12), two-step (double) auxotrophic mutants were isolated for each of the phage types following UV irradiation. Using the prototrophic recovery method that had been developed for *Escherichia coli* (see Section 5.3), attempts to obtain recombination were made. At first, only crosses between mutants of the same strain were made, so-called "self crosses." Crosses between different strains were not made at that time because it was believed that *Escherichia coli* was homothallic. There was even a reasonable basis for this belief, since the successful results had been obtained with mutants that were all derived from the single strain, *Escherichia coli* K-12. (This was before the nature of the F factor controlling mating was understood; see Section 5.6.)

These first *Salmonella* crosses gave completely negative results. Since the selfed crosses were giving negative results, in the summer of 1950 Zinder began making crosses in all possible combinations between the various Lilleengen serotypes. In all, one hundred pairwise combinations of the 20 types were made, including "selfed" crosses and single-culture controls. Fifteen combinations yielded prototrophs, most of which involved strain LT-22, especially when paired with strain LT-2. Subsequently, it would be shown that strain LT-22 produced a phage (initially called PLT-22 and later P22) active against LT-2. Bacterial strain LT-2 and phage strain P22 would become the foundation for the extensive research on *Salmonella* genetics that this initial work would stimulate. However, originally Zinder and Lederberg were unaware that a phage was responsible for the recombination process.

The number of prototroph recombinants recovered in these experiments was quite low, but always considerably above background. The use of double mutants eliminated any confusion that might have arisen from *reversion*. Since genetic recombination in *Escherichia coli* was also low (this was before the discovery of Hfr), the low efficiency of the process did not seem troublesome. However, there was no evidence of genetic linkage, since in any cross only a single marker was transferred.

As noted in Section 5.3, the fact that Lederberg had demonstrated linkage in *Escherichia coli*, including the formation of recombinants for unselected markers, had indicated that mating in this organism was not a "simple" transformation. As in the mating work with *Escherichia coli*, one possible interpretation was that recombination was due to DNA-mediated transformation (see Chapter 9). This was even more likely in the *Salmonella* case since only single markers were being transferred. Employing the Davis (1950) U-tube device (see Section 5.4), Zinder and Lederberg found that, in contrast to the situation in *Escherichia coli*, *Salmonella* prototrophs *did* arise when the two strains were placed on opposite sides of the filter. On the other hand, sterile filtrates of a single strain did not elicit prototrophs, showing that although the two strains did not need to be in contact, "something" was passing back and forth from one to the other. Zinder and Lederberg called this something *filterable agent* (FA).

---

[3]Zinder recalls that Lilleengen's strains were shipped from Sweden in a diplomatic pouch.

Early Interpretations of Transduction

In the early *Salmonella* work there was no reason to believe that a phage was *directly* involved in the recombination process. Because the filterable agent (FA) was resistant to DNase, it was obviously not a transforming principle. Still attempting to equate bacterial recombination processes with those of higher organisms, Lederberg interpreted the new phenomenon in terms of a bacterial life cycle involving a type of structure known as an L form.

L forms, first described by E. Klieneberger-Nobel and L. Dienes in the 1930s (reviewed by Bisset, 1950), were small, swollen forms of bacteria that were able to pass through bacteriological filters. L forms were believed either to be stages in the life cycles of bacteria (see Section 3.6), or separate forms that entered into parasitic or symbiotic relationships with bacteria. They often appeared to arise when bacteria were placed under certain noxious conditions. L forms could only be grown on certain special media and hence often appeared and disappeared from cultures in mysterious ways. It was eventually shown that L forms constituted bacterial cells that had lost their cell walls and grew poorly or not at all in liquid medium, although they could sometimes be made to grow as small colonies on agar. Because of their wall-less nature, they could squeeze through the pores of bacteriological filters. Later it was found that penicillin inhibited cell wall synthesis in bacteria and was an excellent inducer of L-form growth. In one of his side projects, Lederberg carried out several studies on the nature of L-form growth (Lederberg, 1956a; Lederberg and St Clair, 1958).

In the first report of the transduction work, given at the 1951 Cold Spring Harbor symposium,[4] Lederberg stated:

> As a working hypothesis, we suggest that FA can be correlated with the granular phase of L-type colonies...There is an unmistakable suggestion of "filtrable" elements which can regenerate bacteria probably along the lines of the reversion of L-type cultures...The conditions of dormancy and of reversion to bacteria are too poorly understood for more detailed discussion (Lederberg, Lederberg, Zinder, and Lively, 1951).

A number of treatments that were known to cause L-form formation were discussed in this paper, and a photomicrograph was published of an L form of *Salmonella* that had been elicited by exposure to antiserum. Note that this paper mainly concerned genetic recombination in *Escherichia coli*, and the *Salmonella* work was simply appended. Pursuing the L-form analogy further, Lederberg wrote:

> It may be quite obvious that the genetic effects of FA appear, at this time, to duplicate the well-known pneumococcus transformation. While casual published

[4]This paper became famous among those in the bacterial genetics community for its length and obscurity. Several of my correspondents and reviewers who attended the session remarked on this paper. It was given at an evening session and went on until midnight. Zinder (personal communication) has remembered: "In May of 1951 we wrote an overwhelming paper for the Cold Spring Harbor Symposium. This paper was delivered by Josh at the symposium. I don't remember if he took six or nine hours giving it. It is a summary of three years work in the laboratory, everything that was done in the laboratory...I would say that the data in that paper are all really quite good. I would also say that almost every interpretation in that paper is wrong—maybe not every. The pheno-genetics section is a little better than the rest because we ran into real cis-trans effects, which at least were stated properly...This symposium would represent...if you ever want to pick out a symposium that represents what [Kuhn] calls a crisis in revolutionary science, this symposium is it. There are good things, there are bad things; everyone is talking, no one is listening. And no one completely understands what is being said."

statements concerning the size of the transforming principle may rule out the participation of L-granules in that system, the possibility that they play a part in other transformations should be examined carefully.

The idea that phage might be involved in some way had been considered, since in the 1951 Cold Spring Harbor paper the possibility that the filterable agent (FA) might have something to do with bacteriophage was mentioned obliquely:

> FA, then, is not a normal component of [strain LT-2], but is produced under the stimulus of a latent phage (Lederberg, Lederberg, Zinder, and Lively, 1951).

According to Zinder, the Cold Spring Harbor symposium was the turning point in his thinking about the *Salmonella* work. Among others, Harriet Ephrussi-Taylor suggested to him at the Cold Spring Harbor meeting that the phenomenon might be due to a phage with a piece of DNA stuck on it.[5] Returning to Wisconsin, Zinder began to study all of the phages in the laboratory to see whether they had transducing activity. Gratifyingly, one of them did show activity.

One of the first things done was to work out a quantitative assay procedure, using the count of prototrophs as an expression of the activity of FA. If cell number was kept constant, then the fraction of prototrophs increased as the amount of FA increased (up to a limit). A *unit* of FA was defined as that content of filtrate that would elicit a single prototroph from an optimum concentration of cells. Filtrates were generally quite active, containing about 2500 units/ml.

This quantitative assay permitted a careful analysis of the relationships between FA and bacteriophage.

The Filterable Agent is Phage

The paper by Zinder and Lederberg (1952), which is one of the classics of modern biology, describes in detail the experimental steps used to demonstrate that the filterable agent (FA) that participated in *Salmonella* recombination was a bacteriophage. Lysogeny in *Salmonella* was a well-known phenomenon, having been studied by Burnet and McKie (1929) (see Section 7.3) and in detail by Boyd (1951). As noted, bacterial strain LT-22 was lysogenic for phage P22 active against bacterial strain LT-2. Evidence that "forced" attention to phage included studies showing the chemical stability of FA, its resistance to DNase, its relative heat sensitivity, its sedimentability at high speed (but not at low speed), and its size as indicated by filtration through graded ultrafilters. The particles of FA observed under the electron microscope were "phage-like" (although phage P22 has only a very short tail). FA adsorbed to cells in the manner of phage, and both FA and phage had a common specificity of adsorption on *Salmonella* serotypes.

The exact role of the FA/phage as a genetic carrier was unclear, however. The phage particle appeared to be only a *passive* carrier of the donor genetic material, since in serial transfer of one genetic marker after another, the FA obtained in the second round had lost the genotype of the first donor and had acquired the genotype of the secondary donor. FA was not released by cell breakage, but "various deleterious agents elicit its appearance in a way that

---

[5]See Section 9.7 for a discussion of Harriet Ephrussi-Taylor's work on the genetics of transformation, carried out at this same time.

may parallel the action of latent phage." Ideas about lysogeny and the nature of prophage had by this time been clarified by Lwoff's work (see Section 7.4), and lysogeny had been discovered in *Escherichia coli* K-12.

> The following picture appears to be most consistent with the observations to date. An active filtrate contains a population of numerous species of granules, each corresponding to a genetic effect although some may be intrinsically inert. Each bacterium may absorb a limited number of particles, in the possible range of one to perhaps one hundred. Each adsorbed particle has a fixed, independent probability of exerting its particular transductive effect. The low frequency of single, and particularly double transductions, is limited by the total number of particles that may be adsorbed as well, perhaps, as by the low probability that an adsorbed particle will complete its effect.
>
> Two aspects of FA must be carefully distinguished: the biological nature of the particles themselves and their genetic function. There is good reason to identify the particle with bacteriophage. Nevertheless, the phage particle would function as a passive carrier of the genetic material transduced from one bacterium to another. This material corresponds to only a fragment of the bacterial genotype (Zinder and Lederberg, 1952).

8.2 TERMINOLOGY OF TRANSDUCTION

According to Zinder (personal communication, 1989), the term *transduction* was coined by Lederberg in the fall of 1951, after the Cold Spring Harbor meeting but before the nature of the transduction process was clear. The word was derived from "transduce," meaning "to lead across." The term *transduction* was originally used to describe any "genetically unilateral transfer in contrast to the union of equivalent elements in fertilization" (Zinder and Lederberg, 1952; Lederberg, 1956b), but this definition did not stick. With this definition, transformation (see Chapter 9) would also have been considered a transduction. However, it was clear almost immediately that such a broad use of the term *transduction* was of little value, and by the time of the next Cold Spring Harbor symposium, in the summer of 1953, Zinder (1953) was using the term in the more restricted way in which it is used today (phage-mediated gene transfer). Other workers quickly followed this usage, but Lederberg's insistence that transformation was a subclass of transformation was not accepted. The matter was still not resolved at the Oak Ridge symposium of April 1954, when the following discussion took place between Rollin Hotchkiss and Lederberg:

Hotchkiss: As to nomenclature, I think it would be well if we pointed out explicitly that the word "transformation" has disadvantages since it comes from general usage and is adapted for a rather specific sense. But it does have historical value. Many people know what bacterial transformation means. Therefore, I should like to recommend that we retain "transformation" as the generic term, and save "transduction" for the phage-mediated transformations...The term "transforming agent" can be used in reference to a material entity; if it is a phage, it becomes a transducing agent.

Lederberg: No one can deny that in all these experiments cells are being transformed, or rather their properties are being altered. In that context, there is no objection at all.

Hotchkiss: Also, one may have a transformed cell, or "transformant," while you have defined transduction so that it is only the character which is transduced.

Lederberg: Precisely. I hope I kept that straight...Perhaps another term is needed to distinguish phage-mediated transductions (or transformations), though perhaps we ought to learn a little more about them first (Lederberg, 1955).

Lederberg continued to use the word *transduction* in his broader sense for some time (see Lederberg, 1956b), but the restricted use of the term prevailed. Current usage is most clearly stated in the definition of Hartman (1957):

> ...transduction will refer only to those processes in which a fragment of the host nuclear genome, other than the genetic material specific for the transmitting phage itself, is carried by bacteriophage particles from one bacterial cell to another.

Transduction and Lysogeny

Once the connection between transduction and phage had been clearly established, the relationship between transduction and lysogeny became a critical issue. By 1953, the Hershey/Chase experiment had been published (see Section 6.13), providing convincing evidence that the role of the phage coat was in the transfer of the DNA from cell to cell. Zinder (1953, 1955) presented data showing the relationship between FA and phage P22. Of major importance was the observation that although phage and FA behaved the same physically, they were biologically distinct, because a lysate of bacterial strain LT-22 contained a single kind of phage but a *mixture* of transducing particles (depending on which mutant the phage had been grown on). "The phage and the FA may be considered different biological entities sharing a common adsorptive mechanism" (Zinder, 1953).

One reason that the proper interpretation might have been so difficult to develop is that the situation involved a filterable agent that could not be detected in the filtrate of either partner separately. The potential recipient (auxotroph) carried the prophage and rarely released infectious particles. If these particles infected the nonimmune donor, they gave rise to a population of phage particles, some of which contained the wild-type gene for which the recipient was auxotrophic. These particles, returning to the recipient, occasionally transduced it. Even though the recipient, being lysogenic, was immune to infection, it could still be transduced, because immunity to reinfection did not prevent mechanical introduction of the DNA.[6]

Initially, all bacterial transductants isolated were found to be lysogenic, but occasionally transduced cells were found that were nonlysogenic and phage sensitive (Stocker, Zinder, and Lederberg, 1953). The Stocker work involved the transduction of the flagella (and motility) genes. When nonmotile strains were transduced with phage grown on motile strains and then plated out on soft agar, the colonies that developed were surrounded with discrete swarms of motile cells, which spread out from the dense colonial growth of the nonmotile recipient. As Stocker, Zinder, and Lederberg (1953) concluded: "The occurrence of some colonies of non-lysogenic sensitive cells, which however were motile, indicates that some of the offspring of cells which acquire exogenous hereditary material do not contain descendants of the

---

[6]I am indebted to B.D. Davis for contributing the ideas expressed in this paragraph.

phage particle which imported the genetic material." Thus, transduction and lysogenization need not proceed together.

At about this same time it was discovered that *Escherichia coli* K-12 was lysogenic for *lambda* (see Section 7.5), and genetic studies by Esther and Joshua Lederberg and by Elie Wollman showed that *lambda* behaved as a genetic element that was linked (in some ill-defined way) to the bacterial chromosome. It was clear, however, that transduction in *Salmonella* was related to prophage only indirectly:

> The phage does not have the genetic activity conferred upon it at the pro-phage stage. Were this to be so it would restrict the transduction to those characters that were related to the pro-phage and transduction seems to encompass the entire bacterial genome (Zinder, 1953).

Further:

> The role played by the phage is primarily passive. It acts as a vehicle for the genetic material. The phage may have its specific site on a bacterial chromosome, but its transducing potentialities are in no obvious way limited by its presumed position; all markers are transducible.
>
> It is quite apparent that transduction as found in *Salmonella* may be considered a transformation, in the pneumococcus sense, aided by having a vehicle for the penetration of bacteria, a bacteriophage. That is, there is incorporated into phage, during its vegetative growth, fragments of the host's genetic material which have retained their biological specificity (Zinder, 1955).

Since the Hershey/Chase experiment had shown that the DNA of the phage was injected into the cell in the initiation of the infection cycle, it was reasonable for Zinder to conclude that phage-mediated transduction involved the incorporation of bacterial genes within a phage protein coat (see also Hartman, 1957).

However, the soon-to-be-discovered specialized transduction using bacteriophage *lambda* in *Escherichia coli* would make possible a more detailed connection between transduction and lysogeny (see Sections 7.5 and 8.6).

8.3 ABORTIVE TRANSDUCTION

The work on the transduction of motility cited above led to the discovery of an additional phenomenon that was to be important in understanding the mechanism of transduction: *abortive transduction*. In addition to swarms in which all the cells were motile, groups of microcolonies sometimes developed deep in the agar at a distance from the site of inoculation. Subculture from these microcolonies yielded only *nonmotile* forms, similar to the parent, yet because of the distance from the site of inoculation, the cell from which each microcolony developed must have reached the site of the colony by active motility. The phenomenon was first described by Stocker, Zinder, and Lederberg (1953). In explaining the phenomenon, they suggested that these trails resulted when a flagella gene was carried into a nonmotile cell and was expressed but did not replace the homologous gene in the recipient. If it were expressed but not replicated, the recipient would be motile, but at cell division, there would be only one cell containing the foreign gene, and its path through the agar would be marked by a trail of microcolonies growing from the nonmotile daughter cells produced each time it divided.

The idea that abortive transduction was due to the incorporation of the DNA in the absence of integration was later shown to be correct. The transduced fragment remains functionally active but is not replicated. The process is a good example of linear inheritance (Lederberg, 1956c).

It is now known that abortive transduction occurs at a higher rate than generalized transduction. The DNA involved circularizes in the recipient, the ends of the fragment apparently held together by a special protein that is injected along with the DNA (Masters, 1985).

Abortive transduction was subsequently found to also occur for nutritional genes by Ozeki (1956). Because two alleles are present together in the abortively transduced cell, this phenomenon can be used in *cis-trans* (complementation) tests, and was used by the Demerec group in the gene function studies described in Section 8.5. For nutritional markers, abortive transduction is recognized by the presence of minute colonies on minimal agar plates. Such colonies presumably arise because the single abortively transduced cell is able to produce enzyme and grow, but its progeny, which do not receive the unreplicated genetic marker, are not able to produce enzyme. These cells divide a few times from the pool of enzyme they receive from this transduced parent, but this pool of enzyme is gradually diluted out.

8.4 BACTERIOPHAGE P1

Although generalized transduction was first discovered in *Salmonella*, the process was soon extended to *Escherichia coli* using the bacteriophage P1. Bacteriophage P1 was first isolated by Bertani (1951) from the Lisbonne-Carrère strain of *Escherichia coli*. As discussed in Chapter 7, Lisbonne and Carrère had found that *Escherichia coli* produced a phage active against a *Shigella* strain. In addition to *Shigella*, Bertani found that the *Escherichia coli* phage formed plaques on *Escherichia coli* strains B, C, and W. Extending Bertani's work, Lennox (1955) found that bacteriophage P1 was able to transduce a wide range of characters into *Escherichia coli* K-12 (and other *Escherichia coli* strains) as well as into *Shigella*. Since the overall genetic map was known in *Escherichia coli* from mating experiments, it was possible to relate genes defined by mating with genes defined by transduction. Lennox was able to show that closely linked characters could be cotransduced in the same phage particle. P1 was also able to transduce the *lambda* prophage (see Section 7.5 for a discussion of Jacob's work on transduction of prophage).

Bacteriophage P1 quickly became the standard transducing phage for *Escherichia coli* and has been widely used in genetic mapping studies (Masters, 1985). It is interesting, therefore, that in contrast to P22 and *lambda*, phage P1 does *not* become integrated into the genome as a prophage. Instead, its DNA perpetuates itself in the cell as an independent plasmid, although this fact was not known until later (Ikeda and Tomizawa, 1968). (How would the work on the location of the prophage, discussed in Section 7.6, have developed if phage P1 had been used in the first studies?)

8.5 FINE-STRUCTURE GENETIC MAPPING USING TRANSDUCTION

Classical genetic analysis is based on breeding studies, and bacterial and phage genetics followed the same route. It was soon realized that the *resolu-*

tion of genetic analysis using microbes is vastly superior to that of higher organisms because enormous populations can be analyzed (see Section 6.10 and Pontecorvo, 1958). The essential process on which genetic analysis is based is recombination. Recombination can be defined as any process in which individuals arise that contain two or more hereditary determinants in which their ancestors differed. Recombination is recognized by observing the offspring that show a new association of properties, but recombination of properties is only the detectable secondary effect of the reassociation of genetic elements. In eucaryotic genetics, recombination occurs during the crossing-over process during meiosis. Although meiosis in bacteria does not occur, recombination processes do occur, and they can be used to analyze linkage relationships. Eventually, through the work of Benzer and others (see Section 6.10), the recombination processes in bacteria and phage would be defined even at the molecular level.

At the time that genetic analysis in *Salmonella* was being carried out, fine-structure analysis was being done by Benzer on phage T4. One of the important distinctions to come out of Benzer's work was that recombination could take place not only *between* two genes, but also *within* single genes. With the low resolution of classical genetics, intragenic recombination was almost impossible to detect (although it had been seen in a few situations in *Drosophila*), but in microbes such recombination was relatively easy to measure. Studies in the fungi *Neurospora* and *Aspergillus* and in yeast provided some of the early evidence, but it was the work on bacteria and phage that pushed genetic resolution to the limit (Pontecorvo, 1958). A whole new level of sophistication in genetic analysis could be obtained by studies on bacteria and their viruses. Such analysis would lead, eventually, to a merging of genetic analysis with molecular analysis, culminating in the demonstration of the colinearity of gene and protein (see Section 10.14).

Although the molecular details of transduction would not be understood for a number of years, the use of transduction as a *tool* for genetic analysis was quickly taken up, primarily by a group at Cold Spring Harbor under Demerec (see Section 2.9). Demerec's interest in genetic fine structure went back quite a few years. For instance, at the important Missouri symposium of 1945, the following exchange occurred between Demerec and Beadle:

Demerec: Now there are several instances where genes of like action are located in the same chromosome, and we have a number of cases where genes of very similar action are located very close together. I wonder if Dr. Beadle has any indication of instances where genes closely similar are located close together.

Beadle: There are a number of cases [in *Neurospora*] where they seem to be associated in ways that you might not expect. For example, there are two different albino strains that appear to be genetically different, yet they are so close together they practically never cross over...You cannot take the chromosome map as a random sample (Tatum and Beadle, 1945).

Demerec seized on *Salmonella* transduction as a way to analyze the fine structure of the gene and allelic relationships between phenotypically identical mutants. (A moving account of Demerec's switch to bacterial genetics and *Salmonella* transduction can be found in Hartman, 1988.) Benzer's work on the *r*II locus of bacteriophage T4 had shown that recombination occurred between mutants within the same gene. Complementation or functionality tests (*cis-trans* tests) had been used to group mutants into what Benzer called *cistrons*, which were thought to be equivalent to the regions of the DNA controlling the synthesis of single polypeptides. Although Benzer's work elicited great interest, it had one major deficiency: No one knew what the *r*II protein was. It was therefore of considerable interest to extend this type of fine-structure genetic analysis to a bacterial cell, where genes could be related to specific protein molecules (enzymes). Another interest was in determining the molecular mechanism of crossing-over. Demerec recognized that *Salmonella* transduction could be used for such fine-structure genetic mapping and soon had a large group working on this project.

The work of the Demerec group made use of Zinder's strains and phages, principally P22. Two features made *Salmonella* transduction a valuable tool for genetic analysis: (1) Since only a restricted region of the genome was transduced in any single phage particle, recombination between *closely linked* genes (often phenotypic alleles) and sites *within* alleles could be studied. (2) Abortive transduction could be used as a technique for determining allelic (functional) relationships by complementation, since the introduced genetic fragment does not recombine with the recipient gene and hence is held in the *trans* position.

Terminology of Bacterial Genetics

Ever since the early work of Tatum and Lederberg, mutants had been given numerical designations that provided no clue as to their genotypes. The *Salmonella* mapping resulted in the isolation of a large number of mutants and Demerec recognized that a more systematic terminology for designation of mutants had to be devised. As a result, a uniform nomenclature was developed that was to become the standard for all future bacterial genetics. The system devised provided a unique designation for each strain and an indication from the nomenclature itself of the nature of the mutation, its genetic locus, and mutation site. The distinction between *genotype* and *phenotype* was also taken into account. An interesting feature of the nomenclature is that it avoided the use of punctuation, subscripts and superscripts, and Greek letters. (This would be especially valuable when computerization was introduced.) Years later, a formal publication describing the nomenclature appeared (Demerec, Adelberg, Clark, and Hartman, 1966).[7]

Each locus of a given wild-type strain is designated by a three-letter, lowercase italicized symbol. Thus, a tryptophan-requiring mutant would be designated *trp*. Different loci that produce the same gross phenotypic change are distinguished from each other by adding an italicized capital letter immediately following. Thus, *araA*, *araB*, and *araD* affect three different enzymes in the arabinose pathway. Within any one locus, a number of mutants can be isolated independently, each affecting a different site. A mutation site is designated by placing a serial isolation number after the locus symbol. Thus, *ara*-1, *ara*-2, *ara*-3. These numerical designations are given at the time of isolation, even before the enzyme affected by the

---

[7]According to a personal communication from E.A. Adelberg, he and A.J. Clark first put together a publication describing the nomenclature, but realized that their paper was actually just a refinement of that Demerec had developed much earlier. With his permission, Demerec was made senior author of the paper, which was published the year Demerec died.

mutation is known. When the enzyme is identified, the letter designation is added: for example, *araB*-1, *araA*-2, *araC*-3, *araB*-4. The numerical designation does not change, remaining a unique identifier for the mutation.

This terminology proved so useful that it was quickly adopted by the whole bacterial genetics community (Bachmann, 1987). Further enhancements were added for designation of plasmids, episomes, etc. In the *Salmonella* work, a large collection of mutants was quickly isolated, for a variety of nutritional requirements. Certain requirements were found frequently, and those were analyzed in the fine-structure mapping. They included the tryptophan, histidine, and purine loci.

The Basic Transduction Experiment

The basic procedure for genetic analysis in *Salmonella* can be outlined briefly (for a good summary, see Hartman, 1963). The nonlysogenic donor strain is infected with phage P22, which then undergoes lytic multiplication. The lysate that results is then harvested and sterilized to remove donor bacteria (since the phage is resistant to chloroform, this reagent can be used to destroy remaining bacteria). A culture of the recipient is then mixed with phage at a multiplicity of about 5 to 10 and incubated for a few minutes to allow phage adsorption, after which the infected culture is diluted and spread on plates of an appropriate selective agar medium. Because there are no donor bacteria remaining, the counterselection used in mating studies (see Chapter 5) need not be done. The relative efficiency of transduction will depend on several factors: the efficiency of the infection process, the fraction of bacteria that survive the infection, and the linkage relationships of donor and recipient. In this type of transduction, the efficiency is always low, at best around 10^{-5} for any given character, but because of the large population densities that can be plated, and the powerful selection methods, recombinants can be readily obtained, even between closely linked sites. In Demerec's work, about 2×10^7 bacteria were spread on a single plate, and one hundred or so recombinant colonies were obtained. In intragenic recombination, the number of recombinants could be less than 100, but still significantly above any background revertants (Table 8.1).

Initially, large numbers of phenotypically similar mutants were isolated, primarily auxotrophs of amino acids, purines, and pyrimidines. Care was taken to be certain that only mutants of independent origin were obtained. To obtain independent mutants, the culture (generally wild type) was mutagenized and subject to penicillin selection. The mutagenized culture was then diluted into a number of tubes, so that each culture grew up from a small inoculum (this approach is based on the fluctuation test of Luria and Delbrück). The resulting cultures were plated on complete medium and then replica-plated onto minimal medium. Those colonies not growing on minimal medium were then selected and characterized.

In addition to the genetic analysis, the biochemical pathways for tryptophan, histidine, and purines were also studied. The genetics was then related to the enzymes of these pathways. At this time, following up on the *Neurospora* work, a large amount of research on intermediary metabolism was being done in *Escherichia coli* and *Salmonella*, so that the genetic data could be readily correlated with the biochemical data.

Table 8.1   Typical Transduction Experiment with Various *Tryptophan* Mutants

Recipient	Donor										
	try-1	-6	-7	-9	-10	-11	-3	-2	-4	-8	+
try-1.......	0	66	203	104	219	208	291	706	458	418	1264
-6........	141	0	11	60	21	182	188	179	234	100	1617
-7........	21	2	0	10	19	22	444	537	435	107	717
-9........	26	8	41	0	101	66	310	361	247	437	1456
-10.....	4	2	7	12	0	0	270	628	602	206	1822
-11.....	22	1	23	22	0	0	280	240	315	497	1406
-3.......	166	50	30	75	88	107	0	139	111	123	336
-2........	542	375	126	320	295	440	344	0	18	66	3074
-4........	173	120	44	213	145	235	163	20	0	85	2257
-8........	144	123	138	560	345	111	133	125	44	0	3264

About 2×10^7 infected bacteria were used per plate of minimal medium. Each figure represents the number of colonies on 4 plates after 2 days incubation. Multiplicity of infection, 5; absorption, 5–15 min at 37°C. The results show that the mutants can be divided into four groups. (A) try-8; (B) try-2, -4; (C) try-3; (D) try-1, -6, -7, -9, -10, -11. The biochemical blocks of the four groups are shown in Fig. 8.1. (Reprinted, with permission, from Demerec and Hartman, 1956.)

After two-point crosses were done to approximate the genetic relationships (see Table 8.1), three-point crosses were done to determine the *order* of the genes. For a transducing fragment to lead to a viable recombinant, double recombinational events must occur, one on each side of the gene being scored. In the three-point crosses, certain recombinant classes were more common than others, and assuming that double crossover events would be more common than quadruples, the gene order could be determined. The most surprising finding from these studies was that the genes controlling the various enzymes of a single pathway were closely linked, and that their location on the genetic map coincided with the sequence of biochemical steps in the chain of biosynthetic reaction. Thus, for the tryptophan mutants, the order of four genes was obtained: *tryD—tryC—tryB—tryA*, and the enzymes for the tryptophan pathway operated in the order: $A \rightarrow B \rightarrow C \rightarrow D$ (anthranilic acid $\rightarrow$ compound "B" $\rightarrow$ indole $\rightarrow$ tryptophan) (Fig. 8.1) (Demerec and Hartman, 1956). A similar "assembly line" relationship was found for the genes in the histidine pathway (Fig. 8.2) (Hartman, 1956, 1957). As noted by Demerec and Hartman (1956), the same kind of assembly lines had not been found in *Neurospora*. We now know that the linkage relationships reflect the fact that the genes of each of these pathways are linked in a single regulatory unit, an *operon*. Although the reason for the linkage relationship was not known at the time, its discovery played an important role in the development of the operon concept (see Section 10.10).

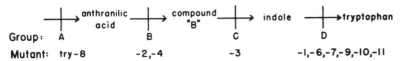

Figure 8.1 Diagram showing the sequence of biochemical steps of tryptophan biosynthesis. The positions of the blocks associated with the different groups of mutants are shown. (Reprinted, with permission, from Demerec and Hartman, 1956.)

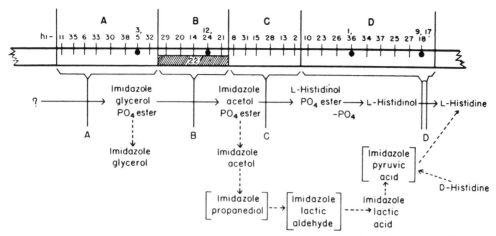

Figure 8.2 Early linkage map of the histidine region of *Salmonella* and the probable relation of genetic blocks to the enzymes. (Reprinted, with permission, from Hartman, 1956.)

Another important discovery from the fine-structure mapping was that mutants of independent origin within the same gene locus were usually at different sites within the locus, a finding analogous to that of Benzer for the *r*II region of phage T4 (Demerec and Hartman, 1959). This is shown by the designations of the individual mutants in the map of the histidine region in Figure 8.2. This finding was made possible by the very high "resolving power" of transductional analysis. Eventually, this work, in the hands of Charles Yanofsky and others, would lead to a clear demonstration that the genetic code and the amino acid sequence were colinear (see Section 10.14).

Because each transducing phage carried such a small part of the bacterial genome, transduction was not useful in determining the overall order of genes on the whole genetic map, for which conjugation was the preferred technique. Although conjugation was eventually adapted to *Salmonella* (see Section 5.13), the use of both techniques in *Escherichia coli* made possible both coarse and fine-structure mapping in this organism.

Mutagenicity Testing

A sideline from the mutation and mapping work was the development of *Salmonella* as a system for testing the activity of chemical mutagens. Such testing developed a broader significance when it was concluded that mutagenicity and carcinogenicity were generally correlated. This field is most closely identified with the work of Bruce Ames (1989), a biochemist who worked closely with Hartman and Demerec in the early days of the development of the *Salmonella* system. The early history of the development of the *Salmonella* mutagen tester strains is given by Hartman (1989).

8.6 TRANSDUCTION IN THE *LAMBDA* SYSTEM

While Zinder was pushing forward the work on *Salmonella* transduction, Lederberg's laboratory was continuing work on all aspects of genetics in

Escherichia coli K-12. As discussed in Section 7.5, phage *lambda* was discovered in this strain by Esther Lederberg in 1951, and its linkage to the *gal* locus was shown by mating (Lederberg and Lederberg, 1953; Wollman, 1953). M.L. Morse, a student of Lederberg, attempted to determine if *lambda* was capable of transduction. Of numerous genetic markers tested, only a cluster of genes for galactose fermentation were transduced by *lambda* lysates (Morse, Lederberg, and Lederberg, 1956a). Also, transducing lysates could only be obtained by *induction* of lysogenic strains; lytic infection by free *lambda* phage particles did not result in transducing lysates. Since *lambda* prophage was linked to *gal* (see Section 7.5), the restriction of transduction to the *gal* region suggested a different and more special situation in *lambda* transduction than that for P22 in *Salmonella*. Because *lambda* transduction was restricted to only a few genes, it came to be called *specialized transduction*.

The gal^+ clones obtained from *lambda* transduction were found to be heterozygous for the *gal* region, segregating gal^- cells about once per 10^3 bacterial divisions. Since these transductants were heterozygous for only a restricted set of genes (the *gal* region), they were called *heterogenotes* and designated gal^-/gal^+. Such heterogenotes could be used in allelism (complementation or *cis-trans*) tests for the various *gal* genes (Morse, Lederberg, and Lederberg, 1956b). In this way, four distinct *gal* genes were recognized, corresponding to four enzymes of the galactose fermentation pathway.

Of greatest interest was the fact that a lysate produced by induction of such a heterogenotic culture contained *lambda* particles that were able to transduce the *gal* genes at very high frequency, whereas a lysate obtained by infection transduced *gal* at a low frequency. The former situation was designated HFT, for *high frequency of transduction* and the latter LFT, for *low frequency of transduction*. The analogy between Hfr and HFT was, of course, obvious, although the mechanisms behind the two phenomena were quite different (see below). The HFT lysates were always characterized by markedly lower numbers of infectious *lambda* particles (5×10^8 instead of 1×10^{10}), but virtually every particle in such lysates was capable of transduction.'

The Defectiveness of Transducing Lambda

In the initial interpretation of the HFT phenomenon, it was imagined that upon induction of the *lambda-gal* prophage, prior to leaving its prophage site, the neighboring *gal* genes were excised and packaged in the mature virus particle (Morse, Lederberg, and Lederberg, 1956b). Such a hybrid virus, infecting a gal^- cell, would attach near the usual chromosomal site and produce the gal^+/gal^- heterogenote. Upon induction of such a heterogenote, the hybrid prophage would give rise to a burst of *lambda* phage all carrying the *gal* genes.

This interpretation was in part incorrect. The transducing particles are actually *defective* and are not able to produce active phage. This was shown by Arber, Kellenberger, and Weigle (1957) and Campbell (1957), following up on the work of Morse and the Lederbergs. These workers found that the consequences of infection with an HFT lysate depended on the multiplicity of infection. At high multiplicities of infection, gal^+ transductants contained the *lambda* prophage, but at low multiplicities (10^{-3}), almost all the transduced bacteria were *defective lysogens*. These defective heterogenotes were immune to *lambda* and were inducible, but upon lysis they never liberated any active phage, and their lysates were incapable of transduction. "This led us to think that HFT lysates contained two kinds of phages: normal λ phages not carrying Gal, active and capable of forming plaques on sensitive indicator strains; and defective phages, incapable of forming plaques, carrying Gal and transducing this character into the bacteria which they lysogenize" (translated from Arber, Kellenberger, and Weigle, 1957).

Arber, Kellenberger, and Weigle (1957) concluded that heterogenotes were doubly lysogenic, for normal *lambda* and for defective *lambda*, and that the defectiveness was due to the inclusion of the *gal* locus in the prophage. When such a heterogenote was induced, two kinds of phage particles would be produced, in approximately equal numbers, and only those that were defective could carry *gal*. Once a gal^+ was obtained from such a defective particle, it would no longer be able to produce active phage, so that transducing particles could not be obtained from the defective genome alone. The active phage would have to supply the function lacking in the defective phage. These defective phage, which were called λ*dg*, could only be produced when a "helper" phage was also present (Campbell, 1958). Genetic analysis of λ*dg* showed that these particles arose by genetic exchange of the *gal* region for about one-third to one-fourth of the whole phage genome and that gal^+/gal^- heterogenotes were really polylysogenic, carrying both defective λ*dg* and nondefective λ prophages.

An important linkage between genetics and chemistry was obtained by a study of λ*dg*. The technique of cesium chloride buoyant density centrifugation permitted a *physical* separation of λ*dg* from normal λ (Fig. 8.3) (Weigle,

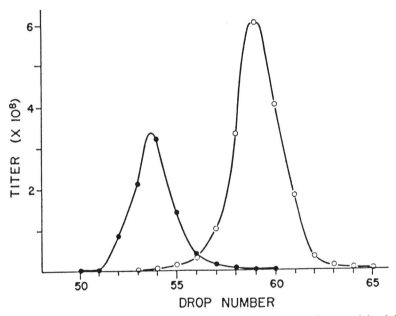

Figure 8.3 Separation of defective transducing phage and normal *lambda* by cesium chloride density-gradient centrifugation. An HFT lysate was centrifuged and droplets were collected and assayed for plaque-forming units (*lambda*) and transducing phages (*gal*). Increasing tube number represents decreasing density. (●) Transducing phages; (○) plaque-forming units. (Reprinted, with permission, from Weigle, Meselson, and Paigen, 1959.)

Figure 8.4 Densities of ten independent λ*dg*s. The numbers indicate the specific λ*dg*
preparations. The densities given below the numbers should be multiplied by 10^2.
(Reprinted, with permission, from Weigle, Meselson, and Paigen, 1959.)

Meselson, and Paigen, 1959). If a number of independent λ*dg* isolates were
studied, some were found to be "lighter" than λ, others "heavier" (Fig. 8.4).
Presumably, this density difference reflected primarily the DNA content of
the phage particles. Thus, the segment of the bacterial genome that was
incorporated into the λ*dg* was variable in length, the differences due to
recombinational events arising at the time when each λ*dg* first developed.

The transducing phage could thus be considered as a genetic recombinant
between the bacterial chromosome and *lambda*, but not a conventional recombinant. Conventional recombination was *reciprocal*, with each of the two
mating partners exchanging equal amounts of genetic material. In the case of
λ*dg*, nonreciprocal recombination occurred, by so-called *illegitimate recombination* between points within the prophage and points on the bacterial
chromosome (Franklin, 1971). Since heterologous breaks were involved,
there was no predictable relationship between the sites undergoing recombination, hence the wide variety of λ*dg* formed.

8.7 DEFECTIVE PARTICLES IN GENERAL TRANSDUCTION

The discovery that transducing particles in *lambda* were defective was quickly followed by the discovery of defective particles in generalized transduction. This work was done with P1, which, as we have seen, is a generalized
transducing phage for *Escherichia coli*. By controlling multiplicity of infection
and by isolating transductants and determining their lysogenicity, Adams
and Luria (1958) were able to show that transducing particles did not carry
the phage genome. Moreover, cells infected with such particles did not yield
HFT lysates. The origin of transducing particles in P1 is as follows: As a result
of induction, bacterial genes become packaged within phage coats, and these
particles are able to transduce but, of course, not lysogenize. Transducing
particles and infective phage could be separated by density-gradient centrifugation (Ikeda and Tomizawa, 1965). Because of the limits on the amount
of DNA that could be packaged into a phage head, only a restricted set of
genes could be transferred by a single phage particle, but any genetic region
of the chromosome could be packaged. Thus, the origin of transducing
particles and the mechanism of generalized transduction are completely
different from those of specialized transduction.

Significance for Recombinant DNA Technology

Although generalized transduction has been used primarily as a tool for
fine-structure mapping, it should be noted that an understanding of its

mechanism and that of specialized transduction has been of considerable importance in the development of the recombinant DNA technology (see Chapter 11). Phage *lambda* in particular has found wide use as a vector for transferring recombinant DNA molecules. *In vitro* packaging systems that make use of phage proteins and recombinant DNA permit the construction of phage particles containing virtually any piece of DNA desired. The knowledge of how the transduction process worked has made the development of such phage vectors possible.[8]

8.8 MEROMIXIS

Recognizing the fragmentary nature of the DNA transferred in the three processes of transduction, conjugation, and transformation, Jacob and Wollman (1961) concluded that *all* recombination mechanisms in bacteria led to the formation of incomplete zygotes, made up "essentially of a recipient cell containing only a segment of genetic material from the donor cell. For this reason, it has seemed useful to unify under the general term of *meromixis* (from *mero*: part and *mixis*: mixture) those processes of partial genetic transfer, or partial fecundation, which result in the formation of incomplete zygotes or *merozygotes*."

As defined, *meromixis* had almost the same meaning as Lederberg's original definition of *transduction* (Lederberg, 1956b). In the years between 1952, when the term "transduction" was coined, and 1961, when Jacob and Wollman defined "meromixis," it had been recognized that mating (conjugation) in bacteria was also a fragmentary process. In the meantime, the term transduction had acquired the more specific meaning of *phage-mediated genetic recombination*. Therefore, a term for all fragmentary processes seemed like a good idea. During the 1960s, the terms meromixis and merozygote were commonly used in bacterial genetics and molecular genetics textbooks, but by the 1970s and 1980s they had been essentially abandoned. There is a Darwinian element to scientific terms: If they do not prove useful, they die out!

8.9 LYSOGENIC OR PHAGE CONVERSION

By *lysogenic conversion*, sometimes called *phage conversion*, is meant the acquisition of new traits (other than those of the prophage itself) by the bacterial cell as a consequence of lysogenization or phage infection. Lysogenic conversion differs from either specialized or generalized transduction in that the new traits are apparently part of the normal phage genome rather than being donor cell traits that have become incorporated into defective phage particles.

---

[8]According to a personal communication from Norton Zinder, when he first reported transduction at the Cold Spring Harbor symposium of 1953, Leo Szilard suggested that he should patent the phenomenon. According to Zinder, Szilard recognized immediately the practical significance of a method for moving genes around on viruses.

Conversion in Corynebacterium diphtheriae

The first and still one of the best-known cases of conversion is that involving toxin production by the pathogenic bacterium *Corynebacterium diphtheriae*. It is of historical interest that this conversion process was actually discovered and published *before* that of Zinder and Lederberg on *Salmonella* transduction, although a genetic interpretation was not made.

Conversion was discovered by Freeman (1951) while he was using temperate phage to type virulent and nonvirulent *Corynebacterium diphtheriae*. Freeman was an epidemiologist at the University of Washington (Seattle) who was using bacteriophages to type various natural isolates of *Corynebacterium diphtheriae*.[9] Virulence in this organism is due to the production of a highly specific toxin, the diphtheria toxin. Although most of Freeman's work was with virulent strains, he found several phages that were active against *avirulent* strains of *Corynebacterium diphtheriae*. Freeman decided to determine whether the phage-induced lysis of an avirulent strain might cause the production of a nonspecific *dermal necrotic endotoxin*, and tested phage lysates of avirulent strains for toxin activity. He did indeed find evidence of toxin production, but it was diphtheria toxin itself rather than a dermal endotoxin. There were two phages active against the avirulent strains, called A and B. Only phage B converted *Corynebacterium diphtheriae* from the nontoxigenic to the toxigenic condition. Natural isolates of *Corynebacterium diphtheriae* that were already virulent were *resistant* to infection with phage B and could be shown to be already lysogenic for B. "Apparently, exposure of the phage-susceptible avirulent cultures to phage B resulted in the production of virulent lysogenic strains of *C. diphtheriae*" (Freeman, 1951). Freeman also was unable to obtain another extracellular product from the virulent culture, other than phage, that could cause the change in avirulent cultures. "These results would tend to rule against any extracellular product of virulent organisms as the agent involved in rendering avirulent strains virulent and suggest that the phage per se or phage by-products were implicated" (Freeman, 1951).

Careful purification tests of the cultures showed that the avirulent strains were not contaminated with small numbers of virulent cells. Freeman's initial hypothesis was that virulent mutants were arising spontaneously in the avirulent cultures, and the phage, lysing the avirulent cells, was selecting out virulents. However, in a second paper (Freeman and Morse, 1952), single-cell isolations with a micromanipulator were used to rule out the presence of rare spontaneous virulent mutants in the avirulent cultures. Therefore: "It would seem logical that the simultaneous acquisition of lysogenicity and virulence in the same bacterial cell is a related, rather than coincidental phenomenon. The change to toxin production might well be interpreted as being due directly to the acquired lysogenicity" (Freeman and Morse, 1952).

It should be noted again that the above work was done without knowledge of either Zinder and Lederberg's paper on transduction (which had not yet been published) or (apparently) Lwoff's work on lysogeny (which had been only published in French; the English-language paper on lysogeny was not

[9]I am grateful to Joshua Lederberg for providing me with information on Freeman's background.

published until 1953). Freeman's phenomenon elicited considerable interest in the medical bacteriology community because of its obvious practical importance. However, it remained a curiosity to bacterial geneticists, since the mechanism was not known.

Although Freeman had attempted to rule out phage selection of toxigenic mutants, the evidence was indirect. Neal Groman, using Freeman's cultures, performed the detailed population analysis needed to establish that toxigenicity and lysogenicity were acquired *simultaneously* upon phage infection (Groman, 1953). By following the kinetics of toxin production and lysogenicity after infection, Groman showed that within one to two hours after infection, lysogenic and toxigenic cells appeared and that a relatively high proportion of the cells was converted simultaneously to the two characteristics. The term "conversion" to describe this phenomenon was first used casually in Groman's paper and then formalized by Barksdale and Pappenheimer (1954). Groman noted the possible relevance of this lysogenic conversion to the transduction studies that had been recently published by Zinder and Lederberg. Although the filterable agent (FA) of Zinder and Lederberg was at that time thought probably to be a phage, it was clear that there was a distinct difference between transduction and conversion. The relative numbers of altered cells was much higher in the *Corynebacterium diphtheriae* case than in the *Salmonella* case, making it unlikely that the two phenomena had similar bases.

Within the year, the fundamental work of Lwoff (1953) on the prophage would become widely known, and the implications for lysogenic conversion became obvious. As Lwoff discussed (see Section 7.4), cells containing prophage differed from nonlysogenic cells by *immunity* to infection with the corresponding phage. Although the mechanism of immunity was not at this time known, it was known that immune cells were able to adsorb phage, the infecting phage persisting for a short while and then disappearing. Although the immunity was specific for the prophage (or closely related phages), it was obviously a phenotypic change brought about by the phage itself. In this respect, it resembled conversion. Furthermore, in *Bacillus megaterium*, Ionesco (1953) had shown that normally smooth colonies became rough once they were lysogenized with a certain phage. In another conversion, which was to play a major role in phage genetics, *Escherichia coli* K-12 cells lysogenized by phage *lambda* acquired resistance to *r*II mutants of phage T4 (see Section 6.10).

Following up on the work of Freeman and Groman, Lane Barksdale determined that the phage preparations used had consisted of two phages, a virulent phage that he called B and a temperate phage that he called β (Barksdale and Pappenheimer, 1954). Infection with phage B resulted in a transient system that produced toxin until the culture lysed, whereas infection with phage β resulted in a stable lysogenic system that continued to produce toxin. Toxigenicity thus became the prototype of the phenomenon of lysogenic conversion (Barksdale, 1959).

The final link between prophage and toxigenicity was forged when Groman (1955) showed that elimination of prophage β from toxigenic *Corynebacterium diphtheriae* resulted in loss of ability to produce toxin. Moreover, mixed infection between phage capable of causing conversion and related nonconverting phages resulted in the formation of recombinant particles that were no longer able to confer toxigenicity (Groman and Eaton, 1955). However, the host was not passive in the process, since toxin production depended on physiological and nutritional factors as well.

Conversion in Salmonella

Independently of the *Corynebacterium diphtheriae* work, a group of Japanese workers was studying changes in antigenicity of *Salmonella* strains as a result of lysogenization. In the basic observation, cells of *Salmonella anatum* containing antigen 10 were converted to antigen 15 upon infection with the temperate phage ϵ^{15} or its virulent mutants. Uetake, Luria, and Burrous (1958) showed that the antigenic conversion, which was actually due to a change in properties of the bacterial surface, was a rapid process, occurring within a few minutes after infection. The conversion process in *Salmonella* therefore clearly had some relationship to the formation of enzymes that were being shown about this time to occur early after infection (see Cohen's work discussed in Section 6.12).

Terminology of Conversion

As noted, Groman (1953) and Barksdale and Pappenheimer (1954) had first used the term "conversion" to describe the phage-induced change in *Corynebacterium diphtheriae*. Subsequently, Lederberg (1955) used the more specific term "lysogenic conversion": "...these lysogenic conversions resemble...transduction...but the alterations are inseparable from lysogenicity, i.e., the genetic quality is specifically associated with the phage nucleus, not a desultory companion." However, Barksdale (1959) and Uetake, Luria, and Burrous (1958) showed that conversion occurred even when virulent phage was used, so that the lysogenization process itself was not necessary. Thus, the term "lysogenic conversion" seemed no longer apt since any phage-induced change would become a conversion. At about this time, the term lysogenic conversion was replaced by the term "phage conversion" to refer to any phage-induced change that is unrelated to the phage genome itself (see discussion in Hayes, 1964). However, operationally, such a broad definition presents difficulties, not only because most temperate phages cause changes in host physiology, but also because it is almost impossible to prove that a change is "unrelated" to phage function. Therefore, in recent years the term lysogenic conversion seems to be reappearing in the microbial genetics literature (Freifelder, 1987).

Phage conversion obviously has some practical implications in medical bacteriology. The *Corynebacterium diphtheriae* case stimulated a search for phage-related toxigenicity in other bacterial pathogens. The erythrogenic toxin of hemolytic streptococci and the fibrinolysin of staphylococci were also found to be associated with lysogenization by specific phage (reviewed by Hayes, 1968). However, conversion has never seemed an especially important part of bacterial genetics research.

8.10 SIGNIFICANCE OF TRANSDUCTION

The discovery of phage-mediated genetic exchange has had far-reaching implications in modern molecular biology. The early importance of this discovery was recognized when Joshua Lederberg shared (with Beadle and Tatum) the Nobel Prize in 1958 for his work on bacterial genetics. The concept of transduction markedly broadened ideas about the nature of viruses, ideas that have become applied not only to procaryotes, but also to eucaryotes, with implications for cancer. Vast areas of viral pathogenesis and immunology also have developed from the knowledge of the transduction process.

In procaryotic genetics, transduction has provided an extremely important tool for genetic analysis. Fine-structure genetic mapping depends on the ability to look at closely linked genes and to study recombination within a single gene, and this became possible once transduction became available. The discovery of close linkage between genes coding for related enzymes in a biosynthetic pathway was made by transductional analysis and played an important role in the development of the operon concept (see Section 10.10).

Another important consequence of transduction is in the area of recombinant DNA research. One of the principal techniques for bringing DNA into a cell is by packaging this DNA inside the phage head; the idea that foreign DNA could be packaged arose from transduction research (see Chapter 11).

REFERENCES

Adams, J.N. and Luria, S.E. 1958. Transduction by bacteriophage P1: abnormal phage function of the transducing particles. *Proceedings of the National Academy of Sciences* 44: 590–594.

Ames, B.N. 1989. Mutagenesis and carcinogenesis: endogenous and exogenous factors. *Environmental and Molecular Mutagenesis* 14 (Supplement 16): 66–77.

Arber, W., Kellenberger, G., and Weigle, J. 1957. La défectuosité du phage lambda transducteur. *Schweizerische Zeitschrift für Allgemeine Pathologie und Bakteriologie* 20: 659–665.

Bachmann, B.J. 1987. Linkage map of *Escherichia coli* K-12 pp. 807–876 in Neidhardt, F.C. (ed.), Escherichia coli *and* Salmonella typhimurium. *Cellular and Molecular Biology*, 7th. edition. American Society for Microbiology, Washington, D.C.

Barksdale, L. 1959. Lysogenic conversions in bacteria. *Bacteriological Reviews* 23: 202–212.

Barksdale, W.L. and Pappenheimer, A.M., Jr. 1954. Phage-host relationships in toxigenic and nontoxigenic diphtheria bacilli. *Journal of Bacteriology* 67: 220–232.

Bertani, G. 1951. Studies on lysogenesis. I. The mode of phage liberation by lysogenic *Escherichia coli*. *Journal of Bacteriology* 62: 293–300.

Bisset, K.A. 1950. *The Cytology and Life-History of Bacteria*. Williams and Wilkins, Baltimore. 136 pp.

Boyd, J.S.K. 1951. Observations on the relationship of symbiotic and lytic bacteriophage. *Journal of Pathology and Bacteriology* 63: 445–457.

Burnet, F.M. and McKie, M. 1929. Observations on a permanently lysogenic strain of *B. enteritidis*. *Australian Journal of Experimental Biology and Medical Science* 6: 276–284.

Campbell, A. 1957. Transduction and segregation in *Escherichia coli* K12. *Virology* 4: 366–381.

Campbell, A. 1958. The different kinds of transducing particles in the λ-gal system. *Cold Spring Harbor Symposia on Quantitative Biology* 23: 83–84.

Davis, B.D. 1950. Nonfiltrability of the agents of recombination in *Escherichia coli*. *Journal of Bacteriology* 60: 507–508.

Demerec, M. and Hartman, P.E. 1959. Complex loci in microorganisms. *Annual Review of Microbiology* 13: 377–406.

Demerec, M. and Hartman, Z. 1956. Tryptophan mutants in *Salmonella typhimurium.* pp. 5–33 in *Genetic Studies with Bacteria.* Carnegie Institution of Washington Publication 612, Washington, D.C.

Demerec, M., Adelberg, E.A., Clark, A.J., and Hartman, P.E. 1966. A proposal for a uniform nomenclature in bacterial genetics. *Genetics* 54: 61–76.

Franklin, N.C. 1971. Illegitimate recombination. pp. 175–194 in Hershey, A.D. (ed.), *The Bacteriophage Lambda.* Cold Spring Harbor Laboratory, Cold Spring Harbor, New York.

Freeman, V.J. 1951. Studies on the virulence of bacteriophage-infected strains of *Corynebacterium diphtheriae. Journal of Bacteriology* 61: 675–688.

Freeman, V.J. and Morse, I.U. 1952. Further observations on the change to virulence of bacteriophage-infected avirulent strains of *Corynebacterium diphtheriae. Journal of Bacteriology* 63: 407–414.

Freifelder, D. 1987. *Microbial Genetics.* Jones and Bartlett, Boston. 601 pp.

Groman, N.B. 1953. Evidence for the induced nature of the change from nontoxigenicity to toxigenicity in *Corynebacterium diphtheriae* as a result of exposure to specific bacteriophage. *Journal of Bacteriology* 66: 184–191.

Groman, N.B. 1955. Evidence for the active role of bacteriophage in the conversion of nontoxigenic *Corynebacterium diphtheriae* to toxin production. *Journal of Bacteriology* 69: 9–15.

Groman, N.B. and Eaton, M. 1955. Genetic factors in *Corynebacterium diphtheriae* conversion. *Journal of Bacteriology* 70: 637–640.

Hartman, P.E. 1956. Linked loci in the control of consecutive steps in the primary pathway of histidine synthesis in *Salmonella typhimurium.* pp. 35–61 in *Genetic Studies in Bacteria.* Carnegie Institution of Washington Publication 612, Washington, D.C.

Hartman, P.E. 1957. Transduction: a comparative review. pp. 408–467 in McElroy, W.D. and Glass, B. (ed.), *The Chemical Basis of Heredity.* Johns Hopkins Press, Baltimore.

Hartman, P.E. 1963. Methodology in transduction. pp. 103–128 in Burdette, W.J. (ed.), *Methodology in Basic Genetics.* Holden-Day, San Francisco.

Hartman, P.E. 1988. Between Novembers: Demerec, Cold Spring Harbor and the gene. *Genetics* 120: 615–619.

Hartman, P.E. 1989. Early years of the *Salmonella* mutagen tester strains: lessons from hycanthone. *Environmental and Molecular Mutagenesis* 14 (Supplement 16): 39–45.

Hayes, W. 1964. *The Genetics of Bacteria and their Viruses.* John Wiley, New York. 740 pp.

Hayes, W. 1968. *The Genetics of Bacteria and their Viruses,* 2nd. edition. John Wiley, New York. 925 pp.

Ikeda, H. and Tomizawa, J. 1965. Transducing fragments in generalized transduction by phage P1. I. Molecular origin of the fragments. *Journal of Molecular Biology* 14: 85–109.

Ikeda, H. and Tomizawa, J. 1968. Prophage P1, an extrachromosomal replication unit. *Cold Spring Harbor Symposia on Quantitative Biology* 33: 791–798.

Ionesco, H. 1953. Sur une propriété de *Bacillus megatérium* liée a la présence d'un prophage. *Comptes rendus Académie Sciences* 237: 1794–1795.

Jacob, F. and Wollman, E.L. 1961. *Sexuality and the Genetics of Bacteria.* Academic Press, New York. 374 pp.

Lederberg, E.M. and Lederberg, J. 1953. Genetic studies of lysogenicity in *Escherichia coli. Genetics* 38: 51–64.

Lederberg, J. 1955. Recombination mechanisms in bacteria. *Journal of Cellular and Comparative Physiology* 45 (Supplement 2): 75–107.

Lederberg, J. 1956a. Bacterial protoplasts induced by penicillin. *Proceedings of the National Academy of Sciences* 42: 574–577.

Lederberg, J. 1956b. Genetic transduction. *American Scientist* 44: 264–280.

Lederberg, J. 1956c. Linear inheritance in transductional clones. *Genetics* 41: 845–871.

Lederberg, J. and St Clair, J. 1958. Protoplasts and L-type growth of *Escherichia coli. Journal of Bacteriology* 75: 143–160.

Lederberg, J. and Zinder, N.D. 1948. Concentration of biochemical mutants of bacteria with penicillin. *Journal of the American Chemical Society* 70: 4267.

Lederberg, J., Lederberg, E.M., Zinder, N.D., and Lively, E.R. 1951. Recombination analysis of bacterial heredity. *Cold Spring Harbor Symposia on Quantitative Biology* 16: 413–443.

Lennox, E.S. 1955. Transduction of linked genetic characters of the host by bacteriophage P1. *Virology* 1: 190–206.

Lwoff, A. 1953. Lysogeny. *Bacteriological Reviews* 17: 269–337.

Masters, M. 1985. Generalized transduction. pp. 197–215 in Scaife, J., Leach, D., and Galizzi, A. (ed.), *Prokaryotic Molecular Biology*. Academic Press, New York.

Morse, M.L., Lederberg, E.M., and Lederberg, J. 1956a. Transduction in *Escherichia coli* K-12. *Genetics* 41: 142–156.

Morse, M.L., Lederberg, E.M., and Lederberg, J. 1956b. Transductional heterogenotes in *Escherichia coli. Genetics* 41: 758–779.

Ozeki, H. 1956. Abortive transduction in purine-requiring mutants of *Salmonella typhimurium*. pp. 97–106 in *Genetic Studies with Bacteria*. Carnegie Institution of Washington Publication 612, Washington, D.C.

Pontecorvo, G. 1958. *Trends in Genetic Analysis*. Columbia University Press, New York. 145 pp.

Stocker, B.A.D., Zinder, N.D., and Lederberg, J. 1953. Transduction of flagellar characters in *Salmonella. Journal of General Microbiology* 9: 410–433.

Tatum, E.L. and Beadle, G.W. 1945. Biochemical genetics of *Neurospora. Annals of the Missouri Botanical Garden* 32: 125–129 and discussion, pp. 252–253.

Uetake, H., Luria, S.E., and Burrous, J.W. 1958. Conversion of somatic antigens in *Salmonella* by phage infection leading to lysis or lysogeny. *Virology* 5: 68–91.

Weigle, J., Meselson, M., and Paigen, K. 1959. Density alterations associated with transducing ability in the bacteriophage λ. *Journal of Molecular Biology* 1: 379–386.

Wollman, E.L. 1953. Sur le déterminisme génétique de la lysogénie. *Annales de l'Institut Pasteur* 87: 281–294.

Zinder, N.D. 1953. Infective heredity in bacteria. *Cold Spring Harbor Symposia on Quantitative Biology* 18: 261–269.

Zinder, N.D. 1955. Bacterial transduction. *Journal of Cellular and Comparative Physiology* 45 (Supplement 2): 23–49.

Zinder, N.D. and Lederberg, J. 1952. Genetic exchange in *Salmonella. Journal of Bacteriology* 64: 679–699.

9
TRANSFORMATION

Bacterial transformation, the process by which genetic material is transferred from one cell to another via free DNA, would seem one of the most unlikely processes of gene transfer in living organisms. Discovered at a time when bacterial genetics itself was still virtually a complete mystery, transformation would not even be considered as a genetic process for many years. A priori, it should have seemed highly improbable that such large macromolecules as DNA could survive intact outside the cell, let alone actually be taken up and integrated into the genomes of recipient cells. Indeed, if transformation were first discovered today, even with the sophisticated knowledge we have of molecular genetics, it would probably be considered a cause for wonder.

In the bacterial transformation process, free DNA becomes transferred from one cell to another and brings about genetic change in the recipient cell. The free DNA can arise spontaneously, as a result of lysis of the donor cells, or it can be prepared experimentally. In either case, the DNA, which must be double-stranded, binds specifically to the recipient cells and is taken up. For the recipient cell to be transformed, it must be capable of binding and incorporating DNA, a state called *competent*. After entry, one strand of the transforming DNA pairs at the homologous region in the recipient and becomes integrated into the genome of the recipient via a genetic exchange process involving breakage and reunion. Thus, unlike the *addition* of DNA that occurs in the integration of a phage or plasmid, the incoming DNA in transformation *replaces* the homologous segment in the recipient. The newly incorporated DNA, now part of a double-stranded molecule in the recipient, is replicated during division of the transformed cell. Eventually, a clone of genetically different cells can develop. If the newly introduced DNA differs genetically from that of the recipient, and the introduced gene is expressed, then the resulting clone will be phenotypically different. If the new phenotype has a selective advantage, it may replace the parental phenotype.

For many years, the only bacterial species known to exhibit the transformation process was *Streptococcus pneumoniae*, a human pathogen with complex nutritional requirements. After the demonstration in 1944 that the active agent in transformation was DNA (Avery, MacLeod, and McCarty, 1944), the process was sought in other bacteria. An early report of the process in *Escherichia coli* could not be confirmed because the strains were lost. (As discussed later, certain technical "tricks," developed in the 1970s, would make transformation possible in *Escherichia coli*.) In 1953, transformation was detected in *Hemophilus influenzae*. Although this organism was also a human pathogen, it was easier in some ways to work with and provided some important knowledge of the transformation process. Soon after its discovery in *Hemophilus*, the process was also detected in *Neisseria meningitidis*, and later in *Bacillus subtilis*.

Research on bacterial transformation became a major milestone in the development of molecular genetics because it focused on DNA as the genetic material. It had been known from cytochemical and cytogenetic studies in higher organisms that chromosomes consisted principally of nucleic acid and protein, and the gene was generally considered to consist of nucleoprotein. However, early (erroneous) ideas of the chemistry of DNA (see Section 9.8) had suggested that the specificity of the gene could not reside in the nucleic acid portion of the nucleoprotein. The discovery of transformation suggested that genetic specificity could reside in DNA. About this same time, cyto-chemical work showed that the DNA content of a nucleus doubled when the chromosome content doubled. Numerous other incidental observations about DNA also were understandable if DNA was the genetic material. James D. Watson and Francis H. Crick were led to study DNA primarily because of a realization of the genetic significance of this molecule. Their discovery of the structure of DNA explained how DNA could function as the genetic material and increased the acceptability of studies on bacterial transformation.

The term "transformation" has acquired different meanings in different fields. In the cancer field, transformation means the conversion of a normal cell to a cancerous cell. Fred Griffith, the discoverer of the bacterial trans-formation process being discussed in this chapter, used the term *transforma-tion* without actually defining it. The usage here is that currently accepted in bacterial genetics.

9.1 BIOLOGY OF PNEUMOCOCCI

The pneumococci are a relatively homogeneous group of bacteria that are recognized by the following characteristics: (1) Gram-positive cocci, usually arranged in pairs or short chains; (2) biochemically related to the other members of the genus *Streptococcus* by lack of a cytochrome-based respirato-ry system and by the generation of energy via a conventional Embden-Meyerhoff lactic acid fermentation; (3) rapid lysis induced by bile salts or other surface-active agents; (4) generally associated with a distinct human disease, *lobar pneumonia*; (5) association of virulence with the formation of a polysaccharide capsule that can be easily demonstrated microscopically.

The nomenclature of the pneumococci has been somewhat variable. Al-though closely related to other members of the genus *Streptococcus*, the pneumococci appeared sufficiently distinct so that they were at one time classified in a separate genus, *Diplococcus*, but today the accepted classifica-tion places these organisms within the genus *Streptococcus* with the species name *pneumoniae*. However, the medical bacteriologists who have been responsible for most of the research on this group have been little concerned with the taxonomic distinctions and have preferred the trivial name "pneu-mococcus." For simplicity, this name will be used in the present discussion.

Pathogenesis

The pneumococci are often normal inhabitants of the upper respiratory tract of humans, but in compromised hosts they can cause a variety of infections of the respiratory tract and adjacent structures. The most dramatic infection is that of the lungs, in the condition called *lobar pneumonia*. Although this disease can also be caused by other organisms, it is the major type of pneumococcus infection, and in the days before antibiotic therapy, was a major cause of human mortality.

A characteristic of the pneumococcus is its high invasiveness, linked directly to the presence on the surface of the cells of a capsule, which makes the cells resistant to phagocytosis. However, if the host contains an antibody specific for the capsule of the invading strain, this antibody promotes the process of phagocytosis (a condition called *opsonization*), and the virulence of the pathogen is overcome. Pneumococcus is a classical example of a pathogen whose virulence is principally connected with its ability to initiate a massive invasion of vital tissues. In the absence of host defenses, pneumococci can grow in large numbers in the lungs, leading to severe respiratory stress and death.

Although pneumococcus infection can be readily controlled by antibiotics, during the time when the early work on transformation was taking place, antibiotics were not available and control of the disease was by means of antibody therapy. Pneumococcal pneumonia was one of the few infectious diseases for which specific serum therapy was developed. Serum therapy was based on the fundamental studies carried out at the Rockefeller Institute for Medical Research by Oswald T. Avery, who was eventually to play the key role in the transformation work (see later and Dubos, 1976). Avery, Chickering, Cole, and Dochez (1917) showed that if a specific antiserum was administered early in the course of the disease, full recovery could occur. To prepare the antiserum, heat-killed pneumococci were injected into horses or rabbits, and the antibody globulins in the crude serum were concentrated and purified.

Among laboratory animals, the mouse is a highly susceptible species, and the virulence of pneumococci is most readily tested by intraperitoneal or subcutaneous injection. With many strains, as few as one to five cells inoculated intraperitoneally is sufficient to initiate an infection that is fatal within about 48 hours. Such infections are systemic, and the bacteria are found disseminated throughout the body, including the blood. The mouse can also be used as a selective agent for virulent mutants. If a large population of avirulent pneumococci is injected, growth of capsulated virulent mutants present in the inoculum will be favored.

Type Specificity

A major contributor to knowledge of the pneumococci was Fred Neufeld at the Robert Koch Institute in Berlin. Neufeld first demonstrated the existence of antigenically different types of pneumococci by studying the protective effect of different antipneumococcal sera in mice. He found that a serum would protect against a homologous strain but not necessarily against a heterologous strain. Avery and co-workers followed up on Neufeld's work and showed, using agglutination methods, that the pneumococci could be divided into several antigenic types. Initially Types I, II, and III were recognized, leaving a large heterogeneous unclassified group that came to be called Group IV (Avery, Chickering, Cole, and Dochez, 1917). Eventually, more than 80 pneumococcus types were recognized, but during the early transformation work only the first three types were known.

In the 1920s, Michael Heidelberger, in Avery's laboratory, initiated extensive chemical studies on the pneumococcal antigens involved in type specificity (Heidelberger and Avery, 1923). It had been discovered by A.R. Dochez in Avery's laboratory that pneumococcus cultures liberated soluble substances that precipitated in antiserum prepared against pneumococcus cells. Because these soluble substances were released early during the growth phase of the culture, they were clearly not a product of cell lysis and were hence called "specific soluble substances" (abbreviated SSS) by Avery. By use of chemical extraction methods followed by fractional precipitation, it was possible to show that the soluble materials that determined type specificity were complex polysaccharides that were identical with the capsular materials on the surfaces of the cells. The purified materials were able to precipitate with specific antibody even in high dilution but were not themselves antigenic. The demonstration of immunological specificity in a polysaccharide was at first taken very skeptically by the bacteriological community, since the orthodox view was that only proteins were of sufficient complexity to possess immunological specificity. The sheer weight of evidence from Avery's laboratory eventually won over opponents and Avery's reputation as an outstanding immunochemist was firmly established. For his discovery and characterization of these antigens, Avery received numerous honors, including election to the National Academy of Sciences and the award of the Paul Ehrlich Gold Medal (Dubos, 1976).

Although numerous other antigenic components were eventually identified in the pneumococci (including a nucleoprotein), the polysaccharides that determined type specificity were of the greatest interest because they played the major role in antiserum therapy. Although the detailed chemistry of the specific soluble substances was not known for many years, early immunochemical studies in Avery's laboratory showed that each type had a different complement of sugars. It would eventually be shown that Type I consists of galacturonic acid and *N*-acetyl glucosamine, Type II contains glucose, glucuronic acid, and rhamnose, and Type III contains an alternating glucose/glucuronic acid copolymer (see also Section 9.3). Eventually, knowledge of the chemistry of the type substances would be of value in interpreting the results of transformation experiments, but in the early days the focus was on the immunochemistry itself. It was realized only much later that the existence of various antigenic types had genetic significance, and that each type could be considered analogous to an allele or group of related alleles.

Other Characteristics of Pneumococci

The pneumococci have complex nutrition and are generally grown in meat extract broth with small amounts of glucose. Although they had a reputation as "fastidious" organisms, they are not actually difficult to culture, provided one ensures that the pH of the medium does not become acid from accumulated lactic acid. Although not an obligate anaerobe, the organism is sensitive to hydrogen peroxide, a common product of its own metabolism in the presence of air. If large inocula are used, the organism through its own metabolic activities lowers the redox potential and makes the medium favorable for growth; but if small inocula are used, growth is better if the medium is reduced by heating to drive off oxygen, or by addition of a reducing agent such as cysteine or thioglycollic acid. Media containing blood promote growth partly because blood contains the enzyme catalase that destroys hydrogen peroxide. Colony growth is thus best if a blood agar medium is used. The difficulty in achieving reproducible colony growth made genetic experiments with pneumococci difficult and probably discouraged some workers from studying this system. As late as 1965, a completely synthetic medium for pneumococci was not available, although MacLeod (1965) described a partially defined medium containing numerous amino acids, vitamins, adenine, uracil, glutamine, asparagine, choline, glucose, buffering agents, thioglycollic acid, and salts. Isolation of auxotrophic mutants, a stimulus to much early genetics work on *Escherichia coli*, was thus not possible with pneumococcus.

Another significant characteristic of pneumococcus is its so-called "bile solubility," which results from an autolytic process that is enhanced by the presence of surface-active agents such as whole bile or bile salts. If a bile salt is added to a fully grown culture, the turbidity clears within a matter of minutes due to the activation of the autolytic enzymes. If the culture is first heated to a temperature of 65°C for 30 minutes, autolysis no longer takes place because the autolytic enzymes have been inactivated. Bile solubility is virtually a taxonomic signature for pneumococci, since the other streptococci are resistant to this phenomenon. One significance of this property is that it provides a convenient way of obtaining cell-free extracts of pneumococci.

9.2 THE SMOOTH/ROUGH TRANSFORMATION

A type of variation that has had a long history in bacteriology is the one which Arkwright (1921) called *smooth/rough* variation (see Section 4.5). Although the terms "smooth" and "rough" subsequently acquired various meanings, depending on which group of bacteria was under study, in Arkwright's first work, smooth (S) referred to strains of enteric bacteria that made stable emulsions when suspended in physiological saline, whereas rough (R) strains agglutinated spontaneously under the same conditions. On solid culture media, an S-form colony has a smooth and glistening surface, whereas the R-form colony is flat, thin, and has a jagged or irregular margin. The S forms differ serologically from R forms, the R variation in general consisting of the loss of a somatic antigen that characterizes the surface of S cells.

Although the S/R variation was first described by Arkwright (1921) for enteric bacteria, S/R variants of pneumococci were soon described by Griffith (1923). Griffith also showed that the R form of this organism was nonvirulent. He thought that the R form was thus a "dissociated" or "degraded" variant of the pathogenic type-specific S form, arising during the course of pneumonia infection. Hobart Reimann, a former associate of Avery working in China, confirmed Griffith's observations on the S/R conversion and showed that the R form had lost its capsule (for a review of this early work, see Arkwright, 1930). As pointed out by Austrian (1953), Griffith's use of S and R did not exactly conform to Arkwright's terminology, since colonies of R pneumococci are not actually rough. In the case of R pneumococci, marked agglutination in salt solution did not occur; the Griffith R form is characterized by an altered dullness of the surface of the colonies and an altered agglutination reaction with specific serum. However, the R form was seen to have lost the specific capsular polysaccharide that determined type specificity. Indeed, true rough strains of pneumococci (called "extreme rough," ER) were later isolated, leading to the suggestion that the terminology be changed to conform to that of Arkwright (Austrian, 1953). However, because Griffith's S/R terminology had by this time been well established (and widely used in the transformation literature), it was maintained to avoid confusion.

9.3 THE DISCOVERY OF PNEUMOCOCCUS TRANSFORMATION

Fred Griffith's observations on type transformation in pneumococci are embedded in a lengthy paper on characteristics of pneumococcus types (Griffith, 1928), the only paper Griffith published on transformation. There are numerous brief textbook accounts of Griffith's discovery of type transformation (some containing errors), as well as more extensive presentations in Dubos (1976) and McCarty (1985), but no recent review presents Griffith's study in the broader context of bacterial genetics. Griffith was a medical officer working in the Ministry of Health in London. His principal job was to carry out epidemiological studies of a serological nature, and in the excitement over lobar pneumonia following the great influenza pandemic of 1918, he began studies on pneumococci.

As noted above, Griffith was the first to describe the S/R conversion in pneumococci, but his principal interest was in the diversity of pneumococcus types found in individual pneumonia patients. Pneumococci were routinely isolated from the sputum of pneumonia patients. Local health officers would send sputum samples to Griffith for characterization of the pneumococcal types. He observed changes in prevalence of particular types during particular epidemics, and in some cases he noted the replacement of a particular type with untypable cultures that were called Group IV (later to become divided into numerous additional types) (Griffith, 1928).

During such studies, Griffith noted a number of instances where more than one serological type was cultured from the sputum of a single pneumonia patient. He considered three hypotheses to explain infection with multiple types: (1) The patient had previously been a healthy carrier of the second type, but became infected with the first type, so that the sputum had bacteria of both types. (2) The patient was a normal carrier of the second type, which mutated to the first type, the latter then initiating the infection. "On this hypothesis, the different serological types would be evidence of the progressive evolution" (Griffith, 1928). (3) The third hypothesis was the most interesting one: "On the other hand, [one type] might be derived from [another type] in the course of successful resistance against the latter strain. With the increase of immune substances or tissue resistance the [one type] would be finally eliminated, and there would remain only the [other types] which are almost certainly of lower infectivity and perhaps of less complex antigenic structure" (Griffith, 1928). Strangely, Griffith did not see his third (somewhat Lamarckian) hypothesis, interchangeability of type, as particularly radical. "On a balance of probabilities interchangeability of type seems a

no more unlikely hypothesis than multiple infection with four or five different and unalterable serological varieties of pneumococci" (Griffith, 1928).

During his studies, numerous R strains also arose in his cultures, and
Griffith became interested in the origin of such strains. He began by using
antiserum to bring about the S/R conversion, since S cells of a particular type
were agglutinated by antiserum against that type. During growth in liquid
culture, this agglutination favored the growth of R cells (which remained in
suspension). Generally after several passages in broth containing the appropriate antiserum, stable R strains were isolated. Attenuated R strains could
also be obtained by simple plating on heated blood agar (chocolate agar) and
examining the resultant colonies for the characteristic R appearance. In some
cases, "rough foci" were seen in otherwise normal S colonies, and R strains
could be isolated from such foci.

Once R strains had been isolated from S cultures, Griffith attempted to
obtain reversion to the original type. It is clear that his intention here was to
test (in some undefined way) the hypothesis of conversion of type. By
injecting a large inoculum of R cells into a mouse, or by passing such an R
culture through several mice in succession, highly virulent S cultures were
obtained of the same type as that from which the R strain had been derived.
When using large doses of R pneumococci to isolate S cultures, Griffith
injected the living bacteria subcutaneously, then isolated pneumococci from
the mice after they had died (Fig. 9.1).

How did an S culture arise from an R culture? Like many bacteriologists of
this period (see Section 4.6), Griffith had no concept of mutation and selection. Instead, he believed that "the subcutaneous inoculation of a mass of
culture under the skin furnishes a nidus in which the R pneumococcus is
able to develop into the virulent capsulated form and thence invade the
blood stream" (Griffith, 1928).

Some R strains reverted to type more readily than others, however, and it
was possible "that such strains may have retained in their structure a
remnant of the original S antigen insufficient in ordinary circumstances to
enable them to exert a pathogenic effect in the animal body. When a strain of
this character is inoculated in a considerable mass under the skin, the
majority of the cocci break up and the liberated S antigen may furnish a

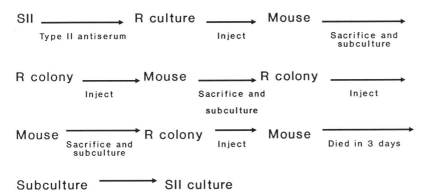

Figure 9.1 Griffith's procedure for conversion of smooth to rough pneumococci of the
same antigenic type, and selection of a virulent smooth culture by passage through
mice.

pabulum which the viable R pneumococci can utilise to build up their rudimentary S structure."

If this hypothesis was correct, how could it be tested? "It appeared possible that suitable conditions could be arranged if the mass of culture was derived from killed virulent pneumococci, while the living R culture was reduced to an amount which, unaided, was invariably ineffective. There would be thus provided a nidus and a high concentration of S antigen to serve as a stimulus or a food, as the case may be" (Griffith, 1928).

It was from studies of this type that Griffith proceeded to the experiments that were to make him famous. Apparently, Griffith had no broader purpose in mind than to determine how several pneumococcus types could be present in the same patient. Because so many "enhancements" of Griffith's work have been made through the years by authors who have not read the original, I quote the above passage to indicate the precise basis of Griffith's work.

Homologous and Heterologous Transformation

On the basis of his hypothesis that S cells provided "a nidus and a high concentration of S antigen to serve as a stimulus or a food," Griffith tested to see whether dead S cells might promote the conversion of R to S. He was pleased to discover that he could carry out the R→S conversion more readily in this way than by inoculating larger numbers of R cells alone (Fig. 9.2). Cultures obtained from the blood of dead mice were Type II by agglutination. Thus, Griffith had achieved transformation, but since it was of *homologous* type, he was not especially surprised.

Griffith's first attempts at *heterologous* transformation (transformation of type) were done essentially as controls, with the anticipation that the results

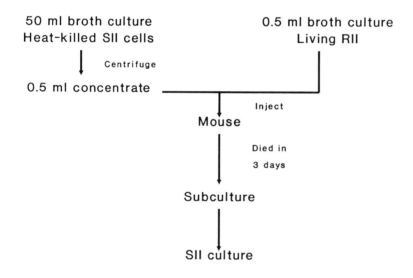

Homologous transformation

Figure 9.2 Griffith's homologous transformation experiment, using heat-killed smooth cells and living rough cells of the same antigenic type.

would be negative. Living R cells derived from Type II were injected together with heat-killed S I cells (Fig. 9.3). Griffith at first used 100°C to kill the S cells, but some erratic results prompted him to use the lower temperature of 60°C, a temperature at which the bacteria were still rapidly killed. In retrospect, the use of a lower heating temperature may have been crucial for success. If he had continued to use 100°C he would have obtained negative results, since pneumococcus DNA melts at that temperature. When cells heated at 60°C were used (the DNA does not melt), the heterologous transformation was readily obtained (Table 9.1). According to Griffith:

> ...the shorter the period during which the culture is heated to 60°C...the more powerful and the less confined to its own type is the effect of the virulent antigen on the attenuated R form when the two are injected simultaneously into mice. For example, virulent cultures of Type I heated to 60°C may cause the attenuated R strain of Type II to assume the capsulated S form in the animal body (Griffith, 1928).

This is the first direct statement of bacterial transformation of type.

In his discussion of his initial experiments, Griffith used the terms "conversion" or "reversion" to refer to the change from one type to another, but when he began to examine the mechanism of the process, he used the word "transformation." Griffith never actually defines the word transformation, although it is clear from usage that he means the change from one antigen type to another.

Griffith proceeded to carry out a large number of experiments, including numerous controls. The obvious explanation, that viable S cells still remained in the heated cultures, was ruled out by numerous sterility controls. The second explanation, that the R Type I had in some way reverted spontaneously to S II, was ruled out by the demonstrated stability of the R strain in

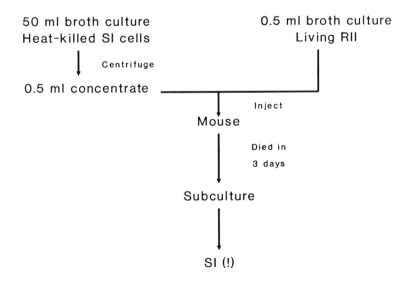

Heterologous transformation

Figure 9.3 Griffith's heterologous transformation experiment, using heat-killed smooth cells of Type I and living rough cells of Type II. A temperature of 60°C was used for heating.

Table 9.1   Transformation of Pneumococcus Type: Conversion of R Type II into S Type I

Killed S cells	Living R cells	Result	Type of culture obtained
Heated Type I	none	mouse survived	none
Heated Type I	none	mouse survived	none
Heated Type I	none	mouse survived	none
Heated Type I	none	mouse survived	none
Heated Type I	0.25 ml R Type II blood culture	mouse died in 3 days	S colonies, Type I
Heated Type I	0.25 ml R Type II blood culture	mouse survived	R Type II
Heated Type I	0.25 ml R Type II blood culture	mouse survived	R Type II
Heated Type I	0.36 ml R Type II grown in heated Type II	mouse survived	R Type II
Heated Type I	0.36 ml R Type II grown in heated Type II	mouse died in 4 days	S colonies, Type I
Heated Type I	0.36 ml R Type II grown in heated Type II	mouse survived	none R Type II

(Modified from Table VII of Griffith, 1928.) The stated mixtures were injected into mice. Each line represents a separate mouse. *Conclusion*: "The rough attenuated Type II Culture has apparently in two instances been changed into a virulent Type I."

numerous studies both in vitro and in vivo. He tried to obtain transformation in vitro, using a concentrated heated suspension of S cells as a culture medium, inoculated with R cells. The whole of each culture was inoculated into mice, which remained unaffected. These in vitro experiments were somewhat limited, but they implied to Griffith that some interaction with the living host was necessary for the transformation process to occur.

Another question of interest to Griffith was whether R I might be an intermediate stage in the conversion of R II to S I; in other words, R II→ R I→S I. However, it was difficult to assess directly whether R cells were potentially of Type I or Type II. The only way to differentiate between the R strains derived from Type I or Type II was to produce reversion to the S form without the use of heated S cells. An R colony was therefore isolated from a mouse that had died of Type I infection after in vivo transformation, and a culture was prepared. Large numbers of cells from this culture were injected into mice; if S cells (presumably mutants) were present, the mice would die and could serve as a source of S cells for type testing. All S colonies isolated from the mice were Type II, like the source of the heated cells. Thus, there was no evidence of an intermediate R I stage.

One of the most interesting experiments Griffith did, and the only one that brings his work toward the realm of genetics, was the use of heated cells of Type III to convert either R I or R II cells. The experiment showed that the capsulated type obtained from the mice was the same as that from which the S (capsulated) cells were derived, rather than that of the original R cells (Table 9.2).

Table 9.2 Transformation of Either R Type I or R Type II to Type III by Use of Heated Type III Cells

Killed S cells	Living R cells	Result	Type of culture obtained from mouse
Heated Type III	RII	mouse survived	none
Heated Type III	RII	mouse survived	none
Heated Type III	RII	mouse died, 7 days	S colonies, Type III
Heated Type III	RII	mouse died, 10 days	S colonies, Type III
Heated Type III	RII	negative	S colonies, Type III
Heated Type III	RII	negative	none
Heated Type III	RII	mouse died, 2 days	S colonies, Type III
Heated Type III	RII	mouse died, 8 days	S colonies, Type III
Heated Type III	RI	mouse survived	none
Heated Type III	RI	mouse survived	none
Heated Type III	RI	mouse survived	S colonies, Type III
Heated Type III	RI	mouse survived	none
Heated Type III	RI	mouse survived	none
Heated Type III	RI	mouse survived	none
Heated Type III	RI	mouse survived	none
Heated Type III	RI	mouse survived	none

(Modified from Table XII of Griffith, 1928.) Each line represents a separate mouse. *Conclusion*: "The injection of S culture of Type III . . . along with living R strains derived from Type I or Type II results in the appearance of an S pneumococcus of Type III. This transformation of type occurs more frequently with the R form of Type II than with the R form of Type I."

In interpreting his transformation results, Griffith certainly had no previous publications to start from. However, as a medical person, his main concern was with the epidemiological and medical significance of the phenomenon. The fact that it appeared to occur only in vivo relegated it to the epiphenomena of pathogenesis, rather than to any fancied bacterial genetics. It is doubtful whether Griffith had ever studied genetics (it was not part of the medical curriculum until many years later), and he certainly did not use any genetic interpretations:

> When the R form of either type is furnished under suitable experimental conditions with a mass of the S form of the other type, it appears to utilise that antigen as a pabulum from which to build up a similar antigen and thus to develop into an S strain of that type (Griffith, 1928).

As McCarty (1985) has noted, Griffith made no mention of the remarkable fact that the change from one type to the other was permanent, or that once the R form had acquired the new S type capsule, it continued to produce it indefinitely in subculture.

> Thus, something had happened to perpetuate the change. But bacteriology had developed as a science almost as if unrelated to the rest of biology. It was too early

in its history to expect genetic interpretations of any phenomena encountered in the laboratory (McCarty, 1985).

Fred Griffith's paper is a classical example of serendipity. The steps he took to discover the transformation of pneumococcus types were certainly logical within the framework of his own research, but the results obtained were quite unexpected. Griffith was hardly prepared to interpret these experiments correctly, let alone to follow them up with more experiments. Indeed, he apparently did no more on the phenomenon, and his name is now linked primarily to its discovery and to the first use of the term *transformation*. Griffith himself would miss the exciting denouement that was to come with the identification of the transforming principle as DNA, since he was killed in a German air raid early in World War II.

Biochemistry of the Capsular Acquisition

In understanding the history of pneumococcus transformation, it is useful to consider the nature of the biochemical events involved, even though these events were not known in the early period of transformation research. We now know that the inability of R cells to produce a capsular polysaccharide is due to mutation of a gene for the production of an enzyme involved in polysaccharide synthesis. Only a very few enzymes are involved specifically in the production of each type of polysaccharide (Mills and Smith, 1962). R strains are mutants that have lost the ability to produce one or more of these "type-specific" enzymes, and restoration of capsule production by transformation is due to the introduction of the gene for the missing enzyme.

In the case of Type III capsular polysaccharide, some of the biochemistry is known (Fig. 9.4). The Type III capsular polysaccharide is a copolymer of

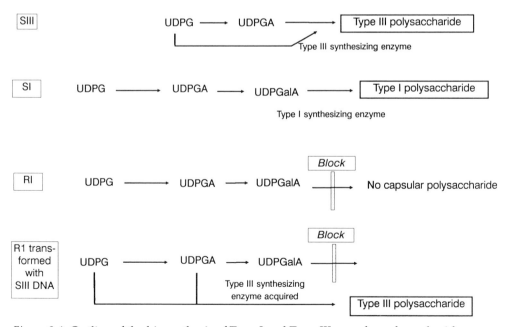

Figure 9.4 Outline of the biosynthesis of Type I and Type III capsular polysaccharides and an explanation of how heterologous transformation might occur. (Modified from Mills and Smith, 1962.)

glucose and glucuronic acid. One R strain has been found to lack the enzyme UDPG dehydrogenase, which catalyzes the formation of uridine diphos-phoglucuronic acid (UDPGA) from UDP glucose. Since UDPGA is used only for Type III polysaccharide biosynthesis, the loss of this enzyme is not lethal but results in the S→R conversion. In a homologous transformation, the gene coding for the enzyme specifically involved in Type III biosynthesis is introduced and the function is restored.

When Type I DNA is used in a heterologous transformation with R cells derived from Type III, the transformed strain is S I. As seen in Figure 9.4, Type I polysaccharide contains galacturonic acid instead of glucuronic acid. One building block of Type I polysaccharide is UDP galacturonic acid (UD-PGalA), synthesized from UDPGA by a racemase. Thus, wild-type I cells have enzymes for the synthesis of both UDPGA and UDPGalA. If a block occurs in Type III in the enzyme that synthesizes UDPGA, then this function could be restored by Type I DNA, but the resulting S strain would be Type I, not the heterologous Type III. The only way the heterologous transformation could occur is if the transformation brings in the gene for the enzyme that does the *final* polymerization step. Thus, if an R strain derived from a Type I strain was blocked at this final step, introduction of Type III DNA could result in Type III cells.

The reverse transformation, from R III to S I, might not actually be possible, depending on the specific mutation that led to R and the linkage relation-ships of the genes involved. If the R III strain was blocked in the final step, the introduction of the gene for formation of Type I polysaccharide would probably not be successful, since one of the building blocks of Type I polysaccharide, UDPGalA, would not be formed by the Type III cells. Since transforming DNA is always in fragments containing only a few genes (see later), it is unlikely that the genes for all the enzymes for a large biosynthetic pathway could be introduced into the cell in a single transformation event.

The point is that, depending on the direction of the transformation being sought, success might or might not have occurred. In fact, under some conditions one might even expect to obtain strains that have acquired the ability to produce *both* polysaccharides, and such binary encapsulated strains have actually been found (Griffith, 1928; Neufeld and Levinthal, 1928; Austrian, 1953; Mills and Smith, 1962). Of course, none of the biochemical points were known, or even vaguely considered, by Griffith, Avery, or other early workers on bacterial transformation.

Confirmation of Griffith's Work

The startling nature of Griffith's results elicited considerable interest in the small "pneumococcus community," and his experiments were quickly con-firmed by others. First to report agreement was Fred Neufeld, the dis-tinguished German bacteriologist who had been the discoverer of pneumo-coccal type specificity. During a visit to London, Neufeld had learned of Griffith's work even before publication, and on his return to Berlin immedi-ately set out to repeat the experiments (Neufeld and Levinthal, 1928). In order to be certain of the background of his strains, Neufeld began with single-cell cultures. He also took great care to ensure that there were no viable cells in the heated cultures. He was able to achieve transformation with one R I strain to Type II, but another R variant could not be transformed. Following Griffith, Neufeld concluded that the S antigen of the dead cells served as a "stimulus" or "fodder" (*"Reiz oder Futter"*) for the R strain.

Another early confirmation of Griffith's work was by Reimann (1929), a former associate of Avery at the Rockefeller Institute who was working in Peking, China. Reimann's interest was in the possible medical significance of the S → R dissociation. He first confirmed Griffith's observations that certain R strains would revert to S spontaneously, either in the animal or in culture. Then, using Griffith's methods, he was able to transform an R strain derived from Type I to S III (but not to S II).

Discussing Griffith's work in his major review on bacterial variation, Arkwright (1930) noted that the surprising results had been confirmed and that "These results, which suggest new and very important processes concerned with variation of bacteria, were carefully controlled and repeated."

9.4 EARLY WORK AT THE ROCKEFELLER INSTITUTE

Oswald T. Avery

Bacterial transformation, and indeed the whole field of molecular genetics, will be ever linked with the name of Oswald T. Avery, whose crowning achievement was the proof that transformation was due exclusively to DNA. Avery's background and work have been extensively documented in books (Dubos, 1976; McCarty, 1985), memorial articles (Chesney, 1957; Hotchkiss, 1965), and scholarly papers (Russell, 1988).

Avery was born in Halifax, Nova Scotia in 1877, but moved with his family to New York City when he was 10 years old. He received his undergraduate training at Colgate University (Hamilton, New York) and studied medicine at Columbia University (College of Physicians and Surgeons), receiving his M.D. in 1904. After a short period in medical practice in New York City, Avery joined the Hoagland Laboratory in Brooklyn, New York, as a research investigator. The Hoagland Laboratory was the first privately endowed laboratory for bacteriological research in the United States (Dubos, 1976). At Hoagland, Avery carried out a variety of studies in medical bacteriology, mostly of a prosaic nature. It was at Hoagland that he first began to work on pneumococci. At this time, pneumococcus pneumonia was the major cause of death in the United States. One of Avery's studies caught the attention of Rufus Cole, the director of the Hospital of the Rockefeller Institute for Medical Research, and in 1913 Avery joined the Rockefeller Institute, where he remained throughout his career. A lifelong bachelor, Avery retired in 1948 and died in 1955. From 1913 on, all of his research dealt with pneumococci and related bacteria. His pioneering work on the chemistry of pneumococcus antigens was discussed in Section 9.1.

Confirmation of Griffith's Work

When Griffith's paper on the transformation of pneumococcus types was published, the paper was widely discussed in Avery's laboratory, but Avery was reluctant to accept its results (Dubos, 1976). He found it impossible to believe that pneumococci could be made to change their immunological specificity. The transformation problem was eventually taken up in Avery's laboratory by Dawson on his own initiative. According to Dubos (1976), Dawson believed that work done in the British Ministry of Health almost *had* to be right, and that therefore Griffith's conclusions were valid.

Martin Dawson was a young Canadian physician in Avery's laboratory. Dawson had earlier shown that R strains were also antigenic and that anti-R sera could be used in vitro to select for S revertants. He initiated studies on the transformation of pneumococcal types at Rockefeller and continued them when he moved to Columbia University (Dawson, 1930a,b). Some of Dawson's experiments, including the first successful in vitro transformations, were done in association with Richard H.P. Sia, who was on leave from the Peiping Union Medical College in China (Dawson and Sia, 1929, 1931; Sia and Dawson, 1931).

At first, Dawson simply repeated Griffith's experiments, but with the goal of deriving a simpler in vitro system for transformation. In his first paper, he studied the reversion of R cells to S cells of the homologous type, either by simple selection in the mouse, or by Griffith's technique of injecting large numbers of heat-killed homologous S cells with small numbers of living R cells. To assure himself that he was working with a homogeneous population of R cells, he prepared "pure lines" by single-cell isolations.

Dawson's initial work was with homologous transformation and was done in vivo (Dawson, 1930a). In this work, he paid special attention to proper preparation of the heat-killed S cells (which he called "vaccines"). He grew large batches of S cells in broth, then subjected them to a preliminary heating at 60°C to prevent subsequent autolysis. One of Dawson's key experiments was a study of the appropriate temperature for preparing active heated cells for homologous transformation. He showed that if the S cells were heated to 60°C, they were active in transformation, but that heating to 100°C destroyed the activity. Attempts to achieve transformation in vitro were initially unsuccessful, but Dawson thought that his lack of success might indicate that the heated S cells had to be "digested" in the subcutaneous tissues of the mouse first, the R cells then utilizing the digested products "to build up their own S substance." Although unsuccessful, these early attempts led eventually to successful in vitro transformations.

Building on his studies on homologous transformation, Dawson then proceeded to study the heterologous transformation process in the mouse, using an R strain derived from a Type II S pneumococcus (Dawson, 1930b). He noted that not all R strains could be successfully transformed, and that the best R strains for transformation were those that could be readily converted to homologous S type either by animal passage or by growth in media containing anti-R serum.

The best R II strain Dawson used could be transformed in the mouse into either an S III, an S I, or a Group IV S, depending on the source of the heated cells. Dawson showed that the temperature to which the donor cells were heated was critical for successful transformation. Heating at any temperature between 60°C and 80°C yielded successful preparations, but if a temperature over 80°C was used, the preparations were inactive.

Dawson was unsuccessful in his attempts to convert one S type directly into another without going through the R form, and he felt that the R form must serve as an intermediary. Such intermediary R forms could be readily isolated from S forms (either natural isolates or transformants) by growth in anti-S serum.

Successful In Vitro Transformation

Dawson's success in repeating Griffith's in vivo transformation experiments apparently encouraged him to continue to search for an in vitro transformation procedure, and shortly afterward he and Sia reported a suitable technique. Initially, Dawson and Sia (1931) attempted an in vitro transformation by injecting heated S cells intraperitoneally in mice, waiting a while, and then washing the peritoneal cavity with saline and removing the peritoneal washings to see whether this would bring about transformation. The hypothesis (erroneous) was that the heated S cells had to first be "digested" in the animal before they were active in transformation. In several experiments using this combined *in vivo/in vitro* procedure, successful homologous and heterologous transformations were obtained. The active material in the peritoneal washings was not filterable.

In the process of doing these in vivo/in vitro experiments, Dawson and Sia (1931) realized that the transformation procedure being used was different from that used in the earlier unsuccessful in vitro experiments. In particular, the R culture was being grown from a small inoculum in broth to which the peritoneal washings had been added. The same procedure was then used, but instead of peritoneal washings, 0.5 ml of heated S III cells was added. Successful transformation was obtained, the first real in vitro transformation.

The experimental procedure was readily repeated, and after further studies, it was found that the key to successful in vitro transformation was the use of a low inoculum of R cells in a culture medium containing blood or serum to which the heated S cells were added. The active ingredient in blood or serum was anti-R antibody, and the R culture therefore had to grow from a small inoculum in the presence of this antibody in order for transformation to be detected. Alloway (1933) would subsequently show that the role of anti-R was to cause R cells to agglutinate and settle to the bottom of the tube, the transformed S cells then growing throughout the culture as a uniform suspension.

Another contribution of Dawson and Sia in this original work was the development of a suitable in vitro method for assessing the success of the transformation process. After the transformation process had occurred, the cultures were streaked on blood agar and the colonies were examined with a low-power microscope. As noted earlier, S colonies are morphologically distinct from R colonies. If S colonies were observed, subcultures were made, and the type specificity was determined by agglutination tests.

The key conditions for successful in vitro transformation were therefore as follows: (1) use of a low inoculum of R cells; (2) incubation for at least 24 hours and preferably longer; (3) addition of anti-R serum; (4) addition of a small amount of blood broth; and (5) use of S cells that had not been heated to a temperature above 60°C. Furthermore, the transformation of type could be induced with very small quantities of heated S cells, the equivalent of 0.1 ml of culture. Although filtrates of heated cells were not effective, suspensions of heated S organisms broken up by freezing and thawing were highly active. Although Dawson and Sia did not obtain an active solution, their experiments served as the starting point for all subsequent work on the transformation system.

Successful Soluble Preparations

After Dawson moved to Columbia University Medical School, he did very little additional work on transformation due to the pressures of his clinical duties. However, by this time, Avery was convinced that pneumococcal transformation was possible and encouraged J.L. Alloway, Dawson's replacement at Rockefeller, to pursue the work. Using Dawson's experiments as a starting point, Alloway was able to obtain successful extracts and even to carry out some preliminary purification procedures (Alloway, 1932).

Those were the days before high-speed blenders and mechanical ball mills. Alloway prepared cell-free extracts by the simple procedure of repeated freezing and thawing in a dry ice/alcohol mixture. The preparations were then immediately heated to 60°C and subjected to high-speed centrifugation. The supernatant, which contained the active ingredient, was then filtered through a bacteriological filter and concentrated at low temperature by vacuum distillation. The final extract, which was about 100 times as concentrated as the original culture, was then heated again to 60°C to ensure sterility and used in the transformation experiments.

In the first successful cell-free transformations, R II cells were transformed with S III extract. In another experiment, R II cells were transformed into S I. Alloway (1933) then modified the preparation procedure so that the cells were lysed with sodium deoxycholate first, then heated. This procedure resulted in preparations of much higher viscosity. The highly viscous extract obtained was then added to cold absolute alcohol (to remove the deoxycholate, which is alcohol-soluble). "A thick, stringy precipitate formed which slowly settled out on standing" (Alloway, 1933). Although this preparation was probably high in DNA, Alloway of course did not know this, nor did he know that the activity of the preparation would be markedly reduced by drying, since the next thing he did was to dry the preparation in vacuo. Even so, these preparations were active in transformation and served as the starting point for further work.

Although Alloway's work was a major advance, his preparations were still a long way from purified DNA. The anti-R serum assay procedure that Alloway developed was convenient, but it provided no quantitative measure of potency. The knowledge that certain R strains were more successfully transformed than others was important, but no understanding of the critical importance of the culture medium or growth phase had been obtained.

In attempting to explain the transformation phenomenon, Alloway (1932) suggested that R cells had the capacity to develop any of the specific capsular polysaccharides, and that this capacity remained "latent until activated by special environmental conditions. The fact that an R strain derived from one type of Pneumococcus...may be caused to acquire the specific characters of the S form of a type other than that from which it was originally derived implies that the activating stimulus is of a specific nature" (Alloway, 1933). Later, he hypothesized that "...the active material in these extracts...acts as a specific stimulus to the R cells which have potentially the capacity of elaborating the capsular polysaccharides of any one of the several types of pneumococci..." (Alloway, 1933).

However, Alloway's procedure for purification of the transforming principle was extremely unreliable. It would be 10 more years before the chemical nature of the transforming principle would be clarified.

9.5 THE TRANSFORMING PRINCIPLE IS DNA

The steps that led Oswald T. Avery and co-workers to discover that the transforming principle is DNA have been presented in some detail in Dubos (1976) and McCarty (1985). The present account is based primarily on a close reading of the paper that described the DNA work (Avery, MacLeod, and McCarty, 1944). Through the years since 1944, Avery's work has been "enhanced" by numerous interpretations, either by defenders or detractors. It is important to separate actual fact from later fancy. Many writers, bringing to their interpretations the knowledge of the importance of DNA, have given Avery more credit than he deserves as a founder of molecular genetics (Russell, 1988). At the same time, it is clear that without Avery's tenacity and insistence on clear and well-designed experiments, the identification of the transforming principle as DNA would not have occurred.

Pre-DNA Interpretations of Pneumococcal Transformation

The transformation work became widely known even before the connection between the transforming principle and DNA was established. Because the Rockefeller Institute was a major research center in the leading U.S. city, work that took place there was widely known. It is generally the case that scientists bring to their reading of the literature the biases that they have formed through their own specialized research. Due to its ill-defined character, transformation was one of those phenomena that could be "interpreted" from numerous points of view. Geneticists, virologists, biochemists, bacteriologists, and physiologists all saw their own significance in the transformation story.

The distinguished Columbia University geneticist Theodosius Dobzhansky discussed the genetic significance of pneumococcal transformation in the second edition of his book on genetics and evolution (Dobzhansky, 1941). After describing the facts of pneumococcal S → R variation and transformation in some detail, Dobzhansky stated:

> The strains "transformed" from one type to another...acquire not merely a temporary polysaccharide envelope of a kind different from that which their ancestors have had, but are able to synthesize the new polysaccharide indefinitely. If this transformation is described as a genetic mutation—and it is difficult to avoid so describing it—we are dealing with authentic cases of induction of specific mutations by specific treatments—a feat which geneticists have vainly tried to accomplish in higher organisms...geneticists may profit by devising experiments along the lines suggested by the results of the pneumococcus studies (Dobzhansky, 1941).

Dobzhansky thus gave the transformation work a genetic interpretation, although calling it a "directed mutation" may have confused the issue. Among others influenced by statements such as Dobzhansky's was Joshua

Lederberg, who began studies on microbial genetics because of his reading of the transformation work (see Section 5.3).

Avery begins the famous 1944 paper with a sentence that is almost a direct quotation from Dobzhansky:

> Biologists have long attempted by chemical means to induce in higher organisms predictable and specific changes which thereafter could be transmitted in series as hereditary characters. Among microorganisms the most striking example of inheritable and specific alterations in cell structure and function that can be experimentally induced and are reproducible under well defined and adequately controlled conditions is the transformation of specific types of Pneumococcus (Avery, MacLeod, and McCarty, 1944).

The virologist Wendell Stanley was another prominent "interpreter" of pneumococcal transformation. Stanley had gained prominence from his work on virus chemistry (Stanley, 1935). There was certainly no biochemical discovery of the 1930s that elicited more excitement than Stanley's crystallization of tobacco mosaic virus (TMV). Stanley was a member of the Princeton branch of the Rockefeller Institute and was, of course, quite aware of the work in Avery's laboratory. In a major review on the biochemistry and biophysics of viruses, Stanley considered the pneumococcal transforming agent in some detail:

> ...there is a factor which may be obtained from any one of the S type of organisms that is normally absent from R type cells, but that when added to such cells induces their conversion into the same type of S organisms from which the factor was derived, with the very important result that more of the factor is produced in the induced S cells. This phenomenon is virus-like, and it is because of this and the fact that it may become important from the standpoint of the chemistry of viruses that a discussion is included here. The various type-specific pneumococci may be regarded as cells infected with different "virus" strains and only the R organisms as healthy. The R organisms may be converted into any one of what we refer to as type-specific organisms by "infection" with any one of the different "viruses." By appropriate treatment it is again possible to free the pneumococci of "virus" and secure the healthy R type. It is of interest, therefore, to examine the nature of this factor or "virus." The type-specificity of the pneumococcus is determined by its capsular polysaccharide, hence it might be assumed that the type of soluble specific substance or polysaccharide...might be responsible for this conversion. However, Dawson and Sia found that the specific capsular polysaccharide in chemically pure form would not induce the transformation in type...It is to be hoped that the study of this agent will be continued because of its virus-like nature (Stanley, 1938).

Another early interpreter of pneumococcal transformation was Ross A. Gortner, a distinguished Minnesota biochemist whose textbook *Outlines of Biochemistry* was a principal reference book in the 1930s (Gortner, 1938). Gortner concluded a chapter containing a thorough discussion of immunochemistry with a brief discussion of the work of Dawson and Alloway, emphasizing that

> ...the type R Pneumococci changed to the particular type characteristic of the material from which the active extract was derived...It appears, therefore, as though some complex chemical substance had induced a change in the hereditary properties and definitely transformed one organism into a different but a closely related organism. This, if it is confirmed, is the first successful experiment of inducing a specifically directed genetic change in an organism by a chemical compound...Since these experiments are of such importance in their genetic impli-

cations, they should be repeated, using cultures from single-cell isolates of the Type R organism. If such experiments confirm the reports, we would have the first demonstrated proof of a directed genetic change (Gortner, 1938).

Another "pre-DNA" consideration of pneumococcal transformation was that of Tracy M. Sonneborn, the distinguished Indiana geneticist and a principal proponent of the role of the cytoplasm in heredity (Sapp, 1987). Sonneborn's studies on a cytoplasmic "particle" in *Paramecium* (Sonneborn, 1943a) led him to a more general consideration of cytoplasmic inheritance, of which he saw pneumococcal transformation as an important aspect (Sonneborn, 1943b). Presumably the reason the pneumococcus phenomenon was considered "cytoplasmic" was because at this time the genetics of bacteria was still considered to be "different" from that of higher organisms.

> The Pneumococcus situation may be compared with the killer situation in Paramecium. In the latter, transfer of a cytoplasmic substance from a killer cell to a sensitive cell containing gene *K* will result in the continued production of this substance and the killer phenotype which depends on it. Possibly all or most strains of Pneumococcus contain the gene or genes (or gene-like materials) required to control the continued production of each of the 50 different polysaccharides, and, as in Paramecium, when an appropriate substance essential for the synthesis of the polysaccharide is added to a cell that lacked it, the "gene" will determine its continued production. As in Paramecium, the gene seems to be unable to initiate its production, but can continue it when the proper substance is provided to start the gene going. The fact that any strain produces only one of these 50 polysaccharides would lead one to suppose that the same gene is involved in all 50 cases and that there are more than 50 cytoplasmic materials which the gene can act upon (Sonneborn, 1943b).

Sonneborn is clearly assuming that the transforming principle is some sort of cytoplasmic particle whose maintenance in the cell is dependent on the presence of "real" genes. (Later work of Sonneborn and others showed that the killer entity in *Paramecium* was actually a symbiotic bacterium.) In Sonneborn's view, the potentiality for production of the pneumococcus antigen is controlled by a gene, but the specificity is controlled by the transforming principle. (For a later view of transforming principle as cytoplasmic particle, see the discussion of Lederberg's interpretation in Section 9.10.)

These statements from prominent American scientists show that the broader significance of pneumococcal transformation was recognized by the biochemical and genetic community even before the connection of the transforming principle to DNA was made. Although it has been emphasized by both Dubos (1976) and McCarty (1985) that Avery strictly avoided broader interpretations of pneumococcal transformation, disinterested onlookers obviously were quite willing to see the phenomenon in a biological context. Avery's letter to his brother Roy has been widely quoted as evidence that Avery understood the significance of his work ("sounds like a virus—may be a gene," see later), but neither of these interpretations was obviously original with Avery. According to Hotchkiss (1965), Avery collected all comments and conjectures regarding the nature of pneumococcal transformation. He clearly did not speculate himself, and even viewed much of the speculation of others with "amusement." For Avery, the principal question was extremely narrow: "what is the substance responsible?" (Hotchkiss, 1965).

The Purification of Transforming Principle

The work of Alloway on the nature of the transforming principle had been taken up by Colin M. MacLeod, a Canadian physician with little research experience who joined Avery's group in the summer of 1934. For unknown reasons, MacLeod elected to work on pneumococcal transformation and, after repeating the Dawson and Alloway work, proceeded to isolate a new R strain that was extremely favorable for transformation studies. This strain, R36A, was selected after 36 subcultures of a parent Type II strain in the presence of specific Type II antiserum. It became the principal "assay" organism during purification of the transforming principle. Not only was R36A stable, but it could also be readily transformed to any of a number of other specific types, thus setting it apart from most of the other R strains that had been isolated.

Holding the limited view of bacterial genetics then current, Avery, MacLeod, and McCarty (1944) described their R36A strain as attenuated rather than as a specific mutant: "The strain R36A has lost all the specific and distinguishing characteristics of the parent S organisms and consists only of attenuated and non-encapsulated R variants. The change S→R is often a reversible one provided the R cells are not too far 'degraded'...Strain R36A has become relatively fixed in the R phase and has never spontaneously reverted to the Type II S form." This statement shows how limited was the grasp of genetics of the Avery group. Note that I am not trying to insist that Avery and co-workers *should* have understood bacterial genetics, but that the impoverished ideas of genetics by bacteriologists of that day prevented them from understanding what was going on. The statement of Avery, MacLeod, and McCarty (1944) was written before the work of Luria and Delbrück clarified the nature of the mutation process in bacteria.

Except for the R36A strain, the other conditions of the transforming principle assay had already been developed by Dawson and/or Alloway but were improved by MacLeod: (1) use of anti-R serum to provide selective conditions for S cells; (2) recognition of successful transformation by the appearance of uniform turbidity in the culture tubes because of growth of transformed S cells after the R cells had agglutinated; (3) confirmation of S by streaking on blood agar and looking for characteristic S colonies; and (4) confirmation of type by picking colonies and testing with type-specific antiserum.

To follow the purification process itself, the transforming principle was assayed by endpoint dilution, a standard immunological procedure. The transforming principle was added to the assay tubes at the beginning of the experiment and allowed to remain throughout the whole incubation period. The specific activity of a preparation was estimated from the highest dilution that gave successful transformation. It is clear from the discussions of Dubos (1976) and McCarty (1985) that the assay system was often "out of control." During the years when attempts were being made to purify the transforming principle, Avery frequently considered dropping the problem completely because of the irreproducible results. An understanding of the critical importance of "competence" for a successful transformation was not obtained until after the DNA work was completed.

After numerous studies during the period of 1934–1938 on the transforming principle, Avery and MacLeod dropped the problem completely, apparently disheartened by the lack of success. Among other things, sulfonamide therapy for bacterial pneumonia was introduced at this time, and an understanding of the transformation of pneumococcal types may have thus seemed less urgent. The problem was not taken up again until the fall of 1940, when work was renewed. (McCarty is uncertain why MacLeod and Avery resumed work on the transforming principle in the fall of 1940. McCarty indicated that the main reason work stopped was because MacLeod had been unsuccessful in obtaining publishable results and turned to other research so that he could obtain a reasonable publication list before going off on his own; see McCarty, 1985.)

Purification was improved by the availability now of numerous enzymes that could be used for removing type-specific polysaccharide, ribonucleic acid, and protein. Additionally, large-volume cultures were now being used, so that much larger amounts of "purified" material could be obtained. The Sharples continuous-flow centrifuge, introduced about this time, permitted harvesting bacteria from large volumes of culture medium relatively easily. Because pneumococci do not require oxygen for growth, the cultures could be readily grown in unstirred carboys or large flasks.

The Sevag Procedure

An important technique developed in the late 1930s was the Sevag procedure for removing protein (Sevag, 1934). M.G. Sevag had been a chemist in the laboratory of Fred Neufeld in Berlin and had developed a deproteinizing procedure that avoided the use of "harsh" chemicals. After coming to the United States, Sevag republished his procedure in an English-language journal (Sevag, Lackman, and Smolens, 1938). Because of Sevag's connection with Neufeld, it was natural that Avery should be aware of the Sevag procedure and use it in the purification of the transforming principle.

In the Sevag procedure, a gel is formed by shaking the protein-containing solution together with chloroform. The chloroform-protein gel separates from the aqueous phase, and the water-soluble nucleic acid remains behind. Among other things, Sevag showed that this procedure could be used to recover protein-free nucleic acid.

The Transforming Principle Is DNA

In 1941, MacLeod left the Rockefeller Institute to become professor at New York University Medical School, and Maclyn McCarty joined Avery's group. By this time, purified preparations had been obtained that were extremely low in protein but highly active in transformation. A type-polysaccharide-splitting enzyme that had been prepared by René Dubos from a soil bacterium was used to eliminate all traces of specific soluble substance, and RNA was eliminated by treatment with pancreatic ribonuclease. (John Northrop's laboratory at the Princeton branch of the Rockefeller Institute had pioneered the purification of pancreatic enzymes.) McCarty describes the next step:

> This material was...subjected to further purification...using both the SIII enzyme [Dubos' enzyme] and ribonuclease. After repeating the Sevag process to remove

the added enzyme protein and dialyzing the solution to eliminate split products, it was precipitated with alcohol. To our surprise, before one volume of alcohol had been added, a stringy mass of fibrous precipitation separated out...there is no indication in the notes that we equated the fibrous alcohol precipitate with DNA. Curiously, we did not even record any diphenylamine tests for deoxyribose on the fraction. The emphasis of our tests was solely on protein, RNA, and SSSIII—the components of the extract that we were trying to get rid of...Nevertheless, this experiment marks the beginning of a period when our attention was focused with increasing sharpness on the possibility that the transforming principle might indeed be DNA (McCarty, 1985).

One of the most important aspects of the purification procedure for transforming principle was the inactivation of endogenous DNases that could destroy the activity. When pneumococcal cells are lysed, both transforming principle and DNase are released. After initial harvest, the cells were suspended in concentrated form and heated rapidly to 65°C, a procedure worked out empirically by Dawson and Alloway, which served to inactivate DNase. After washing, the heated cells were then extracted with deoxycholate, and this extract was added to absolute ethanol to obtain the fibrous DNA precipitate. The precipitate was removed with a spatula, resuspended in saline, and deproteinized by the Sevag procedure. This was followed by treatment with the S III-digesting enzyme, the elimination of all SSS being monitored immunologically. It is noteworthy that at no time during purification was the transforming principle dried. Avery had found that drying caused inactivation. It is now known that drying is detrimental to DNA.

Various chemical tests were used to monitor the purification process. The final purified transforming principle gave negative reactions for protein (biuret and Millon), type-specific polysaccharide (precipitation assay), and lipid (Table 9.3). It gave only a weakly positive test for RNA (orcinol test) but was strongly positive for DNA (the Dische diphenylamine test). Chemical analysis showed a nitrogen/phosphorus ratio of 1.67, in close agreement with that calculated for DNA (from the assumption that DNA was a tetranucleotide).

Properties of the Transforming Principle

The highly purified transforming principle was tested to see whether its biological activity was affected by various enzyme treatments. Neither tryp-

Table 9.3 Chemical Analyses of Purified Transforming Principle

Preparation number	Carbon (%)	Hydrogen (%)	Nitrogen (%)	Phosphorous (%)	N/P ratio
37	34.27	3.89	14.21	8.57	1.66
38B	—	—	15.93	9.09	1.75
42	35.50	3.76	15.36	9.04	1.69
44	—	—	13.40	8.45	1.58
Theoretical for DNA	34.20	3.21	15.32	9.05	1.69

The theoretical values are for sodium desoxyribonucleate (tetranucleotide). (Modified from Avery, MacLeod, and McCarty 1944.)

sin nor chymotrypsin had any effect, nor did RNase, but DNase rapidly destroyed activity. Studies on the effect of DNase on transforming principle were carried out viscosimetrically, and it could be shown that loss of biological activity correlated with loss of viscosity. MacLeod had shown in early experiments that fluoride stabilized the transforming principle during initial extraction, in agreement with the fact that it inhibits DNase. Serological analyses with highly active antibody showed that type-specific SSS had been completely removed.

Electrophoresis studies in the Tiselius apparatus, and molecular weight determinations by diffusion, suggested a molecular weight for transforming principle of the order of 500,000 (actually, the molecular weight of active material is likely to have been much higher). The purified preparations showed an absorption maximum in the region of 2600 Angstroms, an absorption characteristic of nucleic acids.

One of the principal findings of Avery, MacLeod, and McCarty (1944) was the extremely high activity of their purified preparations (Table 9.4). Dilution titrations showed that a $10^{-4.5}$ dilution of the transforming principle was active at a concentration of 0.003 ml per 2.25 ml of culture medium, which represented a final concentration of the purified substance of 1 part in 600,000,000.

The 1944 Paper

Both Dubos (1976) and McCarty (1985) reported that Avery was extremely cautious in writing up his results for publication. Most of the experimental work had been completed by the beginning of 1943, yet the final paper was not submitted until late 1943 and was published in 1944. By this time, considerable discussion had already appeared in the literature about the significance of the transforming principle (see earlier), and Avery commented on this discussion and added a few cautious conclusions of his own.

> Equally striking is the fact that the substance evoking the reaction and the capsular substance produced in response to it are chemically distinct, each belonging to a

Table 9.4   Titration of Transforming Activity

Dilution of TP	Amount added	Trial 1		Trial 2	
		diffuse growth	colony form	diffuse growth	colony form
10^{-2}	1.0	+	SIII	+	SIII
$10^{-2.5}$	0.3	+	SIII	+	SIII
10^{-3}	0.1	+	SIII	+	SIII
$10^{-3.5}$	0.03	+	SIII	+	SIII
10^{-4}	0.01	+	SIII	+	SIII
$10^{-4.5}$	0.003	−	R	+	SIII
10^{-5}	0.001	−	R	−	R
Control	none	−	R	−	R

The experiment was done in quadruplicate; only two trials are shown here. The initial solution contained 0.5 mg/ml of transforming principle (TP), and 0.2 ml of each dilution was added to 2 ml of standard serum broth, and 0.05 ml of a 10^{-4} dilution of a blood broth culture of R36A (derived from Type II) was added. (Modified from Avery, MacLeod, and McCarty, 1944.)

wholly different class of chemical compounds...The experimental data...strongly suggest that nucleic acids, at least those of the desoxyribose type, possess different specificities as evidenced by the selective action of the transforming principle. Indeed, the possibility of the existence of specific differences in biological behavior of nucleic acids has previously been suggested but has never been experimentally demonstrated owing in part at least to the lack of suitable biological methods...If it is ultimately proved...that the transforming activity...is actually an inherent property of the nucleic acid, one must still account on a chemical basis for the biological specificity of its action...The biochemical events underlying the phenomenon suggest that the transforming principle interacts with the R cell giving rise to a coordinated series of enzymatic reactions that culminate in the synthesis of the Type III capsular antigen...the induced alterations are not random changes but are predictable, always corresponding in type specificity to that of the encapsulated cells from which the transforming substance was isolated. Once transformation has occurred, the newly acquired characteristics are thereafter transmitted in series through innumerable transfers in artificial media without any further addition of the transforming agent. Moreover, from the transformed cells themselves, a substance of identical activity can again be recovered in amounts far in excess of that originally added to induce the change. It is evident, therefore, that not only is the capsular material reproduced in successive generations but that the primary factor, which controls the occurrence and specificity of capsular development, is also reduplicated in the daughter cells. The induced changes are not temporary modifications but are permanent alterations which persist...these changes are predictable, type-specific, and heritable.

If...the biologically active substance isolated in highly purified form as...desoxyribonucleic acid actually proves to be the transforming principle...then nucleic acids of this type must be regarded not merely as structurally important but as functionally active in determining the biochemical activities and specific characteristics of pneumococcal cells...If the results of the present study...are confirmed, then nucleic acids must be regarded as possessing biological specificity the chemical basis of which is as yet undetermined (Avery, MacLeod, and McCarty, 1944).

Although this discussion is reasonably to the point, as much of the discussion of the paper deals with the immunological significance of the findings (Avery's primary interest, of course) as with the genetics. It would be almost 10 years before the genetic role of DNA would be universally accepted, but the Avery paper provided the starting point for numerous studies on both the chemistry and genetic significance of DNA.

In May 1943, Avery wrote a letter to his brother Roy, who was on the faculty of Vanderbilt University. In this letter, Avery described the transformation work and its possible significance (Dubos, 1976). The key passage in this letter is

If we are right, and of course that's not yet proven, then it means that nucleic acids are *not* merely structurally important but functionally active substances in determining the biochemical activities and specific characteristics of cells—and that by means of a known chemical substance it is possible to induce *predictable* and *hereditary* changes in cells...after innumerable transfers and without further addition of the inducing agent, the same active and specific transforming substance can be recovered far in excess of the amount originally used to induce the reaction. Sounds like a virus—may be a gene. But with mechanisms I am not now concerned—One step at a time—and the first is, what is the chemical nature of the transforming principle? Someone else can work out the rest. Of course, the problem bristles with implications (Dubos, 1976).

Max Delbrück was at Vanderbilt at this time (see Section 6.3) and was shown the letter by Roy Avery. Although Delbrück remembered the letter later and played a role in its resurrection, there is no evidence that it had any impact on his research or thinking about genetics.

9.6 EARLY ACCEPTANCE OF THE DNA WORK

It is important to note that the significant point of Avery's work is not that the transforming principle was DNA, but that the process of transformation provided a means for determining the material basis of heredity. Until this time, genes were recognized *only* through their effects on phenotype and through transmission from one generation to the next. With transformation, a new paradigm for genetic research became established. This would have been so even if the active substance had been protein or polysaccharide.

A number of authors have insisted that Avery's work was "premature" and hence had little influence on the genetic community (Stent, 1972; Wyatt, 1975). However, although Delbrück and the "phage church" may have resisted biochemistry and DNA (see Sections 6.12 and 7.4), numerous biologists and biochemists accepted it immediately.

As discussed in Section 5.3, it was Avery's transformation work that motivated Lederberg to begin his work on microbial genetics. The distinguished biologist G. Evelyn Hutchinson discussed Avery's work in some detail in his *Marginalia* column in the "American Scientist," a widely distributed publication (Hutchinson, 1945). He noted the activity of purified transforming principle at extreme dilution and used some preliminary size estimates to calculate that relatively few molecules per cell were needed to produce transformation. Although Hutchinson was somewhat confused about what the transforming principle did, he noted that the substance "seems to be at least a fragment of a genetic system."

In an extensive review on physiological genetics, Sewall Wright, a very well known geneticist, reviewed Avery's work in relation to the chemistry of the chromosome and gene (Wright, 1945). Another important reviewer was George Beadle, who covered the transformation work extensively in his lengthy review on biochemical genetics (Beadle, 1945). However, Beadle misinterpreted the basic phenomenon as a "directed mutation." He hypothesized that the pneumococcus cell was mutating from R to S and that if this mutation occurred in the presence of the transforming principle, a new type arose, otherwise the original type developed. However, he did admit to other possibilities:

> This may mean that a specific gene has been mutated, or that the type-specific nucleic acid is capable of autocatalytic reproduction. The latter alternative amounts to essentially the same thing as the suggestion by Wright that the nucleic acid itself may be the gene and that the transformation of type is brought about not by mutation but instead by actual transfer of the gene. Unfortunately, pneumococci do not reproduce sexually, so no direct test can be made (Beadle, 1945).

Another important early supporter was the Harvard microbiologist J. Howard Mueller, who made this clear statement:

> ...it appears that a polymer of a nucleic acid may be incorporated into a living, degraded cell, and will endow the cell with a property never previously possessed ...When thus induced the function is permanent, and the nucleic acid itself is also reproduced in cell division. The importance of these observations can scarcely be overestimated and stimulates speculation concerning such matters as the chemical basis for specificity in nucleic acids, and the genetic implications presented by the ability to induce permanent mutation in a cell by the introduction of a chemical substance. Such speculation may well include consideration of the relation of this phenomenon to the sequence of events following the introduction of a filterable virus (or a bacteriophage particle) into a susceptible cell (Mueller, 1945).

That nucleic acids were perceived as important in genetics is indicated by the fact that the 1947 Cold Spring Harbor Symposium on Quantitative Biology was entitled "Nucleic Acids and Nucleoproteins," and numerous important workers, including Boivin, Brachet, Chargaff, Seymour Cohen, Davidson, Mazia, Mirsky, Ris, Schultz, Spiegelman, and Hollaender, presented papers. The principal critic of Avery's work at the 1947 Cold Spring Harbor meeting was A.E. Mirsky, a colleague of Avery at the Rockefeller Institute. In a discussion of Boivin's paper on transformation in *Escherichia coli*, Mirsky laid out the reasons to doubt the central role of DNA in the transforming principle: (1) The protein tests available were not sensitive enough to detect small amounts of protein; (2) many native proteins were not acted upon by proteases unless denatured; (3) if the amount of material necessary for transformation was calculated on the basis of the number of "particles" instead of weight, it appeared that a large number of particles were needed; (4) the sensitivity of the transforming principle to DNase might merely indicate that the transforming principle was a deoxyribonucleoprotein with the protein part being critical for activity. "In the present state of knowledge it would be going beyond the experimental facts to assert that the specific agent in transforming bacterial types is a desoxyribonucleic acid" (Mirsky, 1947). It is interesting to note that the stringent requirements for purity advanced by Mirsky (and others) did not extend to the Hershey/Chase experiment a few years later. As discussed in Section 6.13, Hershey and Chase found that 10–20% of the phage phosphorus did not enter the cell, and that considerable sulfur entered as well. From a biochemical point of view, the Hershey/Chase experiment was decidedly inferior, as Hershey himself noted (Hershey, 1966).

In an important review paper on the gene, Hermann J. Muller, the principal theoretician of classical genetics, accepted the Avery work and gave it a decidedly genetic twist (Muller, 1947):

> Avery, MacLeod and McCarty (1944) have...given evidence which they believe points to the conclusion that the effective substance...is the nucleic acid itself...If this conclusion is accepted, their finding is revolutionary, no matter whether, with these authors, one adopts the radical interpretation that a transformation of the genetic material in the treated organisms has been induced, converting it into material like that used in the treatment (cf. 'Kappa substances'?), or whether it is supposed that genetic material of the donor strain actually becomes implanted within the treated strain and multiplies there, or whether the material used in the treatment is regarded as merely exerting a selective action so as to favour the survival of such exceedingly rare spontaneous mutants as happen by accident to agree with the other variety (Muller, 1947).

Muller referred to Mirsky's conviction that tiny amounts of protein contaminating the transforming agent might be responsible for its activity.

> ...the extracted 'transforming agent' may really have had its genetic proteins still tightly bound to the polymerized nucleic acid; that is, there were, in effect, still viable bacterial 'chromosomes' or parts of chromosomes floating free in the medium used. These might, in my opinion, have penetrated the capsuleless bacteria and in part at least taken root there, perhaps after having undergone a kind of crossing-over with the chromosomes of the host. [Thus,] viable chromosome threads could also be obtained from these lower forms for in vitro observation, chemical analysis, and determination of the genetic effects of treatment (Muller, 1947).

It is clear from Muller's paper, and from other publications (reviewed by

Ephrussi-Taylor, 1950) that the important point was not that the transforming agent was DNA, but that it could be manipulated outside the cell. Some hesitation among geneticists may have arisen because of "doubts" about the relevance of work on such a specialized property (antigenic polysaccharide) to genetics of higher organisms, although the Luria/Delbrück experiments (see Section 4.8) were by 1947 widely known and accepted. The Australian microbiologist Macfarlane Burnet, an early contributor to bacteriophage research (see Section 6.2) who had turned to animal viruses, made a clear connection between genetics of bacteria and higher organisms:

> ...hereditable variations in bacteria and viruses arise by a process of discontinuous mutation essentially similar to gene mutation in higher forms...Avery's recent work...opens up an important new approach and provides significant evidence of the essential continuity of genetic processes throughout living organisms...this genetic mechanism is like the chromosomes of the higher forms essentially of nucleoprotein structure, and...the "genes" or equivalent units have their specific activity determined by the chemical pattern...of their nucleic acid constituents. The change of type is induced by the filling of the gap...with an appropriate but specifically distinct nucleic acid unit (Burnet, 1945).

Wyatt (1972) has described how Burnet learned of the Avery work during a visit to the United States in December 1943 and wrote to his wife: "Avery...has just made an extremely exciting discovery, which, put rather crudely, is nothing less than the isolation of a pure gene in the form of desoxyribonucleic acid."

It is clear from these (and other) papers that a direct line of descent exists between the Avery paper and the Watson and Crick model for DNA. Although the chemistry of DNA would have eventually been worked out without the Avery work, bacterial transformation focused early attention on this molecule.

9.7 POST-WORLD-WAR-II WORK AT ROCKEFELLER

Although Avery himself was close to retirement at the time the 1944 paper was published, work on the transforming principle continued at the Rockefeller under his direction, as well as after he retired. McCarty continued to work on the project while in the U.S. Navy until 1946, when he was offered a permanent position at the Rockefeller Institute for work on streptococci and rheumatic fever. Harriett Taylor (later Ephrussi-Taylor) joined the Avery laboratory after receiving her Ph.D. in genetics from Columbia University. Rollin Hotchkiss, who had been in Avery's laboratory or that of his student Dubos since the mid-1930s, began work on the transforming principle after McCarty's move. Hotchkiss and his own students were responsible for many of the most important advances in knowledge of the transforming principle in the post-World-War-II period.

Although a number of studies were done on the transforming principle, the essential nature of the transformation process remained unclear, so that appropriate models were not available to aid in experimental design. Numerous background studies were undertaken, such as studies on the effect of serum on transformation efficiency (McCarty, Taylor, and Avery, 1946). The paper on effect of serum was given at the important 1946 Cold Spring Harbor symposium, the same symposium where the Lederberg paper on bacterial mating and the Hershey and Delbrück papers on phage recombination were given. Unfortunately, the transformation paper concerned primarily some "technical" details of the process, so that the geneticists present missed some of the overall significance of the transformation process. McCarty (1985) has described his disappointment with how poorly his paper was received. These serum studies did help in eventually understanding the competence phenomenon. Another study of this period involved measurement of the sensitivity of the transforming principle to oxidation (McCarty, 1945), but the main contribution was McCarty's important study on the effect of DNase on the transforming principle (McCarty and Avery, 1946a). In the 1944 paper, only crude DNase preparations were available, so that McCarty set out to purify a DNase from the beef pancreas. One result of these DNase studies was the discovery that magnesium ions were required for DNase activity and that DNase activity could therefore be inhibited by addition of sodium citrate, a magnesium chelator. McCarty's highly purified DNase had 100,000 times more DNase than protease activity and, since the transforming principle was sensitive to extremely tiny amounts of this DNase, it seemed unlikely that biological activity was due to protein.

Studies on DNase led to the discovery that the pneumococci themselves possessed an active DNase that was released when the cells lysed. Alloway's heating procedure had served to inactivate, at least partially, the pneumococcal DNase, but the discovery that citrate could be used to specifically inhibit the enzyme made it possible to isolate the transforming principle without the use of heat (McCarty and Avery, 1946b). In this way, a fivefold greater yield of transforming principle was obtained, and isolations of active material from other pneumococcal types were achieved.

In 1946, Maclyn McCarty was the recipient of the Eli Lilly Award, presented annually for meritorious research by an investigator under the age of 35. As a result of this award, McCarty had to give an address at the annual meeting of the Society of American Bacteriologists, and this address was published (McCarty, 1946). In earlier papers, McCarty had been coauthor with Avery, who had discouraged speculation, but in the Lilly address McCarty was able to discuss the broader biological significance of the transformation phenomenon. In addition to a reiteration of the possible analogy of transforming principle to a virus, and the clear evidence that transforming principle undergoes self-reproduction, McCarty considered the molecular nature of the transformation process.

> ...it should be pointed out that in all probability only a small portion of the total molecules are endowed with transforming activity. A desoxyribonucleic acid fraction can be extracted from unencapsulated R pneumococci which is similar in all respects to the Type III preparations, except that it is wholly inactive in the transforming system. It is possible that the nucleic acid of R pneumococci is concerned with innumerable other functions of the bacterial cell in a way similar to that in which capsular development is controlled by the transforming substance. The desoxyribonucleic acid from Type III pneumococci would then necessarily comprise not only molecules endowed with transforming activity, but in addition a variety of others which determine structural and metabolic activities possessed in common by both the encapsulated and unencapsulated forms. This implies that any given desoxyribonucleic acid preparation represents a complex mixture of a large number of entities of diverse specificity (McCarty, 1946).

The genetic implications of this statement are clear, although it would be several years before studies of a truly genetic nature would be undertaken.

Table 9.5 Purine and Pyrimidine Content of Hydrolysates of the Transforming Principle of Pneumococcus

Weight of base found in the fraction (mg)	State of purification				
	DNA extract	RNase treated	Sevag purification	alcohol precipitation	calf thymus DNA
Cystosine	9.8	1.7	1.8	1.7	—
Adenine	10.5	1.9	1.8	1.9	—
Thymine	2.3	2.7	2.7	2.7	—
Uracil	5.0	0.0	0.0	0.0	—
Ratio of deoxyribose to 260 nm absorption	0.013	0.074	0.081	0.086	0.093
Fraction of initial activity recovered	1.00	0.87	0.66	0.46	—

Guanine data not given. (Data from Hotchkiss, 1949, 1952.)

Hotchkiss (1979) quotes various post-1944 references to transformation and shows that although many people thought the phenomenon was significant for genetics, many others did not, either not understanding it or calling it something else such as directed mutation, environmentally induced mutation, etc. The idea of a specific mutator was perhaps understandable, since formal genetics had not yet developed the concept of a unit transfer process. The real point is that most of the people discussing transformation were not so convinced of its importance that they began to work on it.

9.8 CHEMISTRY OF DNA

At the time that the Avery, MacLeod, and McCarty paper was published, the structure of DNA was considered to consist of alternating tetranucleotide units, each consisting of one member of each of the four bases, adenine, thymine, guanine, and cytosine (Portugal and Cohen, 1977). This model was developed by Phoebus A.T. Levene, a member of the Rockefeller Institute. Although Levene and Avery were both members of the Rockefeller, they apparently had little contact. The tetranucleotide model also influenced Alfred Mirsky, who, as we have seen, was skeptical that the transforming principle could be DNA, because "nucleic acids are all alike" (McCarty, 1985).

Rollin D. Hotchkiss, in Avery's laboratory, used paper chromatography to characterize the constituent bases of DNA. Since the base compositions of DNA preparations from different organisms varied, Hotchkiss concluded that the tetranucleotide hypothesis was incorrect (Hotchkiss, 1949). (An extensive review of this early work, with facsimiles of some of the key published data, can be found in Hotchkiss [1979].) At a French genetic colloquium organized in 1949 by André Lwoff,[1] Hotchkiss presented the first results of his studies on the base composition of pneumococcal DNA (Table 9.5). According to Hotchkiss:

[1]This colloquium was noteworthy because of the papers by Delbrück, Monod, Lwoff, and numerous others.

If a nucleic acid preparation contains different molecular proportions of the various nitrogen bases, it cannot consist of multiple repetitions of a tetranucleotide, at least in most of the cases...Once this idea is accepted, it is possible to conceive of innumerable arrangements of the different purine and pyrimidine bases along the lengthy molecular chain...These observations, taken together with the knowledge of the biological potentialities of transforming extracts from different strains of pneumococci, emphasize the conclusion that DNA molecules of high molecular weight are able to adopt particular specific structures which would in certain cases be the basis for their specific hereditary properties (Hotchkiss, 1949).

Hotchkiss (1966) has written about how his paper, containing important new data but "translated into flowing French," was buried forever because it was neither read nor abstracted properly. He would subsequently publish the same data elsewhere (Hotchkiss, 1952).

One scientist who apparently was convinced of the importance of DNA by the Avery paper was Erwin Chargaff, a Viennese biochemist on the faculty of Columbia University Medical School. Independently of Hotchkiss, Chargaff began to study the chemistry of DNA. According to Chargaff's later recollections, he decided as a result of the Avery paper to switch fields and devote all his research efforts to a study of DNA (Chargaff, 1978, 1979). Chargaff hypothesized that if DNA was the active agent in transformation, it must carry species specificity. He employed paper chromatography to separate the DNA nucleotide bases and determined the base compositions of DNA preparations from a variety of animal tissues as well as microorganisms. He showed (Chargaff, 1950, 1951) that the molar ratios of the bases of DNA preparations varied markedly (Table 9.6). However, there were several consistencies: Purines generally equaled pyrimidines, and the ratios of adenine to thymine and guanine to cytosine generally were about one. These ratios, subsequently to be called "Chargaff's rules," were to play a significant role in the development of the Watson and Crick model for DNA (Portugal and Cohen, 1977).

It is not necessary to repeat the history of the discovery of the structure of DNA, which has been well covered in other sources (Portugal and Cohen, 1977; Judson, 1979). In the present context, we should merely note that it was the knowledge of the genetic significance of DNA, derived primarily from bacterial transformation studies, that encouraged structural studies on the

Table 9.6   Molar Ratios of Bases in Different DNA Preparations

Source of DNA	Ratios				
	A/G	T/C	A/T	G/C	Pu/Py
Ox	1.29	1.43	1.04	1.00	1.1
Human	1.56	1.75	1.00	1.0	1.0
Chicken	1.45	1.29	1.06	0.91	0.99
Yeast	1.67	1.92	1.03	1.20	1.0
Hemophilus	1.74	1.54	1.07	0.91	1.0
Escherichia	1.05	0.95	1.09	1.08	1.1
Mycobacterium	0.4	0.4	1.09	1.08	1.1
Serratia	0.7	0.7	0.95	0.86	0.9

The *Hemophilus* DNA was active in transformation. (Data from Chargaff, 1951.)

DNA molecule.[2] The Watson and Crick structure was so obviously "right," and explained so much about heredity, that it quickly became accepted. The Watson and Crick structure made the transformation phenomenon explainable in chemical terms, and thus brought transformation into the main stream of molecular genetics.

9.9 TRANSFORMATION IN OTHER BACTERIA

The discovery of transformation in pneumococcus elicited considerable interest in the bacteriological community, and it was natural for workers to look for this process in other bacteria. Because transformation in pneumococcus involved the R→S conversion, the process was first sought in other bacteria that exhibited S/R variation. One of the first reports of successful transformation of another organism was that of Boivin (1947) on *Escherichia coli*. Unfortunately, Boivin died shortly after this report, and his strains were lost. Attempts by Lederberg and others to achieve transformation in *Escherichia coli* were unsuccessful, and it was not until the 1970s that transformation with this organism, although never of high efficiency, became a genetic tool (Mandel and Higa, 1970; Cosloy and Oishi, 1973; Hanahan, 1983).

In 1951, transformation was reproducibly obtained in *Hemophilus influenzae*, a pathogen that also showed S/R variation (Alexander and Leidy, 1951). Hattie Alexander and Grace Leidy, at Columbia University Medical School, followed Avery's procedures almost exactly and were able to obtain active transforming principle that could convert R strains to S, either of the same or of a different type. Successful transformation was later obtained with another pathogen, meningococcus (*Neisseria meningitidis*). A review of early work on transformation in these systems is given by Hotchkiss (1952).

Unfortunately, all of these organisms were nutritionally complex. Although a few biochemical markers eventually became available, these organisms did not provide the vast array of genetic characters that were being exploited so successfully in *Escherichia coli* by Lederberg and others. It was not until the late 1950s that successful transformation was obtained in a nutritionally manageable organism, *Bacillus subtilis* (Spizizen, 1959).

9.10 TRANSFORMATION AND INFECTIVE HEREDITY

A "cytoplasmic" or "infectious" interpretation of pneumococcus transformation was first made by Lederberg (1949). Including pneumococcus transformation under a heading in his review called "infectious transmission" along with lysogenicity, Lederberg stated:

> ...from purely mechanical considerations it would seem most likely that the transforming agents are incorporated into a cytoplasmic system like that of kappa...There would also seem to be a parallelism with the phenomenon of induced lysogenicity... (Lederberg, 1949).

As discussed in Sections 7.5 and 8.1, Lederberg had also initially favored a cytoplasmic interpretation for lysogeny and transduction. The discovery by

---

[2]James Watson was a graduate student at Indiana University. Although a student of Luria, he took courses from both Hermann Muller and Tracy Sonneborn, two geneticists who had written about the genetic significance of the Avery work.

Zinder and Lederberg (1952) of generalized transduction in *Salmonella typhimurium* (see Section 8.1) immediately raised the question of the relation between transduction and transformation. Initially, the role of bacteriophage in the Zinder/Lederberg phenomenon was uncertain, and the term transduction was defined broadly as any "genetically unilateral transfer" (Zinder and Lederberg, 1952) or "the transmission of a (nuclear) genetic fragment from a donor cell (which in every case so far is destroyed in the process) to a recipient cell which remains intact" (Lederberg, 1955). For some years Lederberg used the term transduction to include transformation (Lederberg, 1956). One characteristic in which the two processes were similar was in the fragmentary nature of the gene transfer. Both transduction and transformation brought about the transfer only of closely linked traits. However, the size of the transforming principle was markedly smaller than the size of the transducing particle.

In his review of "infective heredity" at the 1953 Cold Spring Harbor symposium, Zinder (1953) was prompted by the newness of the phage-mediated transduction process to contrast it with the DNA-mediated process in some detail. It was obvious from Zinder's discussion that he considered it inadvisable to use Lederberg's broad definition of transduction (for *any* fragmentary process). It was essential to be able to contrast the two processes, and despite Lederberg's continued broader usage of transduction, the usage adopted by Zinder became accepted. At the major Oak Ridge Symposium in 1954, Hotchkiss attempted to turn the situation around and make "transformation" the broader term:

> ...I should like to recommend that we retain "transformation" as the generic term [for any fragmentary transfer process] and save "transduction" for the phage-mediated transformations. It seems inadvisable to use the term "transforming principle" except when talking about an abstract principle, rather than an actual material. The term "transforming agent" can be used in reference to a material entity; if it is a phage, it becomes a transducing agent (see discussion in Lederberg, 1955 [p. 102]) (see also Section 8.2).

However, usage has not followed these suggestions. The term transformation has remained restricted to those processes involving free DNA, whereas the term transduction refers only to phage-mediated processes (Hayes, 1964).

By the time of Zinder's 1953 Cold Spring Harbor paper, the analysis of transformation had proceeded sufficiently so that its basic genetic character was accepted. Although phage-mediated transduction was a new process, its use as a tool for genetic analysis of *Salmonella* was already under way. What do the two processes have in common?

> The systems of bacterial gene transference discussed herein lead fairly directly to the conclusion that bacteria have hereditary determinants differentiable from the bacteria as a whole. These determinants are so analogous to what are called genes in higher forms that it seems unnecessary to distinguish between the two at this time (Zinder, 1953).

How do the two systems differ from conventional genetic exchange? The two processes result in a replacement of a genetic character by the one introduced, suggesting that at any one time the cell has only a single representative of each genetic factor.

> *A priori*, this singleness of kinds of factors and the necessity of distribution of each kind at cell division speaks for some kind of more tightly knit organization than

free distribution in the cell. Again there is evidence that the transferable units have a spatial differentiation, which in turn could be part of a larger structure, a chromosome. It then would become necessary to explain the unit nature of the transfers. Fragmentation of chromosomes could occur when there is a dissolution of pneumococci or *Hemophilus* and phage lysis of *Salmonella*. Also larger fragments may not be able to penetrate the recipient cells in transformation reactions or be incorporated into the phage carrier in transduction reactions (Zinder, 1953).[3]

Why Infective Heredity?

At almost the same time that transduction was discovered, the F factor was discovered in *Escherichia coli*, and its infectious nature was demonstrated (see Section 5.6). Because of the infectious nature of kappa in *Paramecium*, it seemed likely that there were numerous categories of genetic exchange involving "infectious" particles. Since the transducing phage was clearly infectious, it also seemed not unreasonable to consider transformation as an infective process. Indeed, in one sense transformation was infective, since the DNA that was transferred was able to replicate in the recipient cell, and the resulting culture could provide more DNA for another round of transformation. Thus, it did not seem to do violence to the situation when Cavalli, Lederberg, and Lederberg (1953) stated: "Infective inheritance was first described in micro-organisms 25 years ago (Griffith, 1928)." Clearly, however, this was an attempt to homologize processes that were barely comparable. Although it would eventually be shown that pneumococci lysed spontaneously and released transforming DNA (Ottolenghi and Hotchkiss, 1960), pneumococcal transformation could hardly be considered representative of any sort of "life cycle."

The main contribution that these discussions made to the transformation problem was to focus on the fragmentary nature of the genetic transfer. This, of course, had already been appreciated by the Rockefeller group in a vague way, but analogies with transduction (and vague analogies with the F-mediated genetic exchange) emphasized the need for an understanding of the size of the transforming DNA.

9.11 SIZE OF TRANSFORMING DNA

In their initial work, Avery, MacLeod, and McCarty (1944) estimated the molecular weight of the transforming DNA as about 500,000. Subsequently, Fluke, Drew, and Pollard (1952) used X-ray inactivation and calculations based on target theory to estimate the mean molecular weight of the functional genetic unit in pneumococcus as 5×10^6. However, although X-ray inactivation worked quite well for estimating the sizes of virus genomes,

---

[3]Although the 1953 Cold Spring Harbor paper was given by Zinder alone, Hotchkiss's influence is evident throughout. At Rockefeller, Zinder was in Hotchkiss's laboratory. In a personal communication (19 October 1989) Hotchkiss has informed me that he had been originally invited by Demerec to give the paper and had suggested that Zinder be a coauthor. The plan was to present the common view of these two authors that transduction was another way in which DNA elements got transferred, unit by unit. According to Hotchkiss: "[Zinder] soon had his part ready and I then withdrew, as much for lack of time as for generosity toward a young colleague; instead we discussed his presentation in detail. The comments on fragmentation and single factor analysis had to have reflected both of us, as marker separation and recombination became a preoccupation of mine from the 1951 symposium for years. I can claim a share of credit for the emphasis but not for the succinctness or total scope of the Zinder article."

there were numerous problems with using this procedure for estimating the size of free DNA (Ephrussi-Taylor, 1957). Using sedimentation and diffusion measurements, Goodgal and Herriott (1957a, 1961b) estimated the molecular weight of *Hemophilus* transforming DNA at about 15×10^6. By making assumptions about the amount of DNA per bacterial nucleoid, one could calculate that each transforming unit constituted between 1/200 to 1/500 part of the entire bacterial genome.

Throughout most of the early period of transformation research, it was not known that the DNA of a bacterial cell existed in a single extremely long molecule. Even as late as 1960, postulates of "linkers" in bacterial DNA were being made, and it was not until the work of Levinthal, Davidson, Kleinschmidt, and Cairns in 1962 and 1963 (see Section 3.9 and 5.18) that it was appreciated that the bacterial chromosome was a single molecule of about 1400 μm length, molecular weight about 2.5×10^9 (4×10^6 base pairs). In his 1958 review of size limitations on genetic transformation, Hotchkiss (1958) was still assuming that a molecule of transforming DNA was derived more or less intact from the donor cell. Because linked markers were now available (see later), it was possible to study recombination for single or double (in one case, triple) genes in single transformation events. "The reaction of cells with DNA probably involves in its initial phases entire DNA molecules" (Hotchkiss, 1958). Although true, this statement was made with ignorance of the fact that isolation of the DNA from a cell resulted in the fragmentation of that DNA. Since much emphasis was placed on the *purity* of the DNA (to prove that protein was not involved) the DNA was subjected to numerous steps of chemical treatment, each one of which would introduce shearing forces that would cause the DNA to fragment. "It seems, therefore, that transforming DNA is taken up as whole molecules and chemically is completely utilized for cellular DNA, but biologically speaking, only fragments, usually small, survive to produce a genetic change" (Hotchkiss, 1958).

9.12 COMPETENCE

Use of transformation to study genetics required an understanding of the fraction of cells in any population that were potentially transformable. The term "competence" is used to express the quantitative variability in transformability of bacteria. Competence can vary for either genetic or physiological reasons (Hotchkiss (1954). It was recognized very early that R strains varied in ability to be transformed, and the R36A strain selected by MacLeod (see Section 9.4) proved to be especially useful in the purification and chemical characterization of the donor transforming principle.

Hotchkiss (1954) showed that the ability to be transformed varied throughout the growth cycle and that waves of competent cells occurred. By selection of proper growth conditions, and phasing of cultures with temperature shifts, Hotchkiss was able to obtain a much higher percentage of competent cells. In these early experiments, almost 20% of the cells were found to be transformed for a single character (Hotchkiss, 1955a), and when grown under the right conditions, virtually all of the cells were competent. Subsequently, it was discovered that competent cell suspensions could be stored for months in the frozen state, making it possible to carry out detailed studies on the DNA uptake process on uniform populations of cells (Fox and Hotchkiss,

1957). The development of objective criteria for quantitative transformation made the analysis of the DNA uptake process more precise (Hotchkiss, 1957; Goodgal and Herriott, 1961a).

9.13 GENETIC MARKERS

One of the major difficulties with pneumococcal transformation as a genetic system was the difficulty of obtaining a variety of mutants. The inability to culture the bacterium in a simple synthetic medium and the limited biochemical potentialities of this lactic acid bacterium placed serious restrictions on any mutant selection. As we have emphasized, bacterial genetics has depended on the development of *selective* conditions, so that the rare events being studied can be readily detected. Lederberg's prototrophic selection procedure (see Section 5.3), using auxotrophic mutants, permitted detection of genetic recombination when the frequency was less than 10^{-6}. The capsular markers that were used in the original Griffith and Avery work provided only limited selectivity.

When Lederberg published his first work on mating in *Escherichia coli*, he was able to present a fairly substantial genetic map. In pneumococcus, mapping was not possible as long as only the single capsular character was being studied. The need for additional genetic markers was recognized early by Taylor and Hotchkiss at the Rockefeller Institute, and considerable effort was made to obtain more markers. In some of her initial work, Taylor (1949) described the isolation of a strain she called *extremely rough* (ER), which had arisen spontaneously from the original MacLeod strain R36A. The ER character could be readily recognized in colonies on plates, as well as by the aggregative growth in liquid cultures. Although ER reverted to R when transferred repeatedly in liquid culture, it was reasonably stable on plates. Strain ER could be transformed into R by use of either R DNA or S III DNA, and the ER→R transformation occurred at a much higher frequency than the ER→R reverse mutation. Another important discovery was that an R strain derived by transformation from ER could be transformed into S III in a second step, although not from ER to S III directly. In addition, during transformation, S strains with intermediate levels of capsular polysaccharide arose. Stable strains were obtained which agglutinated to varying degrees with type-specific antiserum. Several of these strains, designated S III-2, S III-1, and S III-N, were studied in some detail and used as sources of DNA for various transformation experiments. Two-step transformations could be obtained of R to S III-1 and from S III-1 to S III-N. Because the S III-1 was not fully encapsulated, it could also be transformed into S II.

This was the first transformation work that provided any genetic insight, and Taylor (1949) concluded that transforming agents were concerned with the heredity of pneumococcus in the same fashion that genes were concerned with the heredity of higher organisms. Taylor suggested that transformation might be used as a means of studying the mechanism of heredity in pneumococcus. The discovery of the ER→R transformation showed that more than one character of the bacterium was under the control of a transforming principle. Thus, the DNA extracted from the S III bacterium must contain "a minimum of two transforming principles." In this 1949 paper, Taylor gave a rigidly genetic interpretation of the transformation system, even so far as to discuss the possibility of the existence of alleles. She credits the French geneticist Boris Ephrussi (later her husband) for extensive advice and discussions during the preparation of her paper. This paper, and a subsequent one presented at Cold Spring Harbor (Ephrussi-Taylor, 1951) that presented further data on the two-step transformations, laid out a framework for the use of genetic analysis in the study of the transformation process. All that would be needed would be more genetic markers.

Antibiotic Resistance Markers

In the early postwar period, antibiotics became widely available, and numerous studies on the development of resistance were reported (see Sections 4.9 and 4.10). Since pneumococci were sensitive to most available antibiotics, it was reasonable that resistant mutants might arise and could be used in genetic studies.

The first antibiotic-resistant mutant of pneumococcus, for penicillin resistance, was isolated by Hotchkiss (1951) from a rough (R) strain by the simple procedure of culturing pneumococci in broth containing the antibiotic.[4] Such a strain, designated RP, could still be transformed to S, and the S transformants remained penicillin resistant. Although this fact would not surprise a bacterial geneticist today, it was considered very significant that the S strain had not become penicillin sensitive in the process of acquiring the capsular polysaccharide! "These findings appear to indicate that the factors for smoothness and penicillin resistance, the S and P factors, can be acquired and exhibited independently of each other" (Hotchkiss, 1951).

The next step, of course, was to show that DNA extracted from penicillin-resistant pneumococci, either S or R, could transform R cells to penicillin resistance. (Transformation of S cells to penicillin resistance could not be done because S cells were nontransformable.) The data showed that penicillin resistance and S character were transformed independently in R receptor cells, with almost equal frequencies.

Hotchkiss also showed that the presence of penicillin did not affect the efficiency of transformation, although the selective action of penicillin made the recognition of transformants easy. Penicillin resistance was a quantitative characteristic and strains were isolated with varying degrees of resistance. The first exposure to transforming agent never resulted in transformants with resistance to more than 0.05 units/ml, even though the donor culture had been resistant to at least 0.30 units/ml. However, such low-level transformants could be brought to higher levels by treatment with the same extract, and the second transformant could be brought to an even higher level after a third step. Thus, penicillin resistance appeared to be controlled by several genes. Hotchkiss noted that the studies by Demerec on antibiotic resistance had shown that resistance to high levels of some antibiotics was acquired in stages. Thus, pneumococcus appeared to have "recapitulated the experience of the resistant donor strains in originally acquiring through

---

[4]It should be recalled that at this time there was still controversy about whether antibiotic resistance was a mutational event or a result of adaptation in the presence of the antibiotic. Lederberg's indirect selection technique, which proved that antibiotic resistance was preadaptive, was not published until 1952 (see Section 4.10).

stepwise spontaneous mutations their resistance to penicillin'' (Hotchkiss, 1951).

Hotchkiss also showed in this paper that high-level streptomycin resistance could be transferred to pneumococcus. The next year Alexander and Leidy (1953) described in *Hemophilus* transformation in a single step to high-level streptomycin resistance.

Linked Markers

One of the principal characteristics of genetic systems is the linkage between unrelated genes that can be demonstrated when genetic exchange occurs. Linkage is one of the main criteria by which genes are mapped, and the behavior of linked genes during meiosis permits an experimental association between the gene and the chromosome. In higher organisms, linkage means the appearance in recombinants of the alleles of two genes with greater than random probability (50%). The closer two genes are together, the lower the probability that they will segregate during genetic exchange.

In bacterial transformation, complete genomes (chromosome complements) are not brought together in the same cell, so the demonstration of linkage must be done in a quite different way. Thus, the results of a recombination experiment will depend not only on linkage, but also on two other factors: (1) the extent to which two markers have been fragmented before they are introduced into a cell; and (2) the extent of homology between the incoming DNA and the recipient (mismatch repair can bring about recombination even if the two units are not precisely homologous). In transformation, linkage is observed if two markers are transferred together on the same DNA fragment.

In assessing linkage in transformation, it is not enough to add DNA and score for double transformants. This is because a single cell is able to take up more than one DNA fragment, and there is a definite probability that two *different* DNA fragments will enter the same cell. The manner by which incorporated DNA is recombined into the recipient cell is still not fully understood. However, there is a significant probability that any cell which has incorporated two DNA fragments will become transformed for characters present in each fragment. The probability of a cell taking up independently two separate DNA fragments should be the product of the probabilities of taking up either fragment separately. Thus, if the fraction of transformants for one marker is 0.005 (0.5%) and the fraction of transformants for the other marker is 0.02 (2%), then the expected fraction of double transformants should be 0.0001 (0.01%).

Extending this analysis, what would be the fraction of double transformants if both markers were linked? This fraction should clearly be in excess of that expected from independent events, but there are two aspects of the transformation process that affect the results: (1) competition between DNA fragments for cellular absorption sites; (2) fraction of cells competent to be transformed. It was determined very early by the Avery group that the frequency of transformants was dependent on the concentration of DNA added, but that as the DNA was increased, a saturation effect was seen due to mutual competition for absorption sites. Even nonspecific DNA (for example, calf thymus DNA) was able to inhibit transformation. Thus, whether double transformants develop from independent uptake events depends on the external DNA concentration.

Cell competence plays an even more significant role, since the fraction of transformed cells must be based on the total population present in the inoculum. If all the cells are competent, then if two genes are unlinked, the random frequency of doubles would be the probabilities of the separate transformations based on the whole population present. However, if only a small fraction of the cells are competent (say, 10%), then the frequency of single transformants is actually *higher* than it appears when the population of cells as a whole is taken. Thus, the true frequency of doubles would be 10 times higher *even* if the markers were unlinked. However, since the investigator does not know what fraction of the population is competent, the total population is used in the calculation. The analysis can therefore lead to a spurious conclusion that linkage exists. (The analysis given here is based on Goodgal [1961] and Hayes [1964], although the basic idea was clearly understood by Hotchkiss and co-workers.)

The first example of linked markers in transformation was the report of Hotchkiss and Marmur (1954) for mannitol utilization (M) and streptomycin resistance (S). The analysis showed that the frequency of double transformants was markedly higher for the M/S pair when the DNA was derived from a doubly marked strain than when a mixture of DNA molecules from separate M and S strains was used. In subsequent experiments, DNA taken from a quadruply marked strain (using sulfonamide and penicillin resistance as additional markers) showed double transformants higher than predicted for the S/M pair but not for the streptomycin/sulfonamide or streptomycin/penicillin pairs.

The discovery of linkage provided further evidence that the transformation process should be viewed in a genetic manner. The M and S characters had no obvious biochemical (phenotypic) relationships and had appeared independently in spontaneous or induced mutations. Further evidence that S and M were linked was shown by the use of S as an *unselected* marker. Cells that were streptomycin-resistant and mannitol-negative were transformed with DNA from mannitol-positive cells, and the mannitol-positive transformants were screened for streptomycin sensitivity. The results showed that the streptomycin-resistance factor was occasionally replaced by the streptomycin-sensitive factor, providing strong evidence for linkage. The reverse experiment was also successful, the mannitol-positive trait from a streptomycin-sensitive donor replacing the trait for streptomycin resistance.

Subsequently, Hotchkiss and Evans (1958) showed that the sulfonamide-resistance locus in pneumococcus was complex and could be separated into three distinct loci, designated *a*, *b*, and *d*, all of which were linked. Transformation could be used to resolve each of these loci separately, although the original resistant mutant appeared to have arisen in a single step. A number of capsular polysaccharide loci were also shown to exhibit linkage in pneumococcus (Ravin, 1960, 1961). In *Hemophilus*, linkage between streptomycin and novobiocin resistance was found (Goodgal and Herriott, 1957b; Goodgal, 1961). As these and other data on transformation gradually accumulated, geneticists became more and more convinced that transformation was indeed a legitimate type of genetic process.

9.14 MECHANISM OF TRANSFORMATION

From the beginning of serious research on the transformation process, questions were being raised about how DNA molecules were able to penetrate cells. Studies on the effect of environmental variables on transformability showed that there were marked variations in susceptibility to transformation, depending on how the cells were grown and treated. Gradually it came to be realized that under the best conditions, all the cells in a population were susceptible to transformation (see Section 9.12) and that the low frequencies of transformation achieved were due to the fact that the transforming principle DNA had become fragmented during isolation and purification. Once competence could be rigidly controlled (see earlier), it was possible to think of experiments on the mechanism of DNA uptake and integration.

Even in the initial studies of McCarty, it was shown that the transforming DNA reached a state in which it was no longer sensitive to DNase (McCarty, Taylor, and Avery, 1946). Later, Hotchkiss showed that DNA became resistant to the action of DNase in just a few minutes after it was added to competent cells. Using ^{32}P-labeled DNA, Lerman and Tolmach (1957) studied in detail the binding and incorporation of DNA into pneumococcal cells. They found that the DNA was first bound to cells so that it could not be removed by washing, but was still sensitive to DNase. Within a few minutes the DNA was converted into a permanently bound state, from which it could not be removed, and in this state it was resistant to the action of DNase. Both the transient and permanently bound states were nonspecific, in the sense that *any* high-molecular-weight DNA could be bound or incorporated. It was estimated that although not all molecules became permanently bound, if a molecule of transforming DNA became fixed, it had a probability of close to 1 of producing a transformant (Fox and Hotchkiss, 1960).

Of considerable interest, because of its analogy with the eclipse period of phage infection (see Section 6.12), was the discovery that transforming DNA lost its biological activity shortly after it had become irreversibly bound to the competent cell (Fox and Hotchkiss, 1960; Fox, 1962). The technique was to allow competent bacteria to absorb genetically and isotopically marked DNA irreversibly, and then to isolate the DNA at intervals and assay it for transforming activity in a *second* round of transformation. The results showed a distinct eclipse period when the newly fixed DNA suffered a loss in function, and this function was rapidly recovered upon further incubation. By following linked markers, it was also possible to show that the newly incorporated DNA became linked to the resident DNA, and that this linkage occurred in the absence of significant DNA synthesis. The eclipse period was shown by Lacks (1962) to occur because one of the two strands of the incorporated DNA was discarded. Although single-stranded DNA is not active in transformation, it seems likely that only one of the two incorporated strands becomes integrated into the recipient genome. Discussion of later work in this area would take us out of the realm of history, but a detailed discussion can be found in Hayes (1968), Hotchkiss and Gabor (1970), and Smith, Danner, and Deich (1981).

9.15 TRANSFORMATION AS A TOOL FOR STUDYING THE PHYSICAL PROPERTIES OF DNA

It was realized fairly early that research on bacterial transformation could be used to correlate the physical and biological properties of DNA. After the Watson and Crick structure of DNA became known, this type of study could be given more specific chemical meaning. Numerous studies were reported on the effect of various chemical and physical treatments on transforming DNA, with attempts to correlate measurable changes in its properties with changes in biological activity. Unfortunately, the interpretation of many of these studies was complex, since biological activity was often lost before any chemical changes could be detected. However, the effect of heat on the structure and biological activity of DNA provided some far-reaching implications. Temperature treatments were to become a major tool for the study of complementary base pairing, hybridization of nucleic acids, and genetic engineering studies.

In 1953, Stephen Zamenhof, a colleague of Erwin Chargaff at Columbia University, showed that the biological activity of transforming DNA in the *Hemophilus* transformation system was much more resistant to heat than the enzyme activity of proteins, but that there was a sharp transitional temperature (80°C in *Hemophilus*), above which the viscosity and biological activity dropped sharply (Zamenhof, Alexander, and Leidy, 1953).

In 1958, Meselson and Stahl (1958a,b) showed that heating did not destroy the polynucleotide structure of DNA, but only the complementary base pairing. These results clarified the heating effects observed by Zamenhof and led the way to the experiments of Marmur and Doty. In 1960, Paul Doty, Julius Marmur, and associates published studies on heated DNA that were to

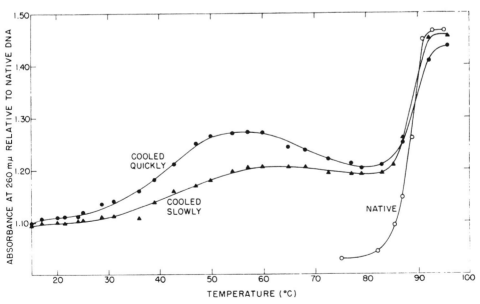

Figure 9.5 Thermal transition of native, slowly cooled, and quickly cooled pneumococcus DNA. Note the sharp change in absorbance for the native DNA. (Reprinted, with permission, from Doty, Marmur, Eigner, and Schildkraut, 1960.)

have a stunning effect on the molecular biology community (Doty, Marmur, Eigner, and Schildkraut, 1960; Marmur and Lane, 1960; Doty, 1961).

Doty was a chemist at Harvard University who had carried out extensive and critical studies on the physical chemistry of polynucleotides, and Marmur was a biologist who had studied bacterial transformation with Hotchkiss at the Rockefeller Institute. They showed that if DNA was slowly heated, there was a dramatic change in its viscosity at a temperature that was characteristic for the DNA. Correlated with the loss of viscosity was a marked *increase* in ultraviolet absorbance (hyperchromicity) of the DNA (Fig. 9.5). The temperature (T_m) at which the transition occurred depended on the percentage of guanine/cytosine base pairs, in agreement with the fact that GC base pairing involves three hydrogen bonds, whereas AT base pairing involves only two (Fig. 9.6) (Doty, Marmur, and Sueoka, 1959).

Even more interesting, under certain conditions the heating effects were *reversible*. If the heated DNA was allowed to cool *slowly* (later called *annealing*), the ultraviolet absorbance decreased again. These effects were not as marked if the DNA was cooled *quickly* after heating (Fig. 9.5).

Marmur and Lane (1960) showed that the loss of biological activity of transforming DNA on heating correlated with the melting temperature of the DNA. Marmur and Lane (1960) also asked the question: Would this biological activity be regained if the heated DNA was cooled slowly? The answer was a dramatic yes. DNA heated to a temperature above the T_m and cooled quickly lost virtually all of its biological activity, but if this heated DNA was allowed to cool slowly, a significant fraction of its activity was restored (Fig. 9.7). The experiment was interpreted to indicate that when the DNA molecules were cooled slowly (annealed), the complementary strands were able to find themselves again, whereas if the DNA was cooled rapidly, reformation

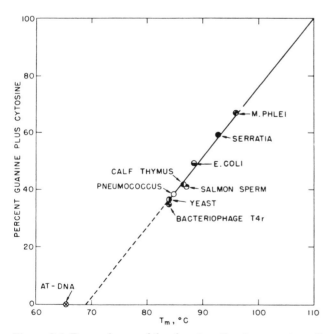

Figure 9.6 Dependence of the denaturation temperature, T_m, on the guanine-cytosine content of the DNA. (Reprinted from Doty, Marmur, and Sueoka, 1959.)

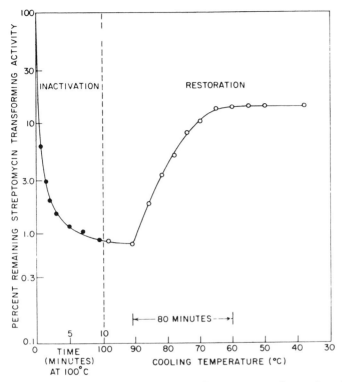

Figure 9.7 Thermal inactivation and restoration of transforming activity of pneumo-coccus DNA. (Reprinted, with permission, from Marmur and Lane, 1960.)

of the double helix could not take place. Although there were numerous technical matters to deal with (ionic strength during heating and cooling, concentration of DNA, rate of cooling, etc.), the technique provided a direct approach to the biological characterization of DNA molecules. It should be noted that although the physical studies could have been done without reference to biological activity, the assay for transforming activity increased greatly the *significance* of the results.

> Bacterial transformation offers a unique approach to [the DNA] problem since the activity of DNA isolated from genetically marked strains can be assayed after being subjected to various treatments...by giving particular attention to the *rate* of cooling from the elevated temperature we have been able to demonstrate a pattern of inactivation and restoration of biological activity consistent with strand separation and specific reunion. It appears that this reunion can take place between complementary strands or between strands of two closely related molecules from mutant strains of *D. pneumoniae* (Marmur and Lane, 1960).

9.16 CONCLUSION

Bacterial transformation certainly provides one of the most fascinating chapters in the history of bacterial genetics. It is also one of the most difficult to assess properly because of the long gestation period and the nongenetic backgrounds of the early scientists working in this field. Transformation was viewed quite disparately by scientists of varying backgrounds. *Geneticists* initially saw it as an example of directed mutation and thought of it as providing an opportunity for carrying out specific genetic change at will. (This promise was eventually to be realized by the development in the recombinant DNA age of *site-directed mutagenesis*.) *Virologists* interpreted the transforming principle as a virus-like structure, emphasizing its ability to replicate independently and its capability for modifying the cell that received it. *Immunochemists* were interested, initially, in the nature of the transforming principle because it brought about a specific change in an antigenic polysaccharide. After the transforming principle was shown to be DNA, *polymer chemists* focused on the basic structure of the polynucleotide chain rather than on its genetic content. *Biologists* viewed transformation in an evolutionary context, seeing it as an example of how one species might emerge from another. *Bacteriologists* saw the transformation process in the context of smooth/rough variation and the implication for the identification of species. Finally, *physicians* were concerned about the epidemiological significance.

Eventually, most of these diverse views melted together under the accumulating evidence of the importance of DNA in genetics. Today, it would be difficult to find a *recent* extensive literature on the epidemiological or medical significance of transformation. The attempts of bacteriologists to use transformation as a taxonomic tool have led to nought, although the nucleic hybridization technique that developed out of the work of Marmur and Doty has had extensive taxonomic uses. Chemists studying the conformation of DNA often do not care especially whether they are working with "biological active" DNA molecules (as witnessed by the fact that calf thymus DNA was for many years the "standard" DNA for chemical research).

It is striking that although transformation is a prominent genetic phenomenon in certain bacteria, the development of bacterial genetics did not depend in any way on the existence of transformation or on the knowledge that the transforming principle was DNA.[5] Bacterial genetics could have developed even if the gene had been protein or polysaccharide. Indeed, even the *chemistry* of the gene could have developed without knowledge of transformation, although more slowly, from studies on bacteriophage and *Escherichia coli* genetics. Transformation studies, in fact, have had little impact on our knowledge of gene structure in bacteria. Even today, there is no detailed knowledge of the genetic map of *Streptococcus pneumoniae*. What transformation did was focus attention early on the possible significance of DNA in the hereditary process. Subsequently, several key steps in our understanding of the chemistry of the gene used transformation as a tool, but even these developments could have developed without the use of transformation.

By 1951 an extensive genetic map of *Escherichia coli* was known, but those working on transformation were still trying to convince others that this process was indeed *genetic*. It was not until 1954 that the first linked markers in pneumococcus were discovered, and by this time even larger genetic maps had been developed in *Escherichia coli*, and the F factor had been discovered. By 1954, the unidirectional nature of gene transfer in bacterial mating was

---

[5]In one sense, this is an overstatement, since Lederberg was stimulated to begin work on bacterial genetics because of the announcement of the chemistry of the transforming principle. Although transformation may have provided the original motivation, it did not help him in his actual work, and the development of genetics of *Escherichia coli* did not directly depend on the transformation work, or even on the knowledge that the transforming principle was DNA.

known, and Jacob and Wollman were beginning to use interrupted mating to study the conjugation process. Transduction had been discovered and was already being used to carry out fine-structure mapping of bacterial genes.

By 1954, biochemical genetics was a well-established field, with numerous pathways being studied in *Neurospora* and *Escherichia coli*. The one-gene/ one-enzyme hypothesis was accepted doctrine, and the connection between genes and enzymes was being widely studied. However, it would not be until the late 1950s that significant information about the enzymology of the pneumococcus transformation process would be known. In 1955, Hotchkiss was able to list about 25 distinct bacterial transformations (Hotchkiss, 1955b) but none of these transformation reactions was especially useful for studying the relationship between genes and proteins.

Stent (1972; see also Wyatt, 1975) has argued that geneticists did not seem to be able to do much with the initial discoveries of transformation, or even with the connection between the transforming principle and DNA. However, many have argued otherwise (Dubos, 1976; McCarty, 1985). Indeed, much of the controversy regarding the acceptance of Avery's work has centered on whether Avery knew its genetic significance. Actually, the critical point is not whether Avery realized the general significance of transforming principle as DNA (it is clear from the 1943 letter to his brother quoted earlier that he did), but whether he emphasized this point in the 1944 paper. Although the biological significance of DNA is mentioned in the discussion of this paper, it is only one of many things mentioned, and much more space is devoted to immunochemistry (Avery's specialty) than to genetics. The standard medical curriculum that Avery had passed through would certainly not equip him for a genetic analysis. It may have been because of the earlier writings of Stanley, Dobzhansky, Burnet, Gortner, and Sonneborn that Avery included a discussion of genetics. As has been emphasized by Dubos and McCarty, Avery strongly discouraged unbridled speculation in his papers. Although pneumococcus transformation could not have been discovered by a geneticist, it would require genetics to "make sense" out of it.

Another point that deserves mention because it is often overlooked is the critical distinction between the experimental approaches used by genetics and biochemistry. Genetics deals primarily with rare events, whether these are mutations or recombinations. Indeed, the reason that bacterial genetics has flourished is because it handles such rare events so much better than classical genetics. Biochemistry, on the other hand, deals primarily with extremely common events, since chemical analyses require large amounts of material. Biochemistry deals, therefore, with the *average* molecules of the cell, whereas genetics deals with the *rare* molecules. This has had important implications in the study of bacterial transformation. Although the basic chemistry of DNA was determined early, the connection between DNA chemistry and gene structure was a much later development. Only in the recombinant DNA era has it been possible to produce biochemically significant amounts of *single* genes for chemical (base sequence) analysis. Arthur Kornberg's extremely important work on the enzymology of DNA replication never made direct use of transformation as a tool. Attempts to produce biologically active DNA by copying transforming principle with the Kornberg polymerase gave variable and not completely convincing results (Litman and Szybalski, 1963). Apparently, the first successful enzymatic synthesis of biologically active DNA was not with transforming principle but with the single-stranded circular phage ϕX174 (Goulian, Kornberg, and Sinsheimer, 1967).

Wyatt (1972) has done a detailed analysis of the impact of Avery's work on the genetics literature. Although geneticists may have been discussing DNA, they were apparently not doing anything about it. *Advances in Genetics*, a major review annual edited by Demerec, contained no mention of DNA in the index through 1956, three years after the Watson and Crick structure became widely known and accepted. This is, of course, reasonable, since *conventional* genetic analysis has nothing *directly* to do with the chemistry of the gene.

Wyatt (1972) has also commented on the reluctance of the phage group under Delbrück and Luria to consider Avery's transformation studies seriously. This idea is implicit also in Stent's discussion of prematurity (Stent, 1972). It is clear that Delbrück and the "phage church" had developed an extremely economical (some would say narrow) approach to research, ignoring or refusing to consider any facts or ideas that did not apply immediately to an analysis of how the T-even phages behaved (see Section 6.6). We have discussed in Section 7.4 the difficulties that Delbrück raised regarding lysogeny. There was also an almost "anti-chemical" bias among the principal members of the phage group. It was not until two of their own members, J.D. Watson and A.D. Hershey, legitimized DNA that it became acceptable to discuss Avery's work seriously. As Hotchkiss has written:

> It is interesting that T-coliphages (by 1952, a support for genetic roles of DNA) were earlier used against transforming DNA. If one phage particle could, as shown by Delbrück and Luria, establish an infection, the gambit ran, why could not one or a few (unknown) protein molecules? Luckily, Alfred Hershey began to ask an almost opposite question; if phage contained DNA—and that was known —why couldn't that be its active component? (Hotchkiss, 1966).

As Hotchkiss notes, the Hershey/Chase experiment that was accepted as strong support for DNA as the genetic material was hardly convincing. Almost 20% of the protein sulfur could not be removed by blending, whereas some of the nucleic acid phosphorus could. Furthermore, some proteins have phosphorus, and there is no reason why some exotic DNA such as phage might not have small amounts of sulfur. "We dealt with DNA of much higher purity...but isolated DNA acts by random mass action laws and we could not approach the high efficiencies of infection obtained with phage" (Hotchkiss, 1966).

The group of researchers studying pneumococcus transformation after World War II remained exceedingly small and, except for Harriett Ephrussi-Taylor and Rollin Hotchkiss, were primarily in bacteriology departments or medical institutes. Despite the encouraging early work by Boivin, there were few attempts to transform *Escherichia coli* (see, e.g., Chargaff, Schulman, and Shapiro, 1957). Lederberg was initially challenged that his mating experiments in *Escherichia coli* might have been a result of transformation (Lederberg, 1972), but once this matter was laid to rest, he seems not to have considered transformation as a possible tool for genetic analysis. Soon, transduction would provide a much more accessible and readily manageable alternative.

Finally, I cannot finish this chapter without making the point that the development of new paradigms will generally come not from the established members of a discipline, but from the "outsiders." No geneticist, and probably no biochemist, would have been able to arrive at the place Avery and co-workers arrived at. It took an association of circumstances to lead to the discovery of transformation in the first place, and only a bacteriologist such as Oswald T. Avery would have been sufficiently interested in the process, in the early days, to pursue its nature (Davis, 1988). Although most "harebrained" ideas come to nought, the few that do have value lead to the real revolutions in science (Cohen, 1985). It is one of the greatest ironies in the history of biology that the phage workers, pursuing the nature of the material of heredity in a logical manner, were "scooped" by a group of medical bacteriologists, self-taught in their biochemistry and innocent of any genetics.

REFERENCES

Alexander, H.E. and Leidy, G. 1951. Determination of inherited traits of *H. influenzae* by desoxyribonucleic acid fractions isolated from type-specific cells. *Journal of Experimental Medicine* 93: 345–359.

Alexander, H.E. and Leidy, G. 1953. Induction of streptomycin resistance in sensitive *Hemophilus influenzae* by extracts containing desoxyribonucleic acid from resistant *Hemophilus influenzae*. *Journal of Experimental Medicine* 97: 17–31.

Alloway, J.L. 1932. The transformation in vitro of R pneumococci into S forms of different specific types by the use of filtered pneumococcus extracts. *Journal of Experimental Medicine* 55: 91–99.

Alloway, J.L. 1933. Further observations on the use of pneumococcus extracts in effecting transformation of type in vitro. *Journal of Experimental Medicine* 57: 265–278.

Arkwright, J.A. 1921. Variation in bacteria in relation to agglutination both by salts and by specific serum. *Journal of Pathology and Bacteriology* 24: 36–61.

Arkwright, J.A. 1930. Variation. pp. 311–374 in *A System of Bacteriology in Relation to Medicine*, volume I. His Majesty's Stationery Office, London.

Austrian, R. 1953. Morphologic variation in pneumococcus. I. An analysis of the bases for morphologic variation in pneumococcus and description of a hitherto undefined morphologic variant. *Journal of Experimental Medicine* 98: 21–37.

Avery, O.T., Chickering, H.T., Cole, R., and Dochez, A.R. 1917. *Acute Lobar Pneumonia. Prevention and Serum Treatment*. Monographs of the Rockefeller Institute for Medical Research, Number 7, New York.

Avery, O.T., MacLeod, C.M., and McCarty, M. 1944. Studies on the chemical nature of the substance inducing transformation of pneumococcal types. Induction of transformation by a desoxyribonucleic acid fraction isolated from pneumococcus type III. *Journal of Experimental Medicine* 79: 137–159.

Beadle, G.W. 1945. Biochemical genetics. *Chemical Reviews* 37: 15–96.

Boivin, A. 1947. Directed mutation in colon bacilli, by an inducing principle of desoxyribonucleic nature: its meaning for the general biochemistry of heredity. *Cold Spring Harbor Symposia on Quantitative Biology* 12: 7–17.

Burnet, F.M. 1945. *Virus as Organism*. Harvard University Press, Cambridge. 134 pp.

Cavalli, L.L., Lederberg, J., and Lederberg, E.M. 1953. An infective factor controlling sex compatibility in *Bacterium coli*. *Journal of General Microbiology* 8: 89–103.

Chargaff, E. 1950. Chemical specificity of nucleic acid and mechanism of their enzymatic degradation. *Experientia* 6: 201–209.

Chargaff, E. 1951. Structure and function of nucleic acids as cell constituents. *Federation Proceedings* 10: 654–659.

Chargaff, E. 1978. *Heraclitean Fire. Sketches from a Life before Nature*. Rockefeller University Press, New York. 252 pp.

Chargaff, E. 1979. How genetics got a chemical education. *Annals of the New York Academy of Sciences* 325: 345–360.

Chargaff, E., Schulman, H.M., and Shapiro, H.S. 1957. Protoplasts of *E. coli* as sources and acceptors of deoxypentose nucleic acid: rehabilitation of a deficient mutant. *Nature* 180: 851–852.

Chesney, A.M. 1957. Oswald Theodore Avery. 1877–1955. *Journal of Pathology and Bacteriology* 74: 451–460.

Cohen, S.B. 1985. *Revolution in Science.* Harvard University Press, Cambridge. 711 pp.

Cosloy, S.D. and Oishi, M. 1973. Genetic transformation in *Escherichia coli* K12. *Proceedings of the National Academy of Sciences* 70: 84–87.

Davis, B.D. 1988. Molecular genetics, microbiology, and prehistory. *BioEssays* 9: 129–130.

Dawson, M.H. 1930a. The transformation of pneumococcal types. I. The conversion of R forms of pneumococcus into S forms of the homologous type. *Journal of Experimental Medicine* 51: 99–122.

Dawson, M.H. 1930b. The transformation of pneumococcal types. II. The interconvertibility of type-specific S pneumococci. *Journal of Experimental Medicine* 51: 123–147.

Dawson, M.H. and Sia, R.H.P. 1929. The transformation of pneumococcal types in vitro. *Proceedings of the Society for Experimental Biology and Medicine* 27: 989–990.

Dawson, M.H. and Sia, R.H.P. 1931. In vitro transformation of pneumococcal types. I. A technique for inducting transformation of pneumococcal types in vitro. *Journal of Experimental Medicine* 54: 681–699.

Dobzhansky, T. 1941. *Genetics and the Origin of Species.* Columbia University Press, New York.

Doty, P. 1961. Inside nucleic acids. *Harvey Lectures* 55: 103–139.

Doty, P., Marmur, J., and Sueoka, N. 1959. The heterogeneity in properties and functioning of deoxyribonucleic acids. *Brookhaven Symposia in Biology* 12: 1–16.

Doty, P., Marmur, J., Eigner, J., and Schildkraut, C. 1960. Strand separation and specific recombination in deoxyribonucleic acids: physical chemical studies. *Proceedings of the National Academy of Sciences* 46: 461–476.

Dubos, R. 1976. *The Professor, the Institute, and DNA.* Rockefeller University Press, New York. 238 pp.

Ephrussi-Taylor, H. 1950. Heredity in pneumococci. *Endeavour* 9: 80–84.

Ephrussi-Taylor, H. 1951. Genetic aspects of transformations of pneumococci. *Cold Spring Harbor Symposia on Quantitative Biology* 16: 445–456.

Ephrussi-Taylor, H. 1957. X-ray inactivation studies on solutions of transforming DNA of pneumococcus. pp. 299–320 in McElroy, W.D. and Glass, B. (ed.), *Chemical Basis of Heredity.* Johns Hopkins Press, Baltimore.

Fluke, D., Drew, R., and Pollard, E. 1952. Ionizing particle evidence for the molecular weight of the pneumococcus transforming principle. *Proceedings of the National Academy of Sciences* 38: 180–187.

Fox, M.S. 1962. The fate of transforming deoxyribonucleate following fixation by transformable bacteria, III. *Proceedings of the National Academy of Sciences* 48: 1043–1048.

Fox, M.S. and Hotchkiss, R.D. 1957. Initiation of bacterial transformation. *Nature* 179: 1322–1325.

Fox, M.S. and Hotchkiss, R.D. 1960. Fate of transforming deoxyribonucleate following fixation by transformable bacteria. *Nature* 187: 1002–1006.

Goodgal, S.H. 1961. Studies on transformations of *Hemophilus influenzae.* IV. Linked and unlinked transformations. *Journal of General Physiology* 45: 205–228.

Goodgal, S.H. and Herriott, R.M. 1957a. Studies on transformation of *Hemophilus influenzae.* pp. 336–340 in McElroy, W.D. and Glass, B. (ed.), *Chemical Basis of Heredity.* Johns Hopkins Press, Baltimore.

Goodgal, S.H. and Herriott, R.M. 1957b. A study of linked transformations in *Hemophilus influenzae. Genetics* 42: 372.

Goodgal, S.H. and Herriott, R.M. 1961a. Studies on transformations of *Hemophilus influenzae.* I. Competence. *Journal of General Physiology* 44: 1201–1227.

Goodgal, S.H. and Herriott, R.M. 1961b. Studies on transformations of *Hemophilus influenzae.* II. The molecular weight of transforming DNA by sedimentation and diffusion measurements. *Journal of General Physiology* 44: 1229–1239.

Gortner, R.A. 1938. *Outlines of Biochemistry*, 2nd. edition. John Wiley, New York.

Goulian, M., Kornberg, A., and Sinsheimer, R.L. 1967. Enzymatic synthesis of DNA, XXIV. Synthesis of infectious phage ϕX174 DNA. *Proceedings of the National Academy of Sciences* 58: 2321–2328.

Griffith, F. 1923. The influence of immune serum on the biological properties of pneumococci. *Reports on Public Health and Medical Subjects, no. 18. Bacteriological Studies.* His Majesty's Stationery Office, London.

Griffith, F. 1928. The significance of pneumococcal types. *Journal of Hygiene* 27: 113–159.

Hanahan, D. 1983. Studies on transformation of *Escherichia coli* with plasmids. *Journal of Molecular Biology* 166: 557–580.

Hayes, W. 1964. *The Genetics of Bacteria and their Viruses*. John Wiley, New York. 740 pp.

Hayes, W. 1968. *The Genetics of Bacteria and their Viruses*, 2nd. edition. John Wiley, New York. 925 pp.

Heidelberger, M. and Avery, O.T. 1923. The soluble specific substance of pneumococcus. *Journal of Experimental Medicine* 38: 73–79.

Hershey, A.D. 1966. The infection of DNA into cells by phage. pp. 100–108 in Cairns, J., Stent, G.S., and Watson, J.D. (ed.), *Phage and the Origins of Molecular Biology*. Cold Spring Harbor Laboratory of Quantitative Biology, Cold Spring Harbor, New York.

Hotchkiss, R.D. 1949. Études chimiques sur le facteur transformant du pneumocoque. pp. 57–65 in *Unités Biologiques Douées de Continuité Génétique*. Publications of the Centre National de la Recherche Scientifique, Paris.

Hotchkiss, R.D. 1951. Transfer of penicillin resistance in pneumococci by the desoxyribonuclease derived from resistant cultures. *Cold Spring Harbor Symposia on Quantitative Biology* 16: 457–461.

Hotchkiss, R.D. 1952. The role of desoxyribonucleates in bacterial transformations. pp. 426–439 in McElroy, W.D. and Glass, B. (ed.), *Phosphorus Metabolism*, volume II. Johns Hopkins Press, Baltimore.

Hotchkiss, R.D. 1954. Cyclical behavior in pneumococcal growth and transformability occasioned by environmental changes. *Proceedings of the National Academy of Sciences* 40: 49–55.

Hotchkiss, R.D. 1955a. Bacterial transformation. *Journal of Cellular and Comparative Physiology* (Supplement 2) 45: 1–22.

Hotchkiss, R.D. 1955b. The biological role of the deoxypentose nucleic acids. pp. 435–473 in Chargaff, E. and Davidson, J.N. (ed.), *The Nucleic Acids*, volume II. Academic Press, New York.

Hotchkiss, R.D. 1957. Criteria for quantitative genetic transformation of bacteria. pp. 321–335 in McElroy, W.D. and Glass, B. (ed.), *The Chemical Basis of Heredity*. Johns Hopkins Press, Baltimore.

Hotchkiss, R.D. 1958. Size limitations governing the incorporation of genetic material in the bacterial transformations and other non-reciprocal recombinations. *Symposium Society for Experimental Biology* 12: 49–59.

Hotchkiss, R.D. 1965. Oswald T. Avery. 1877–1955. *Genetics* 51: 1–10.

Hotchkiss, R.D. 1966. Gene, transforming principle, and DNA. pp. 180–200 in Cairns, J., Stent, G.S., and Watson, J.D. (ed.), *Phage and the Origins of Molecular Biology*. Cold Spring Harbor Laboratory of Quantitative Biology, Cold Spring Harbor, New York.

Hotchkiss, R.D. 1979. The identification of nucleic acids as genetic determinants. *Annals of the New York Academy of Sciences* 325: 321–342.

Hotchkiss, R.D. and Evans, A.H. 1958. Analysis of the complex sulfonamide resistance locus of pneumococcus. *Cold Spring Harbor Symposia on Quantitative Biology* 23: 85–97.

Hotchkiss, R.D. and Gabor, M. 1970. Bacterial transformation, with special reference to the recombination process. *Annual Review of Genetics* 4: 193–224.

Hotchkiss, R.D. and Marmur, J. 1954. Double marker transformation as evidence of linked factors in desoxyribonucleate transforming agents. *Proceedings of the National Academy of Sciences* 40: 55–60.

Hutchinson, G.E. 1945. Marginalia. *American Scientist* 33: 56–57.

Judson, H.F. 1979. *The Eighth Day of Creation*. Simon and Schuster, New York.

Lacks, S. 1962. Molecular fate of DNA in genetic transformation of *Pneumococcus*. *Journal of Molecular Biology* 5: 119–131.

Lederberg, J. 1949. Bacterial variation. *Annual Review of Microbiology* 3: 1-21.

Lederberg, J. 1955. Recombination mechanisms in bacteria. *Journal of Cellular and Comparative Physiology* (Supplement 2) 45: 75–107.

Lederberg, J. 1956. Genetic transduction. *American Scientist* 44: 264–280.

Lederberg, J. 1972. Reply to H.V. Wyatt. *Nature* 239: 234.

Lerman, L.S. and Tolmach, L.J. 1957. Genetic transformation. I. Cellular incorporation of DNA accompanying transformation in *pneumococcus*. *Biochimica et Biophysica Acta* 26: 68–82.

Litman, R.M. and Szybalski, W. 1963. Enzymatic synthesis of transforming DNA. *Biochemical and Biophysical Research Communications* 10: 473–481.

MacLeod, C.M. 1965. The pneumococci. pp. 391–411 in Dubos, R.J. and Hirsch, J.G. (ed.), *Bacterial and Mycotic Infections of Man*. J.B. Lippincott, Philadelphia.

Mandel, M. and Higa, A. 1970. Calcium-dependent bacteriophage DNA infection. *Journal of Molecular Biology* 53: 159–162.

Marmur, J. and Lane, D. 1960. Strand separation and specific recombination in deoxyribonucleic acids: biological studies. *Proceedings of the National Academy of Sciences* 46: 453–461.

McCarty, M. 1945. Reversible inactivation of the substance inducing transformation of pneumococcal types. *Journal of Experimental Medicine* 81: 501–514.

McCarty, M. 1946. Chemical nature and biological specificity of the substance inducing transformation of pneumococcal types. *Bacteriological Reviews* 10: 63–71.

McCarty, M. 1985. *The Transforming Principle*. W.W. Norton, New York.

McCarty, M. and Avery, O.T. 1946a. Studies on the chemical nature of the substance inducing transformation of pneumococcal types. II. Effect of desoxyribonuclease on the biological activity of the transforming substance. *Journal of Experimental Medicine* 83: 89–96.

McCarty, M. and Avery, O.T. 1946b. Studies on the chemical nature of the substance inducing transformation of pneumococcal types. III. An improved method for the isolation of the transforming substance and its application to pneumococcus types II, III, and VI. *Journal of Experimental Medicine* 83: 97–104.

McCarty, M., Taylor, H.E., and Avery, O.T. 1946. Biochemical studies of environmental factors essential in transformation of pneumococcal types. *Cold Spring Harbor Symposia on Quantitative Biology* 11: 177–183.

Meselson, M. and Stahl, F. 1958a. The replication of DNA in *Escherichia coli*. *Proceedings of the National Academy of Sciences* 44: 671–682.

Meselson, M. and Stahl, F. 1958b. The replication of DNA. *Cold Spring Harbor Symposia on Quantitative Biology* 23: 9–12.

Mills, G.T. and Smith, E.E.B. 1962. Biosynthetic aspects of capsule formation in the pneumococcus. *British Medical Bulletin* 18: 27–30.

Mirsky, A.E. 1947. Discussion after paper by A. Boivin. Directed mutation in colon bacilli. *Cold Spring Harbor Symposia on Quantitative Biology* 12: 15–16.

Mueller, J.H. 1945. The chemistry and metabolism of bacteria. *Annual Review of Biochemistry* 14: 733–748.

Muller, H.J. 1947. The gene. *Proceedings of the Royal Society* B134: 1–37.

Neufeld, F. and Levinthal, W. 1928. Beiträge zur Variabilität der Pneumokokken. *Zeitschrift für Immunitätsforschung und experimentelle Therapie* 55: 324–340.

Ottolenghi, E. and Hotchkiss, R.D. 1960. Appearance of genetic transforming activity in pneumococcal cultures. *Science* 132: 1257–1258.

Portugal, F.H. and Cohen, J.S. 1977. *A Century of DNA*. MIT Press, Cambridge, Massachusetts. 384 pp.

Ravin, A.W. 1960. Linked mutations borne by deoxyribonucleic acid controlling the synthesis of capsular polysaccharide in pneumococcus. *Genetics* 45: 1387–1404.

Ravin, A.W. 1961. The genetics of transformation. *Advances in Genetics* 10: 61–163.

Reimann, H.A. 1929. The reversion of R to S pneumococcus. *Journal of Experimental Medicine* 49: 237–249.

Russell, N. 1988. Oswald Avery and the origin of molecular biology. *British Journal for the History of Science* 21: 393–400.

Sapp, J. 1987. *Beyond the Gene*. Oxford University Press, New York. 266 pp.

Sevag, M.G. 1934. Eine neue physikalische Enteiweissungsmethode zur Darstellung biologisch wirksamer Substanzen. *Biochemisches Zeitschrift* 273: 419–429.

Sevag, M.G., Lackman, D.B., and Smolens, J. 1938. The isolation of the components of streptococcal nucleoproteins in serologically active form. *Journal of Biological Chemistry* 124: 425–436.

Sia, R.H.P. and Dawson, M.H. 1931. In vitro transformation of pneumococcal types. II. The nature of the factor responsible for the transformation of pneumococcal types. *Journal of Experimental Medicine* 54: 701–710.

Smith, H.O., Danner, D.B., and Deich, R.A. 1981. Genetic transformation. *Annual Review of Biochemistry* 50: 41–68.

Sonneborn, T.M. 1943a. Gene and cytoplasm. I. The determination and inheritance of the killer character in variety 4 of *Paramecium aurelia*. *Proceedings of the National Academy of Sciences* 29: 329–338.

Sonneborn, T.M. 1943b. Gene and cytoplasm. II. The bearing of the determination and inheritance of characters in *Paramecium aurelia* on the problems of cytoplasmic inheritance, pneumococcus transformations, mutations and development. *Proceedings of the National Academy of Sciences* 29: 338–343.

Spizizen, J. 1959. Genetic activity of deoxyribonucleic acid in the reconstitution of biosynthetic pathways. *Federation Proceedings* 18: 957–965.

Stanley, W.M. 1935. Isolation of a crystalline protein possessing the properties of tobacco mosaic virus. *Science* 81: 644–645.

Stanley, W.M. 1938. Biochemistry and biophysics of viruses. pp. 491–492 in Doerr, R. and Hallauer, C. (ed.), *Handbuch der Virusforschung*. Julius Springer Verlag, Vienna.

Stent, G.S. 1972. Prematurity and uniqueness in scientific discovery. *Scientific American* 227: 84–93.

Taylor, H.E. 1949. Additive effects of certain transforming agents from some variants of pneumococcus. *Journal of Experimental Medicine* 89: 399–425.

Wright, S. 1945. Physiological aspects of genetics. *Annual Review of Physiology* 7: 75–106.

Wyatt, H.V. 1972. When does information become knowledge? *Nature* 235: 86–89.

Wyatt, H.V. 1975. Knowledge and prematurity: the journey from transformation to DNA. *Perspectives in Biology and Medicine* 18: 149–156.

Zamenhof, S., Alexander, H.E., and Leidy, G. 1953. Studies on the chemistry of the transforming activity. I. Resistance to physical and chemical agents. *Journal of Experimental Medicine* 98: 373–397.

Zinder, N.D. 1953. Infective heredity in bacteria. *Cold Spring Harbor Symposia on Quantitative Biology* 18: 261–269.

Zinder, N.D. and Lederberg, J. 1952. Genetic exchange in *Salmonella*. *Journal of Bacteriology* 64: 679–699.

10
GENE EXPRESSION AND REGULATION

Up until the present point in this book, the emphasis has been on genetic phenomena in bacteria, with only occasional glances into biochemistry and physiology. Certain genetic processes, such as transformation, had important biochemical aspects, and enzymology was at the basis of interpretation of many genetic experiments, but for the most part the discussion has proceeded in an almost formal way.

If bacterial genetics had proceeded no further, it might have been interesting primarily for its bacteriological applications, such as in medical microbiology or infectious disease. Bacterial genetics, however, was to have a much broader role to play, a role that was to thrust it into the center of the burgeoning field of molecular biology. It was through an intimate merging of bacterial genetics with biochemistry and physiology that the central processes of cellular metabolism were to be understood. Through this amalgamation, the fundamental dogma of modern biology: DNA → RNA → protein, was to emerge.

In this chapter, we discuss the development of our understanding of the process that is now called *gene expression*, culminating in the development of the messenger RNA concept. The work on adaptive (or what came to be called induced) enzymes in bacteria constitutes one of the most exciting chapters in the whole field of biology; its revelations became the foundation of the young field of molecular biology.

The term gene expression, coined by Monod in the 1960s, focused on an aspect of genetics that became increasingly important during the 1930s. Gene expression is concerned with the manner by which the genotype controls the phenotype. With the development of the one-gene/one-enzyme hypothesis by Beadle and Tatum in the early 1940s (see Section 5.1), the focus became clearer: How do genes bring about the formation of specific proteins (enzymes)? (For a review of early ideas of how this question was formulated, see Spiegelman, 1946.) Induced enzymes seemed to offer interesting material for a study of enzyme synthesis, and in the early post-World-War-II period, because of the work primarily of Jacques Monod, Sol Spiegelman, Roger Stanier, and Martin Pollock (the so-called "Adaptive Enzyme's College of Cardinals"; Cohn, 1978), induced enzymes became an important area of research. Although the study of enzyme induction became a central theme of molecular genetics, it is actually not the central question in gene expression (which is the genetic code). Enzyme induction is really an aspect of the *regulation* of gene expression, and not even, as it turns out, necessarily the most important aspect. However, during the period under discussion, enzyme induction was *viewed* as central, and its solution led, indirectly, to an understanding of gene regulation in the broader sense. The *idea* of messenger RNA arose from studies of enzyme induction, but the direct proof of messenger RNA came from biochemical investigations.

10.1 BACKGROUND ON ENZYMES

Enzymes were first discovered in mammals (pepsin, trypsin, amylase, and other digestive enzymes), but most modern ideas of enzyme function and synthesis depended on biochemical genetic studies in microorganisms. The first work was with yeast (Pasteur, 1860) and led to the first cell-free enzyme preparation by Buchner (1897). Buchner's work led to the development of the discipline of biochemistry.

The early part of the 20th century saw the development of a detailed understanding of the nature of enzymes and the chemical reactions of intermediary metabolism. Many hydrolytic enzymes were purified extensively, and their proteinaceous nature was demonstrated. An important aspect of enzymology, in the present context, was the demonstration of the high specificity of enzyme action, which was made possible by knowledge of the stereospecificity of organic compounds. The physiological properties of organisms came to be seen as determined by their enzymatic makeup. During this same period, the isolation of numerous bacteria capable of carrying out distinct fermentations, primarily by the Delft School of General Microbiology, pointed the way to an understanding of the diversity of microbes in terms of their physiology and enzymology (see Kluyver and van Niel, 1956, for a brief account of the viewpoint of the Delft school). Thus, although yeast served as the primary organism for studies in intermediary metabolism in the early 20th century, fermentative bacteria gradually became of interest. Industrial developments during and after World War I, especially the production of acetone and butanol by anaerobic clostridia, elicited considerable interest.

The details of these early developments, in the context of the present work, can be best obtained from the important book of Stephenson (1930) on bacterial metabolism. However, it is important to emphasize that cell-free enzyme assays were uncommon in those days, and many assays were done using whole washed cell suspensions. Methods for breaking microbial cells were only perfected in the 1940s. Before that, if nonliving preparations were used, they were either crude dried cell preparations or toluenized cell suspensions.

Marjory Stephenson was a member of Sir F.G. Hopkins's biochemistry laboratory at the University of Cambridge, and with her students John Yudkin, Ernest Gale, and L.H. Stickland, carried out numerous important studies on enzyme adaptation during the 1930s (discussed below). Her book covered the complete field of bacterial physiology and was based on sound quantitative experiments. Stephenson's book has an especial interest in the present context because, according to Lwoff (1977), it served as Monod's first introduction to the whole field of adaptive enzymes. Monod (1966) in his

Nobel Prize address also acknowledged Stephenson's early influence. The book went through three editions, and it was the second edition of 1938 that Monod was given (cited in his doctoral thesis of 1942).

10.2 ADAPTIVE AND CONSTITUTIVE ENZYMES

It was discovered fairly early that the levels of enzymes varied depending on growth conditions. Much of this early work was with yeast, the organism that Pasteur and Buchner had first used. Because tools for studying genetics in yeast were developed before those in bacteria, for some years yeast provided the only direct demonstration that genes were involved in enzyme adaptation. Unfortunately, however, enzyme adaptation is not the dramatic phenomenon in yeast that it is in bacteria, and presents some complications due to compartmentation, complex nuclear-cytoplasmic relationships, and uncontrolled permeability changes. Hence, many of the sidelines that developed in enzyme adaptation through studies on yeast will be glossed over in the present account. The early yeast work is covered in some detail by Monod (1947).

The first work on adaptive enzyme formation in yeast was that of Dienert (1900). He was interested in the fermentation of galactose by yeast, a process that had been questioned by earlier workers. Dienert showed that glucose-grown yeast could not ferment galactose directly, but if cells were transferred to a medium containing galactose and a nitrogen source, they acquired the ability to ferment galactose. Dienert called this phenomenon acclimation (*accoutumance* in French). He found that glucose- or sucrose-grown cells would only ferment galactose after a 24-hour lag, whereas galactose-grown cells began to ferment galactose almost immediately. He also showed that disaccharides containing galactose, such as lactose and melibiose, would also bring about acclimation to galactose, whereas disaccharides not containing galactose, such as maltose or sucrose, did not. If cells acclimated to galactose were transferred back to glucose, they *lost* their ability to ferment galactose.

Moreover, Dienert recognized that the acclimation phenomenon he had discovered had a broader biological significance than being simply an interesting aspect of the alcoholic fermentation of yeast:

> This work on acclimation to galactose is only a particular case of the general problem of habituation. Processes also occur in animals that can be considered acclimations. Leukocytes may become acclimated to toxins, acclimation to arsenic has been shown by Besredka, and Bordet has shown acclimation in blood serum. Further, leukocytes secrete antitoxins which are heat-labile like enzymes. Just as in yeast, this ability to acquire immunity sometimes occurs readily, at other times more slowly, depending on the animal and the substances in the blood that are involved...Acclimation, provoked by a carbohydrate very closely related to glucose, constitutes a profound modification of the state of the cell (translated from Dienert, 1900).

This remarkable statement, written at almost the dawn of modern biology, presaged numerous future developments. The relationship between adaptive enzymes and antibodies, hinted at in Dienert's words, was to become an important theme of subsequent research. Strangely, despite superficial resemblances, antibody formation is one adaptive phenomenon that does *not* fit into the negative control model that was to arise out of studies of enzyme regulation.

Although the first work on enzyme adaptation was done on yeast, it was not until Karström's work on bacteria that the field was placed into focus. Henning Karström was a student of the distinguished Finnish microbiologist Artturi I. Virtanen, whose broad interests had carried him into studies on bacterial fermentations and enzymes. For his doctoral research, Karström studied the role of enzymes in carbohydrate metabolism of certain fermentative bacteria. His initial interest was in the *specificity* that different bacteria exhibited toward carbohydrates, but in the process of this work he discovered that the *previous history* of the culture determined, in part, its fermentative ability. For instance, if the enteric bacterium now called *Enterobacter aerogenes* was cultivated on glucose, cell suspensions were unable to ferment galactose, arabinose, maltose, or lactose, but each of these sugars could be fermented if the culture was first *adapted* to the appropriate sugar by growth in its presence. However, glucose could be fermented by *all* cultures, no matter what the previous sugar had been. Another example was the adaptation of *Escherichia coli* to the fermentation of lactose (Table 10.1).

On the basis of numerous observations of this kind, Karström recognized *two* kinds of enzymes, which he called *constitutive* and *adaptive*:

> From these results with coli-aerogenes bacteria it is possible to conclude that the *species-specific enzymes of each cell type* can be divided into two groups based on their manner of formation:
>
> 1. *The constitutive enzymes*, which cells *always* produce, *independently* of the composition of the culture medium in which they have been grown.
> 2. *The adaptive enzymes*, which are produced by cells only *when required*; that is, their formation is linked to the adaptation[1] to the corresponding substrate (translated from Karström, 1930, emphasis in the original).

Karström's later review of the problem of enzyme adaptation, published in a widely available journal, served as useful background for Spiegelman, Monod, and others interested in this question (Karström, 1937).

An important point of Karström's paper, which is historically interesting because it anticipated Monod, was the demonstration that adaptation to lactose by *Escherichia coli* consisted of two components, one genetic, the other physiological. Quoting his own work and that of Hershey and Bronfenbrenner (1936),[2] Karström made the clear statement that certain lactose-negative strains of *Escherichia coli* could mutate to lactose-positive. These were the classical *Escherichia coli mutabile* strains studied by Lewis (see Section 4.4) and by Monod later. The *lactase* produced by such strains is an adaptive enzyme.

However, when it came to presenting a possible explanation for enzyme adaptation, Karström could present only the teleological interpretation that the organism was "forced" to synthesize the enzyme to "adapt" to the changing conditions and that bacteria adapted to glucose did not produce enzymes capable of breaking down disaccharides because they did not need them for growth.

Karström's ideas became quickly adopted by the British, primarily through

---

[1]Karström uses the German word *Gewöhnung* and the English word *Adaptation* interchangeably.

[2]We have discussed Hershey's phage work in detail in other places in this book. The paper by Hershey cited here was one of his earliest, and as in his phage work, was done under Bronfenbrenner's direction. It shows an important "talent" for thinking genetically.

Table 10.1 Adaptive and Constitutive Carbohydrate-degrading Enzymes of *Escherichia coli*

Growth substrate	Sugars fermented										
	Glu	Fru	Man	Gal	Suc	Mal	Lac	Xyl	Ara	Rha	Raf
Glucose	+	+	+	−	+/−	−	+/−	−	−	−	−
Sucrose	+	n.t.	n.t.	n.t.	+	−	−	n.t.	n.t.	n.t.	+/−
Maltose	+	n.t.	n.t.	n.t.	+	+	−	n.t.	n.t.	n.t.	n.t.
Lactose	n.t.	n.t.	n.t.	n.t.	−	+	+	n.t.	n.t.	n.t.	n.t.
Arabinose	+	n.t.	n.t.	n.t.	n.t.	n.t.	n.t.	−	+	−	n.t.
Xylose	+	n.t.	n.t.	n.t.	n.t.	n.t.	n.t.	+	−	−	n.t.

The fermentations were measured on washed cell suspensions. (+) Immediate fermentation, without an induction period; (+/−) fermentation after an induction period of 50–165 min; (−) no fermentation in 225 min. n.t. indicates not tested. (Translated, with modifications, from Karström, 1937.)

the influence of Marjory Stephenson. Stephenson's students John Yudkin and Ernest Gale quickly took up the study of enzyme adaptation (Yudkin, 1932; Stephenson and Yudkin, 1936; Stephenson and Gale, 1937). The main contribution of this early work was the proof that enzyme adaptation was not due to the selection of a "variant" strain, but was a conversion of the *whole* population. As discussed in Section 4.6, the British had advanced the concept of "training," a phenomenon that involved the gradual acquisition of a new function. As we have seen, training was interpreted in a rather Lamarckian manner. The phenomenon of adaptation was seen by the Stephenson group to be different from training, since adaptation was concerned with reactions already normal to the organism that appeared fully upon a single transfer to a new medium (Yudkin, 1932; Gale, 1947).

The Yudkin Hypothesis

In 1937, John Yudkin wrote an important review on enzyme variation in microorganisms and advanced a hypothesis to explain adaptive enzyme formation that was to dominate thinking for many years (Yudkin, 1937). After making a distinction between genetic and physiological change, Yudkin discussed the rationale of enzyme adaptation. Rejecting teleological explanations, Yudkin discussed the facts that needed to be explained. (1) In some cases, adaptation occurred in response not only to the substrate, but also to *products* of the enzyme action. (2) In many cases, unadapted cells were not totally devoid of enzyme, but contained smaller amounts. Even where no enzyme could be found in unadapted cells, Yudkin felt that the technique used to detect enzyme was just not sensitive enough. "If this is so, all examples of enzyme production are cases of increase in enzyme and none are instances of the formation of completely new enzyme." This conclusion became the central core of Yudkin's so-called "mass action theory of enzyme formation":

> Without attempting to postulate any definite compounds or definite sequence of events, it is clear that the adaptive enzyme is produced from a precursor or precursors. It does not matter for the purposes of this discussion whether such precursors are compounds of the enzyme with some inhibiting substances or whether they form part of the cell protoplasm normally concerned in some quite other process. In either case it is assumed that an equilibrium exists between such precursors and the formed enzyme. An immeasurably small amount of enzyme in untreated cells would mean that the equilibrium is on the side of the precursors. The combination of the enzyme with any substance would result in a disturbance of the equilibrium and more enzyme would be formed from precursor in order to restore it. Such a virtual removal of the enzyme could of course be effected by combination with its substrate.
>
> We have here then a simple conception of the process by which enzyme production may be induced by the presence of the substrate of the enzyme. The addition of the substrate means that a part of the enzyme present is at any given moment combined with the substrate and the restoration of the precursor-enzyme equilibrium involves the formation of more enzyme from precursor (Yudkin, 1937).

According to this (erroneous) hypothesis, the equilibrium in the cell between precursor and enzyme was altered by the presence of the substrate (or in some cases, the product), which combined with the enzyme and stabilized it, thus "pulling" the reaction to the right. In this model, the substrate played a major role, since it was only because the substrate combined *specifically* with the enzyme that the stabilization occurred. Another deduction from the theory was that other substances which combined with the enzyme, such as inhibitors, might also affect enzyme synthesis. Subsequently, when Monod discovered diauxie (see later), this idea would return to play a prominent role in theoretical interpretations.

Spiegelman's Work on Yeast

Sol Spiegelman became one of the central figures in the study of enzyme adaptation and, indeed, the whole process of protein synthesis. Spiegelman began as a student of Carl C. Lindegren, the pioneer *Neurospora* geneticist (see Section 5.1) who had turned to yeast when he moved to Washington University in St. Louis. In the late 1930s, Otto Winge in Copenhagen, working with the Carlsberg brewing industry, had developed procedures for doing yeast genetics, and Spiegelman and Lindegren capitalized on these procedures in a study of enzyme adaptation. Although Stephenson in England had studied adaptation primarily because of its biochemical interest, Spiegelman and Lindegren saw the problem in its genetic context. The Beadle/Tatum one-gene/one-enzyme hypothesis had by this time begun to dominate thinking on gene function. After confirming Stephenson and Yudkin's conclusion that adaptation was due not to selection of a mutant, Spiegelman went on to study the genetics of galactose and melibiose fermentation. Although the genetic system was not yet under complete control, the results clearly indicated that ability to adapt was inherited in a Mendelian fashion (Spiegelman, 1945, 1946).

At this time, Yudkin's model was the only focused explanation of the mechanism of enzyme adaptation, and Spiegelman noted that this model led to specific kinetic predictions. The rate of enzyme synthesis should be most rapid immediately after induction commenced and become gradually slower as the hypothetical precursor was used up. However, Spiegelman's experiments showed quite different kinetics: The rate of enzyme synthesis *increased* with time rather than decreased. To explain his results, Spiegelman postulated that a cytoplasmic self-replicating mechanism was involved, such that "once the enzyme is formed, further formation of enzyme molecules can proceed without the intervention of the gene. A mechanism of this sort would result in an autocatalytic transformation of [precursor] into [enzyme]... (Spiegelman, 1946).

It is interesting that this radical idea did not find immediate rejection, but it should be recalled that at this time, research on cytoplasmic inheritance elicited marked excitement. The discoveries of Sonneborn on the *kappa* particle of *Paramecium* (Sapp, 1987) had convinced everyone of the central role of cytoplasmic inheritance, and microorganisms appeared to be the best organisms in which to study this phenomenon. The term *plasmagene* was used to describe these self-replicating cytoplasmic entities. It is not clear whether Spiegelman's kinetic experiments "forced" the self-duplication hypothesis, or whether the hypothesis came first and the kinetic experiments were analyzed quantitatively to see if they fit.

Since Spiegelman's model involved the gene only in the initial events, the enzyme itself (or some derivative) should act as a self-duplicating cytoplasmic particle (plasmagene) once the adaptation process had begun. This idea was tested in the following way: Using the melibiose system, genetic crosses of *mel*$^+$ and *mel*$^-$ were made in both the presence and absence of the substrate melibiose (Spiegelman, Lindegren, and Lindegren, 1945). In the *absence* of melibiose, the conventional Mendelian segregation ratio of 1:1 was obtained, whereas in the *presence* of melibiose, all four spores from a single meiotic event yielded strains with the ability to ferment the sugar, even though two of the four should have been *mel*$^-$. These results, a good example of a "strong inference" experiment, were of course erroneous and were subsequently withdrawn (Lindegren and Lindegren, 1946).[3]

Spiegelman's model for adaptive enzyme formation, based on these erroneous hybridization experiments, postulated that adaptive enzymes were produced by plasmagenes, the role of the gene being to initiate enzyme synthesis by virtue of a low capacity to produce a few of the cytoplasmic self-duplicating units. Once the plasmagene had formed, the gene was no longer needed (Spiegelman, 1946). What was the role of the substrate in Spiegelman's model? The plasmagene-enzyme complex was hypothesized to be unstable, and if substrate was absent, the plasmagene would not be maintained. Once present, however, the plasmagene could replicate and continue to bring about enzyme formation. Thus, once the substrate had initiated plasmagene formation, it was no longer essential for continued enzyme synthesis. This 1946 model is a refinement of Spiegelman's 1945 model in which the enzyme itself had been postulated to be self-replicating. Between 1945 and 1946, a role for nucleoproteins in enzyme formation had come to the fore (Spiegelman and Kamen, 1946), and the idea that enzymes themselves should be self-duplicating seemed less attractive.

Although Sonneborn and most other advocates of cytoplasmic inheritance viewed the phenomenon as ancillary to the nuclear gene, Spiegelman viewed his plasmagene theory as central to *all* gene action:

> I believe that I have proposed, not a plasmagene theory but a theory of gene action which involves plasmagenes...The critical difference between the plasmagene concept which I have used from others...is that...the plasmagene is not a special, or unique, or isolated, cytoplasmic component, in the sense that it is outside the normal physiological processes. On the contrary, it is assumed to be an integral part of the enzyme-synthesizing system and is presumed to be the *normal* link by means of which genes can effect control over protein formation in the cytoplasm.[4]

Although Spiegelman recognized that the experimental verification of this model would be difficult, at the time it had the satisfying strength of bringing together both the genetic and physiological phenomena of enzyme adaptation. However, the data suggesting autocatalytic synthesis of adaptive enzymes could be explained in other ways, and Monod (1947) soon pre-

---

[3]Although Spiegelman himself soon abandoned the plasmagene model, Lindegren continued to insist on variants of it for his cytogene hypothesis (and even a version involving a kind of reverse transcription!). Spiegelman eventually became a respected contributor to molecular biology, but Lindegren's stubborn insistence on radical ideas unsupported by solid experimental evidence led to his eventual ostracism from the genetics community (Sapp, 1987).

[4]From a letter from Sol Spiegelman to Tracy Sonneborn, September 22, 1947, quoted by Sapp (1987). Both Spiegelman and Sonneborn participated in the Cold Spring Harbor Symposium on Quantitative Biology in July 1946.

sented a model that retained the autocatalytic kinetics that Spiegelman had observed but retained strict nuclear gene control. Monod was uncomfortable with self-replicating enzymes or cytoplasmic genes, since such ideas did not conform to the doctrines of Mendelian genetics: "...the mere existence of Mendelian genetics makes it rather obvious that *purely nuclear* inheritance of enzymatic properties must be considered an almost absolute rule." (Monod, 1947). In Monod's 1947 model, the gene brought about the synthesis of "specific building blocks" that were required to form specific active sites of the enzyme. These building blocks would not be enzymatically active until they were arranged into a more complex structure by combination with nonspecific building blocks. If the rate of interaction between the specific and nonspecific building blocks was increased by the presence of other molecules already formed, then autocatalytic enzyme synthesis would be seen. The role of the substrate was to increase the "competitive value of the corresponding pattern or molecular structure by stabilizing it" (Monod, 1947).

It is interesting to note that in this model, the substrate does not play an "instructive" role but merely a "selective" or regulatory role, and hence the model is reminiscent of the operon model ultimately developed. However, at the time, so little was known about protein synthesis that there was no way of testing Monod's model.[5]

The main significance of the work on enzyme adaptation in yeast was the demonstration that the process exhibited Mendelian inheritance. At the time that the yeast work began, mating had not yet been discovered in bacteria. Yeast therefore provided the only well-defined system where genetics and physiology could be combined (Stanier, 1951).

10.3 AN EARLY MODEL OF GENE FUNCTION

The symposium on "Gene Action in Microorganisms" held in St. Louis February 2 and 3, 1945 at the Missouri Botanical Garden was an important highlight for the whole field of microbial genetics. At this meeting, all the notables were present, including Beadle, Tatum, Lindegren, Demerec, Hollaender, Sonneborn, Delbrück, and Luria. In addition, the geneticist Sterling Emerson gave an important theoretical paper on the template theory of gene action (Emerson, 1945). In his paper, Emerson made use of current ideas about the structure of genes and the one-gene/one-enzyme hypothesis of Beadle and Tatum to present a model for gene function. Emerson's model made considerable use of research on the nature of antibody specificity and antibody-antigen reactions. The important research of Karl Landsteiner on synthetic antigens had focused attention on the complementarity of antibody and antigen, and models of antibody formation by Breinl and Haurowitz (1930) and by Pauling (1940) had postulated that the antigen serves as a template upon which the antibody surface was determined, the surfaces of the two molecules thus being mutually complementary in shape. Pursuing this idea, Emerson noted that complementary surfaces had been suggested

---

[5]Spiegelman and Monod competed for many years. Several of my correspondents have informed me of Monod's distaste for Spiegelman and his ideas. Some of this distaste comes across in Francois Jacob's descriptions of a seminar which Spiegelman gave at the Pasteur Institute (Jacob, 1988).

for other biological systems, such as enzymes and their substrates. Although no direct evidence existed for the role of complementarity in gene function, Emerson noted that genes controlled the specificities of enzymes as well as antigens (the one-gene/one-enzyme and the one-gene/one-antigen hypotheses). The gene thus served as a template, a sort of mold, that produced an enzyme with a complementary configuration (Fig. 10.1a). Since only the active site of the enzyme had to combine with the substrate, it was thought that the gene template only had to be specific for the enzyme in the region of

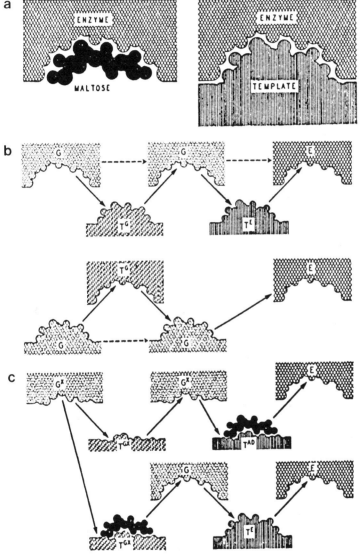

Figure 10.1 Emerson's model for gene action and the role of substrate in enzyme adaptation. (*a*) Complementary surfaces of enzyme/substrate and enzyme/template (gene). (*b*) Possible routes by which specific surface configurations can be transmitted from the gene (G) to gene and from gene to enzyme (E). T^G is a complementary gene template, T^E is an enzyme template. (*c*) Formation of an adaptive enzyme. T^{AD}, adaptive template with substrate attached. T^{GX}, gene template. T^E, enzyme template. (Reprinted, with permission, from Emerson, 1945.)

the active site. (The complexities of protein folding, including the importance of the amino acid sequence, were not understood.) Another critical aspect that Emerson considered was how the gene itself was duplicated. Again complementarity was introduced, but here the gene was envisioned to produce its own complementary template which then produced the gene (Fig. 10.1b). It is clear that in this model the gene and its template are alternate configurations: "If both gene and template are part of the genic material it is purely a matter of convenience which is called gene and which template" (Emerson, 1945). This model bears a striking analogy with the Watson/Crick model for DNA, with the gene and its template being equivalent to the two complementary strands of the DNA polynucleotide.

Emerson also introduced speculations about the role of enzyme substrate in gene function. Although wrong, these speculations were to have an important bearing on thinking of gene expression for many years. It was assumed that the substrate itself was involved in directing enzyme synthesis by combining with and modifying the template (Fig. 10.1c). Such an "instructive" model for the substrate was in keeping with current ideas on the instructive role of antigen in antibody formation. At the time Emerson proposed this model, virtually nothing was known about the details of gene structure, and the central role of DNA had not yet been assimilated into theoretical models. (Emerson's speculations, although cited by Monod and others, have not been considered in the various histories of molecular genetics. Sterling Emerson was an influential member of the faculty at CalTech, and his paper should have been heard at St. Louis by all the key individuals who subsequently participated in the development of molecular biology: Beadle, Tatum, Delbrück, Hershey, and Luria.)

10.4 THE EARLY WORK OF JACQUES MONOD

In the post-World-War-II period, the most influential figure in research on gene expression was Jacques Monod. It is important to emphasize that Monod's gropings toward an understanding of the mechanism of gene expression were often erroneous. After his death, Monod was frequently eulogized by his colleagues and friends, most of whom had short memories or were ignorant of Monod's early work. It is not my point to "debunk" Monod, but in the interest of historical accuracy, it is essential that one not ignore his blunders. In the interests of simplification, and to avoid boring the reader, history is often written as if it moves inexorably in a single direction. However, life does not work this way, and there are numerous false starts and incompletes. Only *after* an event has transpired can we discard all the erroneous ideas; during the exploratory stages, these ideas may not have seemed erroneous at all, and may have affected progress markedly.

The most extensive biographical account of Monod's life and work is that by Lwoff (1977), which also contains a complete bibliography. Monod's key papers have been collected in a useful and admirable volume (Lwoff and Ullmann, 1978), and his associates and colleagues have offered a series of moving tributes (Lwoff and Ullmann, 1979). Extensive background can also be found in his Nobel Prize address (Monod, 1966) and in that by Jacob (1966), as well as in the book by Jacob (1988). However, it is important to emphasize that Monod's work was only part, albeit the most central part, of a large body of genetic, physiological, and biochemical work that came together in the mid to late 1950s to result in the elucidation of the mechanism of protein synthesis and the negative control model of gene expression.

Monod studied biology at the Sorbonne in Paris and did his first research work on protozoa. Among other things, he studied morphogenesis in ciliates and did some early work on the electrical phenomenon of galvanotropism. In 1936, with the support and encouragement of the geneticist Boris Ephrussi, Monod received a Rockefeller Foundation Fellowship to study genetics at CalTech. This stay in Morgan's group resulted in papers on developmental genetics of *Drosophila*, an area of interest to Beadle and Ephrussi (see Section 5.1). Upon returning to Paris in 1937, Monod decided against further work with Ephrussi on *Drosophila* and returned to the Sorbonne to work on the growth of protozoa. At that time, the recognized expert on the growth and nutrition of protozoa was André Lwoff at the Pasteur Institute (see Section 7.4), and Monod came to Lwoff to discuss his work. According to Lwoff: "I told him that ciliates were the worst material to attack the problems of growth, and advised him to use a bacterium able to grow in a synthetic medium, for example *Escherichia coli*. 'Is it pathogenic?' asked Jacques. The answer being satisfactory, Monod began, in 1937, to play with *E. coli* and this was the origin of everything. For it is the systematic analysis of the various parameters of growth of *E. coli* which led to the study of induced enzyme synthesis—at the time enzymatic adaptation—a study which developed into the physiology of the gene and the laws of molecular biology" (Lwoff, 1977).[6]

Monod's personality has been well encapsulated by Melvin Cohn, a close associate of Monod during the early years of his research on gene expression:

> Monod, as well as myself, had an essentially Cartesian view of the world...Monod's resistance to giving up a parsimonious and elegant hypothesis was due to the inability of anyone to present him with a Cartesian argument, conceptual or experimental. When this was done he generally gave up his hypothesis with no personal involvement. He was the most secure scientist I have ever known ...Monod's charisma and leadership derived from his extraordinary breadth of knowledge and his interest in the social and political world around him...He was a cellist, a philosopher, extraordinarily knowledgeable in literature, painting, etc...He had polio and he limped on his left leg which was like a matchstick, yet he was one of the finest mountain climbers in France, he rode horseback and he played tennis. His Protestant background (Hugenot) made him puritanical with respect to himself, and he was driven to overcome any infirmity. Science was only a plaything for him.[7]

Another view of Monod is given by Francois Jacob:

> A personality of extraordinary range, whose many facets included a marvelous intellectual mechanism; an interest in every field; great culture in the arts as well as in science; an absolute rigor in criticism, for others as for himself;...a desire to dominate intellectually, even to terrorize, that led him to sit in the first row at seminars to bombard the speaker with questions and explain to them either what they had done or what they should have done; an enthusiasm in support of any cause he found just. In short, an ardent personality (Jacob, 1988).

---

[6]Lwoff oversimplifies the history. Monod's first work was not with *Escherichia coli* but with *Bacillus subtilis*. Monod did not begin to study adaptive enzyme synthesis in *Escherichia coli* until 1946.

[7]Letter from Melvin Cohn to the author, March 14, 1989.

Monod's doctoral thesis on the growth of bacterial cultures is a classic (Monod, 1942). Published in the darkest period of the German occupation, and while Monod himself was a member of the French resistance movement, the work is a fundamental treatment of how populations of single cells grow. The work is noteworthy for the manner in which quantitative and mathematical analyses are used to illuminate the significance of experimental studies. In the present context, the most significant material deals with the phenomenon of diauxie (see later), but it should be emphasized that Monod's quantitative treatment of microbial growth has played a *major* role in thinking about all microbiological processes. One of the key discoveries from Monod's thesis work was that *growth rate* was a function of *substrate concentration*. Although the thesis itself has remained a rather obscure document, Monod published the essence of the work later in a widely available English-language paper (Monod, 1949).

By the time his thesis was published in 1942, Monod had joined the French resistance. Although still nominally at the Sorbonne, he found time throughout the war to do occasional experiments in Lwoff's laboratory at the Pasteur Institute. When the war ended, Monod returned temporarily to the Sorbonne, but no one in the Laboratoire de Zoologie, where he was working, was interested in the growth of bacterial cultures. In the fall of 1945, he joined Lwoff's group at the Pasteur Institute (the *Service de Physiologie Microbienne*, located in the attic of the Institute), where he quickly established a solid research program on enzymatic adaptation. Contacts with U.S. scientists were reestablished: Lwoff and Monod were invited by Demerec to participate in the 1946 Cold Spring Harbor symposium on heredity in microorganisms, the historic meeting at which so much of the future of microbial genetics was initiated (see Sections 5.3 and 6.9). Soon after the war, U.S. funding for biological research expanded dramatically, and the French group received significant research money from the U.S. National Institutes of Health. Also, many U.S. scientists received financial support to go to Paris and study with Lwoff and Monod.[8] Numerous important collaborations were forged that were to lead to the major research developments in the whole field of gene expression (see below).

Diauxie

The aspect of Monod's thesis research that led him into the study of adaptive enzymes was *diauxie*, the phenomenon of *double growth*. Monod's discovery of diauxie was a logical outgrowth of his studies on the effect of carbohydrate concentration on growth rate and growth yield (Monod, 1941a). Monod found that when *Bacillus subtilis* or *Escherichia coli* was grown on mixtures of certain sugars, *two* growth phases occurred, separated by a short time period when growth either did not occur, or slight lysis took place (Fig. 10.2). Monod initially termed this phenomenon *preferential attack* (*attaque préférentielle*) (Monod, 1941b) but subsequently coined the term *diauxie* (Monod, 1942). A systematic examination of this phenomenon showed that sugars

[8]Another important source of research funds for Monod was the Jane Coffin Childs Memorial Fund of New Haven, Connecticut, a small fund noteworthy for its support of pioneering research. Joshua Lederberg was also supported by this fund during his graduate research at Yale University.

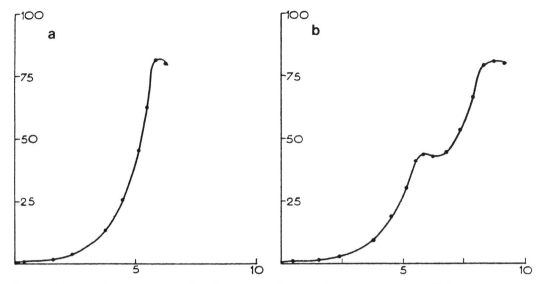

Figure 10.2 The phenomenon of diauxie. (*a*) Growth of a culture of *Bacillus subtilis* in a medium containing two sugars that do not elicit diauxie, sucrose and glucose. (*b*) Growth of a culture in a diauxie-inducing medium, sucrose and arabinose. The amount of growth is expressed in arbitrary units (100 units is equivalent to 80 mg dry weight of bacteria per liter). Time in hours. (Reprinted, with permission, from Monod, 1941b.)

could be divided into two groups, which Monod called A (no diauxie) and B (diauxie). In *Escherichia coli*, the sugars in group A were glucose, mannose, fructose, and mannitol, and in group B were galactose, arabinose, xylose, rhamnose, sorbitol, dulcitol, maltose, and lactose (Monod, 1942).

According to Lwoff, Monod came to the Pasteur Institute with his first diauxie results in December 1940 and asked what they could mean.

> I said it could have something to do with enzymatic adaptation. The answer was "Enzymatic adaptation, what is that?" I told Monod what was known—what I knew—and he objected that the diauxic curve showed an inhibition of growth rather than an "adaptation." We know today that repression and induction are complementary, but I simply repeated that diauxie should be related to adaptation. Anyhow I gave him Emile Duclaux's *Traité de microbiologie*, Marjory Stephenson's *Bacterial Metabolism* and a few reprints I had secured, among them the precious Ph.D. thesis of Karström—which I never saw again (Lwoff, 1977).

According to Monod (1966), his 1940 visit to Lwoff's laboratory was the turning point of his life, and from then on all his scientific activity was focused on a study of the phenomenon of enzymatic adaptation.

In pursuing the phenomenon of diauxie, Monod showed that it was a result of the *inhibition* by the A sugar (glucose) of the utilization of the B sugar. In reviewing Karström's work, Monod was struck by the fact that the sugars which Karström had found to be attacked by constitutive enzymes fell in Monod's A group, whereas those which were attacked by adaptive enzymes fell in his B group. It was this fact which convinced Monod that diauxie had something to do with adaptive enzyme formation, even though there was an inhibitory effect of the A sugar. "Taken together, these results show that diauxie is due to variation in enzymatic constitution, and this variation is linked to the adaptive nature of certain enzymes and to the ability that certain sugars seem to possess of inhibiting these enzymes, or, more precisely, the inhibition of their adaptation. The phenomenon of diauxie will be explained when the precise mechanism of this inhibition is explained" (translated from Monod, 1942).

At that time, the only model for adaptive enzyme formation was that of Yudkin (1937), discussed earlier in this chapter. To encompass diauxie under Yudkin's theory, Monod postulated that a single precursor could give rise to many enzymes, with different equilibrium constants for each reaction. Inhibition of enzyme formation by glucose might then be due to its strong combination with the precursor, preventing the precursor's conversion to any other enzyme. This theory was in keeping with a common model for antibody formation: The antibody was thought to be molded to the antigen (see Section 10.3). One of Monod's attempts to test the precursor hypothesis was to study the *competition* between an A and a B sugar when they were present in different concentrations (varying over a 1000-fold range). When the adaptive sugar was present in much larger amounts than the suppressible sugar, the suppression did not occur as much. From these observations, Monod concluded that the two sugars competed for the same surface in the cell (Monod, 1945).

Unfortunately, diauxie provided misleading signposts. The inhibitory sugar and the inducer do not work at the same site, and the idea of a common target led Monod and co-workers to many inappropriate experiments. It was only when the study turned from diauxie to adaptation itself that the situation became clarified.

10.5 THE EARLY POSTWAR PERIOD: THEORIES AND EXPERIMENTS

Although Monod was trained in a classical biology department, he had the good fortune of having numerous biochemically trained collaborators (many American). In addition, Lederberg's important genetics work not only became central to the experimental analysis of the problem, but Lederberg himself provided some key discoveries on the *lac* system upon which Monod would be quick to capitalize (see below).

The Preenzyme Model

It should be emphasized again that in the immediate postwar period, very little was known about protein structure, and essentially nothing about gene chemistry (DNA was still a rather obscure molecule). Not a single amino acid sequence of a protein was known, and it was not even known whether enzyme specificity was determined uniquely by amino acid sequence: Most workers felt that folding alone was enough. Pauling's model for development of antibody specificity by folding of the antibody globulin in contact with the antigen dominated biochemical thinking (see Section 10.3).

At this time, however, the one-gene/one-enzyme hypothesis was also dominating thinking. If an enzyme was controlled by a gene, what was the role of the substrate in adaptive enzyme formation?

> The basic question is, whence does the molecule derive its specific structure? In the case of "biosynthetic" enzymes, there is no experimental evidence to con-

tradict the hypothesis that a specific gene may be entirely responsible for it. But in the case of adaptive enzymes, there is proof that the substrate plays an important and decisive part in the synthesis, although specific genes are implicated in determining the competence to adapt (Monod, 1950).

One of the first studies that Monod did when he moved in 1945 from the Sorbonne to the Pasteur Institute was to distinguish between mutation and enzyme adaptation in *Escherichia coli mutabile*. We have discussed this organism earlier and have noted that it provided the first evidence of a classical mutation mechanism in bacteria (see Section 4.4). Monod and Audureau (1946) isolated a new strain of this organism[9] and confirmed the earlier work that showed that the transformation of *lac⁻* to *lac⁺* was due to a mutational event. However, Monod went further and showed that the ability to utilize lactose in the resulting *lac⁺* mutants was due to the formation of an adaptive enzyme. The activity of the enzyme, called at that time *lactase*, was measured using cumbersome oxygen uptake assays in a Warburg apparatus. Glucose-grown cells did not respire when suspended in lactose-containing buffer, whereas lactose-grown cells did. In addition, *lac⁺* but not *lac⁻* strains exhibited diauxie when cultured in mixed glucose/lactose medium. Monod and Audureau concluded that these results were in accord with Spiegelman's (erroneous) plasmagene hypothesis and Emerson's model of gene action (see Section 10.3).

This work was reviewed extensively by Lwoff (1946) at the important 1946 Cold Spring Harbor symposium, a symposium that Monod also attended. A key statement of Lwoff was:

> In the case of the L⁻ to L⁺ mutation it is reasonable to assume that the change consists of a structural modification of a precursor, which makes it "adaptable" to lactose...This, however, brings up the problem of the mechanism of enzymatic adaptation, which is obviously of greatest importance to our understanding of enzymatic mutations [after mentioning Yudkin, Lwoff then proceeds to present Monod's hypothesis]. Enzymes allowing attack of carbohydrates by bacteria are all derived from a *common precursor* [italics in original]. This precursor (preenzyme) has a slight general affinity for carbohydrates. Transformation of the precursor into an adapted, specific enzyme occurs as a result of the substrate-preenzyme combination (which would account for the specificity of enzymatic adaptation).
>
> Certain substrates might have greater affinity for the preenzyme and thus be able to displace other substrates from it, which would explain the *nonspecificity* of the inhibiting effect of "constitutive" substrates (Lwoff, 1946).

This preenzyme hypothesis, based on immunology, would dominate Monod's research for the next five or six years.

In 1948, an important American visitor, the immunologist Melvin Cohn, came to Monod's laboratory. Cohn was to remain in Paris for seven years and was to be a major participant in the "new" thinking that was to develop (Cohn, 1989).

In 1948, Monod reviewed the work on enzymatic adaptation at a symposium organized by Lwoff. This symposium, entitled "Biological units capable of genetic continuity" was one of the first postwar genetics meetings held in Europe. The main motivation for the meeting appears to have been the recent excitement about cytoplasmic inheritance, since this was the main thrust of the talks, which included Sonneborn (*kappa*), Rhoades (plant plas-

---

[9]Strain ML became the forerunner of a whole group of strains that played an important role in both enzyme induction and in the discovery of bacterial "permeases" (see below).

tids), Bawden (plant viruses), Darlington (plasmagenes), and Brachet (role of plasmagenes in development). However, numerous "microbial" papers were given, including Taylor and Hotchkiss on pneumococcus transformation, Boivin on transformation in *Escherichia coli*, Delbrück on phage genetics, Rountree on lysogeny, Ephrussi on respiratory deficient yeast, and Monod on adaptive enzymes. It seems evident that these microbial systems were represented at the meeting because they were thought to relate to the question of cytoplasmic inheritance. "Conventional" bacterial genetics, such as Lederberg's work, was not included.

At this meeting, Monod presented work that showed that enzyme adaptation for lactase only occurred in *growing* cells. Cells inhibited by chemicals, or more specifically by bacteriophage infection (work done with Elie Wollman; Monod and Wollman, 1947), did not synthesize the adaptive enzyme. Monod presented an alternative model to Spiegelman's plasmagene hypothesis, a model that was an important step in his thinking. He stated that although the gene controlled the enzyme, it was unlikely that it was responsible for all the reactions involved in the synthesis of a molecule as complex as an enzyme protein. Rather, the gene was considered to have a guiding role, elaborating only a small part of the protein, "carrying out a specific priming reaction which brings about the organization of *nonspecific* substances that play a common role in the synthesis of all enzymes in the cell, causing these to be laid down in specific oriented structures. This specific function would only need to occur once, after which the primary unit would no longer be needed. One can readily imagine that this primary unit can be a partial replica (unendowed with genetic continuity) of the gene itself..." (translated from Monod, 1949). It is possible to read too much into such speculations, but the idea of a partial gene replica unendowed with genetic continuity might be considered an antecedent of the messenger RNA hypothesis of 1961 (see later).

10.6 LEDERBERG'S WORK ON THE β-GALACTOSIDASE OF *ESCHERICHIA COLI*

The enzyme β-galactosidase has played an important role in studies on gene expression. In his initial work, Monod used a strain of *Escherichia coli*, strain ML, that could not be analyzed genetically. However, almost from the beginning of his work on bacterial mating in *Escherichia coli* K-12, Lederberg had studied the genetics of the *lac* character. J. Lederberg (1947) initially used *lac* as an unselected marker in his crosses (see Section 5.3). *lac*⁻ mutants were isolated by plating a large number of cells derived from mutagenized cultures on EMB-lactose indicator agar. In the first genetic map published for *Escherichia coli* (see Section 5.3), the *lac* gene was mapped in a location that is almost exactly where it is figured on current maps (Bachmann, 1987). Following up on this work, J. Lederberg (1948) isolated a large number of *lac*⁻ mutants and recognized phenotypic differences. Esther Lederberg (1948, 1950, 1952) studied these *lac* mutants further, and showed that there were several alleles. She also found a "position effect" in the interaction of two of the *lac* alleles, a discovery that presaged the discovery of polarity and the *cis-trans* effect of Jacob and Monod some years later. Some of these mutants were certainly in other loci than *lac*, since they had marked pleiotropic effects (J. Lederberg, 1948).

The most important consequence of this genetic work is that it focused attention on the nature of the specific enzyme involved in the hydrolysis of lactose. In a paper that was to have far-reaching consequences, Lederberg developed procedures for assaying the enzyme and studying its biosynthesis (J. Lederberg, 1950). Using his contacts with the Department of Biochemistry at the University of Wisconsin, Lederberg obtained the synthesis of a number of synthetic β-galactosides, including the important compound o-nitro-phenyl-β-D-galactoside (ONPG) (Seidman and Link, 1950). When this colorless substrate was hydrolyzed, the yellow-colored o-nitrophenol was released. Thus, ONPG, being chromogenic, could be used for a quick, specific, and extremely sensitive assay for the enzyme β-galactosidase. Using this new assay, Lederberg studied the effect of substrate concentration on the kinetics of the enzyme in both whole cells and cell-free extracts. Because the Lineweaver/Burk plots in whole cells and autolyzed cells were different, he concluded that "permeability" factors influenced the apparent activity of the enzyme. It is interesting that Monod's laboratory "rediscovered" the role of permeability in the activity of the organism some years later (see later and Rickenberg, Cohen, Buttin, and Monod 1956).

But of most importance in the present context was Lederberg's study of enzyme activity in adapted and unadapted cells. He showed that lactose-grown cells had markedly higher activities than cells grown on other substrates, and that glucose-grown cells had lower activity than cells grown on succinate, a further confirmation of the so-called "glucose effect." Another important discovery presented in this paper was that unadapted cells still retained a small amount of enzyme. Monod had used a manometer to assay for the enzyme, but according to Lederberg "Manometric experiments would usually miss activity of the order of a small per cent of adapted cells..." This presence of a basal level of enzyme in unadapted cells should have, but apparently did not, rule out certain of Monod's models for the mechanism of enzymatic adaptation.

In 1950, the Genetics Society of America held a jubilee meeting entitled Genetics in the 20th Century, celebrating the 50th anniversary of the rediscovery of Mendel's laws (Dunn, 1951). At this meeting, Lederberg presented an important paper on bacterial genetics that had an extensive section on *lac* (J. Lederberg, 1951). This paper presented a systematic study of Lederberg's *lac⁻* mutants, using the newly available ONPG chromogenic substrate and some synthetic galactosides as possible inducers. Although many of the *lac⁻* mutants were clearly pleiotropic, several were directly involved with β-galactosidase. One of Lederberg's mutants, designated *lac₁⁻*, was found to synthesize minimal amounts of β-galactosidase when grown on lactose but substantial amounts when grown on an alkyl galactoside. In discussing this result, Lederberg made the following prescient statement:

> This results in the paradox that cells grown on a heterologous substrate are better adapted to lactose than those grown on lactose itself. Since "adaptation" is presumably a physicochemical rather than an entelechist process, such deviations are not surprising but suggest the need for revising "adaptive enzyme formation" in favor of a more general term connoting "enzyme formation under environmental influence" (J. Lederberg, 1951).

Subsequently, Monod expressed this same idea when he adopted the term "enzyme induction" (see later) but did not mention Lederberg's prior work (Monod, Cohen-Bazire, and Cohn, 1951).

Another important announcement in Lederberg's 1951 paper was the isolation of the first *constitutive* mutant (which he designated Cst^+). Because such mutants provided one of the strongest points against the somewhat teleological models of enzyme adaptation that had been advanced earlier, it is significant that Lederberg's constitutive mutant was announced at least a year before those of the French group. The manner in which Lederberg isolated the mutant was also interesting. *Escherichia coli* did not attack the sugar neolactose; Lederberg attempted to isolate a mutant with altered specificity to β-galactosides by growing the cells on this neolactose. Such a mutant was readily isolated; then it was found that β-galactosidase could split neolactose but that the latter was unable to induce the formation of the enzyme. It was then found that the adaptive system had been changed to a constitutive one. Excellent β-galactosidase production was obtained even when the mutant was grown on glucose. Lederberg concluded that substrate-dependent β-galactosidase formation was subject to rather complex genetic control (J. Lederberg, 1951).

In this paper Lederberg thus presented the key ideas that later became the Monod canon: (1) certain β-galactosides are substrates of β-galactosidase but not inducers; (2) certain β-galactosides are inducers and are not substrates; (3) mutants can be isolated that do not form β-galactosidase because of altered regulatory properties, rather than because of alterations in the structural gene of the enzyme (constitutive mutants); (4) an adaptive enzyme can be synthesized in the absence of its substrate, so that the β-galactoside must be viewed as a regulatory substance as well as a substrate.

These ideas were not very strongly presented, and the data to support them were almost "throwaways." Furthermore, much biochemical work would be needed to rule out any alternative explanations,[10] but these results provided the essential leads for the subsequent work of Monod and his collaborators. Why were they ignored by Monod? Although some chauvinism may have been involved, my contention is that the evidence was genetic rather than physiologic, which was what Monod was seeking. In addition, "Jacques did not like to get rid of his theories. He had a strong tendency to stick to his model, sometimes slightly beyond the point of reason" (Jacob, 1979). Only gradually would the old ways of thinking die out and be replaced by the new.

10.7 FROM ENZYMATIC ADAPTATION TO INDUCED ENZYME SYNTHESIS

In Monod's laboratory, the period from about 1950 until 1955 represented a critical transition in thinking on the problem of enzyme adaptation. An overview of this period can perhaps best be obtained from Melvin Cohn's Eli Lilly Award address (Cohn, 1957), whose main points are summarized here.

Constitutive Mutants

The isolation of constitutive mutants placed the problem of the role of the substrate in a new light. With Lederberg's work as a background, Cohen-Bazire and Jolit (1953) set out to isolate constitutive mutants in Monod's laboratory. Instead of neolactose, they used lactose but adopted a cycling

---

[10]It was necessary, above all, to show that constitutive mutants produced the *same* enzyme protein as the wild type.

procedure in which the culture was grown repeatedly in a glucose medium and then transferred into lactose medium. Because the wild type lacked β-galactosidase when grown in glucose, there was a latent period in the lactose medium before growth was resumed. However, constitutive mutants present in the culture would begin growth immediately, so that at the end of the latent period the culture would be enriched in constitutive mutants. This "adapted" culture was then transferred back to glucose medium, where both the wild types and the constitutives grew normally. Constitutive mutants eventually dominated and could be purified by streaking on plates.

Although J. Lederberg (1951) had interpreted constitutive mutants as due to changes in regulatory properties, the Monod laboratory subscribed to a different theory, hypothesizing that these mutants produced an *internal inducer*. The hypothesis of an internal inducer was in line with the strong feeling of Monod (1947) that a unitary hypothesis was necessary to explain adaptive and constitutive enzyme synthesis. Why was an internal inducer needed for *every* adaptive enzyme in the cell? Monod had obviously not yet thought of the more economical (and correct) hypothesis of a negative regulatory system that was abolished in constitutive mutants. A genetic consequence of the internal inducer hypothesis was that if complementation tests were run, constitutivity should be dominant over inducibility. At the time, *Escherichia coli* genetics was not far enough advanced to permit complementation tests to be run. Such tests later would readily eliminate the internal inducer hypothesis (see Section 10.10).

Artificial Galactosides

The synthesis of artificial β-galactosides made it possible to separate their role as enzyme substrates from their role as inducers. As noted earlier, J. Lederberg (1950) had developed synthetic β-galactosides as both substrates and inducers. However, his interest was primarily genetic, and he did not pursue this line of work. Melvin Cohn, in Monod's laboratory, would carry the study of synthetic galactosides to its ultimate conclusion. He synthesized a wide array of β-galactosides, among which were several that were excellent inducers but were not substrates of the enzyme. According to Monod's theory (1947), the substrate was to have a directive influence on the aggregation of protein subunits by combining with the active site of the enzyme, leading to the formation of the final product. However, the existence of nonsubstrate inducers ruled out any direct connection between the β-galactoside as substrate and as inducer. This discovery had a profound philosophical impact on Monod, for it ruled out any teleological explanation of enzyme adaptation. This led to the use of a new term, *induction*, to replace the philosophically "loaded" term *adaptation*. The fact that a β-galactoside could be an inducer but not a substrate led to the use of the term *gratuitous induction*, a term meaning that the kinetics of enzyme formation were being studied under conditions such that neither the enzyme nor its inducer influenced general cellular metabolism. Gratuitous induction made it possible to study enzyme synthesis separately from enzyme function, thus making the study of kinetics more meaningful (Monod, Cohen-Bazire, and Cohn, 1951; Monod and Cohn, 1952; Cohn, Monod, Pollock, Spiegelman, and Stanier, 1953).

Precursor Pz

The most widely held model for adaptive enzyme synthesis postulated a conversion of a precursor protein into the final enzyme under the influence of the inducer. This model was first elaborated in detail by Spiegelman (1945, 1946), but had already been conceived independently by Monod, and was presented by Lwoff (1946) at the 1946 Cold Spring Harbor symposium. Subsequently, the precursor model had received some support from the discovery by Cohn and Torriani (1952) of a protein called Pz, present in uninduced cells, that was immunologically related to β-galactosidase (abbreviated Gz). A study was therefore initiated of the possible role of Pz as a precursor of Gz (Cohn and Torriani, 1952, 1953). Since Pz was related immunologically to Gz, it was reasonable to hypothesize that Pz was the postulated precursor. The Pz protein was found in all strains of enteric bacteria that fermented lactose, but was absent from non-lactose-fermenters. Uninduced cells that contained negligible amounts of Gz (just the basal level) contained Pz. Finally, when a gratuitous inducer was added to a culture, the amount of Pz decreased while Gz increased (although the decrease in Pz appeared to be quantitatively less than the increase in Gz).

Pz, however, was one of those sidelines down which research is often drawn. Radioisotope studies would subsequently show that Gz was not formed from Pz, but *de novo* (see below). Pz turned out to have nothing to do with β-galactosidase, and the fact that it cross-reacts immunologically with β-galactosidase is incidental (Pappenheimer, 1979).

The parallel between the Pz model and Spiegelman's model is illustrated in Figure 10.3b.

a

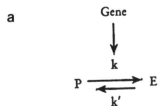

b   Pz + acides aminés --→ Gz

c

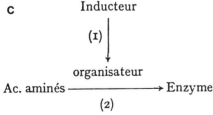

Figure 10.3 Comparison of precursor models. (*a*) Spiegelman's model for adaptive enzyme formation. (Reprinted, with permission, from Spiegelman, 1945.) (*b*) The Pz model of Monod. (Reprinted, with permission, from Monod, Pappenheimer, and Cohen-Bazire, 1952.) (*c*) Pollock's organizer concept, as visualized by Monod, Pappenheimer, and Cohen-Bazire. (Reprinted, with permission, from Monod, Pappenheimer, and Cohen-Bazire, 1952.)

Differential Rate of Enzyme Synthesis

The study of the quantitative relationships between cell growth and enzyme synthesis led to the development of the concept of the *differential rate of synthesis* (Monod, Pappenheimer, and Cohen-Bazire, 1952; see also Pappenheimer, 1979; Stanier [Cohen-Bazire], 1979). This concept arose out of studies by Cohen-Bazire and Pappenheimer on the conversion of Pz→ Gz under conditions of nitrogen starvation. They found that no enzyme was produced if the inducer was added to a culture starved for an essential amino acid, but enzyme synthesis began almost immediately upon adding back the required amino acid. In order to express the results quantitatively, they plotted the kinetics not as a function of time but as a function of increase in bacterial mass. When Monod saw the results, he was struck by the fact that immediately after addition of the amino acid, enzyme was synthesized as a *constant proportion* of the total bacterial protein. The precursor theory would have predicted a lag before enzyme synthesis began at maximal rate. The results thus appeared to be incompatible with a hypothesis that assumed the enzyme was derived, in part, from a preexisting protein.

These studies focused attention on the fact that enzyme synthesis was being measured in *populations* of growing cells; nothing was known about the process in single cells, or whether heterogeneity at the cellular level might exist. This problem was investigated by Seymour Benzer, who was at that time in the laboratory of André Lwoff at the Pasteur Institute, after having spent two years at CalTech with Delbrück (Benzer, 1966). An experienced phage worker (see Section 6.10), Benzer decided to use phage infection to analyze induced enzyme synthesis at the cellular level (Benzer, 1953). The rationale was as follows: Infection of *Escherichia coli* by one of the T-even phages blocks host enzyme synthesis, but reproduction of the phage requires active metabolism on the part of the host. Thus, if a phage-infected cell is placed under conditions where a given intracellular enzyme is required for metabolism (for instance, β-galactosidase when *Escherichia coli* is growing on lactose), the development of the phage in that cell will be dependent on whether the enzyme was present *before* infection. Using his strong mathematical-physical background, and with a careful kinetic analysis, Benzer was able to show that "all the cells participate [in enzyme synthesis] to a similar degree. There is no inherent discreteness, the amount of enzyme in each cell rising gradually" (Benzer, 1953).

De Novo β-Galactosidase Synthesis

One "phantom" that had haunted work on induced enzyme synthesis was the idea that the proteins of a living organism were in a dynamic state and turned over constantly. Protein turnover had been shown to occur in mammals in a classic series of experiments carried out at Columbia University in the 1930s by Schoenheimer and his associates. This work, using the stable isotope of nitrogen, ^{15}N, had provided one of the first approaches using isotopes to the study of metabolism in the living organism. However, the work was done in the whole mammal, and since cell turnover is common in certain tissues of the body, the work did not tell anything about protein turnover at the molecular level.

Although Monod's kinetic analysis provided certain "limits" to the problem, kinetics could not furnish *direct* evidence of whether the induced enzyme was synthesized *de novo*. This required the use of radioisotopes. Studies on the conversion of Pz→Gz (see above) led to the use of radioisotopes to show that β-galactosidase was synthesized *de novo* from amino acids. The results laid to rest the idea of the precursor and put the whole problem of induced enzyme synthesis in a new light. Two groups carried out these experiments, Spiegelman's at the University of Illinois (Rotman and Spiegelman, 1954) and Monod's in Paris (Hogness, Cohn, and Monod, 1955), with equivalent results. The Illinois group used ^{14}C, and the Paris group used ^{35}S to trace protein synthesis during enzyme induction. The experimental procedure was to fully label uninduced cells, wash away excess isotope, and then induce cells in unlabeled medium. At various stages of the induction process, portions of the cells were broken open, and the β-galactosidase was purified and assayed for radioactivity. If any of the enzyme protein was derived from *preformed* cellular proteins (for instance, Pz), then the enzyme should be radioactive, whereas if synthesis was only from amino acids synthesized *after* removal of the label, then the enzyme should be found to be unlabeled. The results showed that essentially all of the enzyme protein was synthesized *de novo* after induction and that if a protein precursor of β-galactosidase existed in uninduced cells, its level was less than 0.04% of that for fully induced cells (Hogness, Cohn, and Monod, 1955).

These experiments were to provide an important basis for all subsequent experiments over the next five years on the use of radioisotopes to study macromolecular synthesis. Together with the phage studies (see later), they lighted the way to the development of the operon-messenger RNA model of protein synthesis.

The Organizer

In attempting to develop a model to account for the results available at this time, Monod adopted the concept of the *organizer* (Cohn and Monod, 1953), an idea first advanced by Martin Pollock. Pollock had originally postulated the hypothetical organizer to explain penicillinase induction in *Bacillus cereus* (Pollock, 1953). He had shown that if the inducer (penicillin) was added to washed cells for a brief period of time, even at 0°C, the cells were changed so that they would subsequently synthesize penicillinase at a rapid rate, even if the inducer was removed. Studies with radioactive penicillin had shown that much more enzyme was formed than could be accounted for by the amount of inducer bound to the cells. The idea was that the inducer programmed a specific enzyme-assembling site called the *organizer*, which was supposed to work catalytically (Fig. 10.3c). The organizer was visualized as "catalyzing the formation of a specific enzyme precursor from a non-specific precursor. It could be regarded as a sort of template directing the moulding of enzyme molecules around it, or (even more vaguely) as a specifically developed catalytic cycle, analogous to the Krebs...cycle...The essential characteristics of the organizer are that it (1) is distinct from, but reflects the specificity of, the inducer, and (2) acts as a catalyst in enzyme formation..." (Pollock, 1953).

According to the Cohn/Monod modification of Pollock's idea, the inducer "forms an *unstable* union with a (presumably macromolecular or particulate) cellular constituent ('apo-organizer'), [and] forms the 'organizer'" (Cohn and Monod, 1953). Glucose was assumed to inhibit induction by preventing entry of the inducer, whereas β-galactoside analogs were assumed to inhibit induction by preventing conversion of the external inducer to the co-organizer; the constitutive state was due to the formation of an internal inducer. The organizer itself controlled the conversion of amino acids to enzyme. Although not specified, the apo-organizer was presumably under gene control.

In retrospect, it is clear that the penicillin system of Pollock was much too complicated to analyze properly and contributed nothing but confusion to the understanding of enzyme induction. The organizer provides an interesting example of how tortuous is the path to truth. I have tried to determine whether any of the "entities" later found to be involved in gene regulation could be analogized to the organizer. Perhaps positive regulator proteins would come the closest. The strangest thing about all the models constructed at this time is that the gene itself played such a minor role.

Permeability and Another Galactoside Entity

At about this time, the Monod laboratory became involved in extensive studies on the permeability of *Escherichia coli* cells to β-galactosides. This led to the discovery of the so-called "galactoside permease," an entity that was to attract considerable attention. Although Monod's studies on galactoside permeability represent, in one sense, a sideline on the path to the repressor and mRNA, they played an important role in the development of the operon concept, since they added an additional inducible protein to the β-galactosidase system. They also clarified a number of confusing kinetic studies on enzyme induction. In addition, permease studies opened up a major new area of bacterial physiology and provided important new insights into how genetics could be used to study a physiological problem. Although Monod has received much of the credit for the development of an attractive model of permeability phenomena in bacteria, he was by no means the first. Early studies on permeability that were equally convincing were those of B.D. Davis on citrate accumulation (Davis, 1956, 1958, 1986) and Arthur B. Pardee on melibiose uptake (Pardee, 1957). Monod cited such work when it suited his purpose but often conveniently ignored it. Because only studies on galactoside permease relate to the broader story being presented here, the present section presents only a narrowly focused view of the history of permeability research.

Most of Monod's models of induced enzyme synthesis in the mid-1950s had a step that involved the "metabolism" of the inducer (Cohn and Monod, 1953). It was natural, therefore, when gratuitous inducers such as thio-β-D-galactoside (TMG) became available, to determine how such compounds were modified during the induction process. ^{35}S-labeled TMG could be prepared at high specific radioactivity, thus providing a way to study the fate of the inducer. The first work on galactoside permease was done by Georges Cohen, who was soon joined by Howard Rickenberg. The original purpose was to see whether radioactive TMG could be found linked to one of the cellular macromolecular components: DNA, RNA, or protein. Cohen soon found a phenomenon more suitable for research: The amount of radioactivity that could be found intracellularly was low in uninduced cultures, but very high in cultures that had been preinduced. It was readily shown that the accumulation of TMG was energy-dependent, reversible, and stereospecific. Furthermore, strains constitutive for β-galactosidase were also constitutive for TMG transport. Other constitutive strains were unable to transport TMG but could make galactosidase, a phenomenon that was called "crypticity." Still other strains could not synthesize galactosidase, but could be induced for TMG uptake. Thus, although related via β-galactosides, the β-galactosidase enzyme and the β-galactoside transport system were independent genetic characteristics.

The amount of TMG incorporated into the cell was very high, reaching 2–4% of the bacterial dry weight, thus excluding the hypothesis that accumulation was due to the presence of stoichiometric receptor sites. This led Monod to the concept of a "permease," an enzyme-like entity that acted as an active pump system (Cohen, 1979).

The first general presentation of the permease data was in a lecture Monod presented at an important symposium on enzymes at the Henry Ford Hospital in Detroit, Michigan (Gaebler, 1956; Monod, 1956).[11] Subsequently, the work would be published in full in French (Rickenberg, Cohen, Buttin, and Monod, 1956), and a virtual repeat of the French data would be published in English (Cohen and Monod, 1957). A key point of all these papers is the evidence for an additional entity in the galactosidase system, which Monod called y. It was significant that the y system and the β-galactosidase (z system) were genetically distinct.

The discovery of the y system showed that the pathway of induction proceeded from i to y to z, and that the so-called "metabolism of the inducer" was connected with the role of y in transporting i to the z-forming system. Since the inducer was bringing about induction of its own uptake, kinetic analyses of enzyme formation immediately after addition of inducer were doomed to misinterpretation. Under some conditions, this phenomenon gave results that resembled a self-replicating entity, analogous to Spiegelman's plasmagene. The discovery of the galactoside permease eliminated for all time any considerations of *autocatalytic* models of induced enzyme formation.

Terminological note: Monod's use of y to designate the permease was logical. For several years, the β-galactosidase enzyme had been designated z (Monod, Cohen-Bazire, and Cohn, 1951), and x was often used to express bacterial mass in the differential equation describing bacterial growth. The term i had been used for the inducer (Monod and Cohn, 1952). In the Ford Hospital paper, Monod formalized these designations: z for β-galactosidase, y for the permeation system, and i for the inducer. It was only later that i was used for the gene, but at the time of the Ford Hospital symposium the existence of an i gene was not suspected.

In the initial work presented at the Ford Hospital symposium, Monod did not use the term *permease*, but he clearly indicated that he considered the y component to be an enzyme. In the full paper describing the y system (Rickenberg, Cohen, Buttin, and Monod, 1956), the term *permease* was used

---

[11]This symposium was noteworthy in that most of the major researchers in what would subsequently be called molecular genetics were present. The published volume is especially valuable historically because all of the discussions are recorded.

for the first time, the -*ase* ending clearly indicating that Monod considered this to be an enzyme. This key permease paper was originally planned for publication in the journal *Biochimica et Biophysica Acta* (see Monod, 1956 [p. 28]), but the editor, Fromageot, objected to Monod's use of the term *permease*, since it had not been shown that the *y* system was indeed an enzyme (H. Rickenberg, personal communication). Rather than rewrite the paper, Monod withdrew it and published it, in French, in the *Annales de l'Institut Pasteur*, where Pasteur Institute workers could publish relatively easily. Fromageot was right: The choice of the term "permease" was unfortunate, because the *y* system is not an enzyme. (The term "permease" has now virtually disappeared from the literature.) It is now known that the *y* gene codes for a transmembrane protein that serves as a *carrier* for β-galactosides (Kennedy, 1970; Beckwith, 1987b). For many years, the Monod group looked diligently for the postulated *y* enzyme and actually found a protein that did enzymatically modify β-galactosides, the *thiogalactoside transacetylase* (Zabin, Kepes, and Monod, 1959). Initially, genetic and physiological evidence strongly suggested that this enzyme was the product of the *y* gene, and in a number of Jacob and Monod papers of the 1959–1961 period it was called the permease, but it was later found that thiogalactoside transacetylase had nothing to do with the transport of galactosides and probably plays a role in the detoxification of certain analogs of lactose (Beckwith, 1987b). The thiogalactoside transacetylase is a product of the *a* gene, closely linked to but separate from the *y* gene.

Despite its insignificance for galactoside transport, the *a* gene also played an important role in the development of the operon concept, since it specified another assayable enzyme closely linked and coordinately regulated with *z*.

10.8 ENZYME REPRESSION

An important factor in the development of the operon (negative control) model of enzyme induction was the recognition of the parallels between enzyme induction and enzyme repression. The term *enzyme repression* was first used by Henry J. Vogel as a result of studies on the pathway for biosynthesis of the amino acid *arginine* (Vogel, 1957). Subsequently, repression was found to occur for a number of other biosynthetic enzymes.

Vogel's initial work was prompted by the prevailing model of Cohn and Monod (1953) that postulated a "unitary" mechanism to explain induced and constitutive enzymes (see Section 10.7). The Monod model at this time was the so-called "generalized induction theory," which postulated that enzymes are constitutive because of the formation of "internal inducers." One important piece of evidence in favor of the internal inducer model was that of Vogel on the arginine system.

The Arginine System

Vogel was studying the pathway for the biosynthesis of the amino acid *arginine* in *Escherichia coli* (Fig. 10.4). In this pathway, seven distinct enzymes had been recognized, including the enzyme *acetylornithinase*, that catalyzed the conversion of acetylornithine to ornithine. Vogel and Davis (1952) had isolated an arginine-requiring mutant that was blocked in the formation of

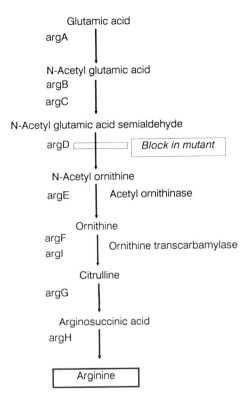

Figure 10.4 The arginine pathway in *Escherichia coli*, as it was known at the time of Vogel's repression experiments (Vogel, 1961). A mutant blocked in the formation of N-acetylornithine would grow on either N-acetylornithine or arginine. Such a mutant was used to obtain evidence for the synthesis of the enzyme acetylornithinase in the absence of the hypothetical internal inducer. The gene designations follow the present terminology (Glansdorff, 1987).

acetylornithine; this mutant could be cultured on either arginine or acetylornithine. Because it was blocked in a step *prior* to that catalyzed by acetylornithinase, this mutant could be employed in experiments designed to exhibit a possible *inducing* effect of an exogenously supplied substrate without interference from endogenously produced substrate. In this way, an induction effect that might be "masked" in the corresponding wild type due to the continuous presence of the substrate (the presumed internal inducer) could be detected.

The results actually showed that rather than acetylornithine being an inducer, arginine was an *inhibitor* of the synthesis of the enzyme. Vogel found that an arginine-grown inoculum had a low level of acetylornithinase activity, whereas a culture grown with acetylornithine had a markedly higher level (Table 10.2). In a mixture of arginine and acetylornithine, the arginine seemed to antagonize acetylornithine. It was first concluded that acetylornithine was an inducer of its enzyme, but since the mutant had to be grown in either arginine or acetylornithine, an alternative interpretation was that the arginine present during the growth of the inoculum lowered enzyme synthesis, and this inhibition was subsequently overcome when the external arginine was exhausted. Vogel's data provided clear evidence for an antagonistic effect of arginine on the synthesis of acetylornithinase. Further-

Table 10.2   Effect of Cultivation Conditions on the Relative Specific Activity of Acetylornithinase in *Escherichia coli*

Growth substrate	Relative specific activity[a]
Arginine, 0.06 mM	1.0
Acetylornithine, 0.30 mM	8.3
Arginine, 0.10 mM	0.6
Arginine, 0.06 mM plus acetylornithine 6.00	1.2

(Data from Vogel, 1957.)
[a]Specific activity relative to arginine-grown cells.

more, arginine (but no other amino acid) also inhibited formation of the enzyme in wild-type *Escherichia coli*. According to Vogel:

> The results under discussion can be explained on the hypothesis that acetylornithinase synthesis is spontaneous (non-induced) and antagonizable by arginine (or by a chemical relative of arginine), and that, following the disappearance of added arginine, the rate of enzyme synthesis accelerates through *release* from the antagonistic influence...The action of arginine on acetylornithinase formation may be regarded as a rather specific antagonistic effect on enzyme synthesis of a substance following in biosynthetic sequence the substrate of the enzyme; in this case, the substance is the "end-product" of the biosynthetic sequence involved. It is readily seen that this action of arginine constitutes a control mechanism that presumably is of great value to the organism; for example, an *E. coli* (wild-type) cell, in an environment that supplies arginine, will utilize this exogenous arginine and at the same time will conserve its resources by sharply curtailing the now unnecessary synthesis of acetylornithine (Vogel, 1957).

Recognizing that end-product inhibition of enzyme synthesis was likely a general phenomenon, Vogel coined the term "enzyme repression" to describe "a relative decrease, resulting from the exposure of cells to a given substance, in the rate of synthesis of a particular apoenzyme..." (Vogel, 1957). Furthermore, the substance that caused repression was termed the "represser." (Vogel also gave "repressor" as an alternate spelling, and this latter has become accepted.)

Vogel was quick to note the general significance of repression and its possible relation to enzyme induction. It was noteworthy that in the case of repression, a structural relationship between the small-molecule effector (arginine, in this case) and the enzyme substrate or its product (acetylornithine or ornithine) did not exist, whereas in the model of Cohn and Monod (1953), a structural relationship between inducer and enzyme was required.

> [According to Cohn and Monod]...the inducer would carry at least part of the information for the structure of the enzymatically active portion of the protein. However, there is a good basis for the belief that the cell has this kind of information prior to induction...It would thus seem that the inducer is dispensable with respect to the structural information required for enzyme synthesis.
>
> Accordingly, there would appear to be no valid theoretical objection to a...hypothesis...[in which] enzyme induction would be primarily that of a control mechanism. This view has been maintained from time to time and has been advocated in a recent paper by Lederberg (1955). Indeed, the picture of induced enzyme formation as a regulatory device is consistent with the available results,

including the following: enzymes of indistinguishable specificity are evoked by different inducers; there is no necessary relationship between the properties of a given substance as inducer, substrate, or complexant; and inducible enzymes usually show a "basal" level in the absence of added inducers (Vogel, 1957).

With these words, Vogel clearly anticipated Monod by at least two years. I am not certain whether these written remarks were also given orally, but it should be noted that Francois Jacob was a participant in this symposium and should have heard Vogel's paper. However, at this time Jacob was working primarily on lysogeny and had not yet begun his close collaboration with Monod. It appears that at this time Monod was still thinking of an "instructive" role for the inducer, and it was only after Leo Szilard suggested a negative repression model (Szilard, 1960; Monod, 1966) that Monod became enthusiastic for it (see later). However, in his Nobel Prize address, Monod stated that he himself had suggested the inhibition (repressor) model to Vogel during a visit of Vogel to the Pasteur Institute (Monod, 1966).

In retrospect, Vogel's paper at the Johns Hopkins symposium was of great importance, but it was not so recognized at the time of the meeting. Although other (forgettable) papers elicited considerable discussion (all printed in the published volume), there is no recorded discussion after Vogel's paper, and the paper is not even mentioned in the extensive executive summary of the meeting!

The Johns Hopkins symposium at which Vogel presented his work took place in June, 1956. Within a year, a further example of repression in the arginine pathway was reported by Gorini and Maas (1957). Using mutants and continuous cultures, these workers showed that the synthesis of the enzyme ornithine transcarbamylase, a key enzyme in the arginine pathway (see Fig. 10.4), was specifically inhibited by arginine. Furthermore, enzyme synthesis responded to *endogenous* as well as exogenous arginine, showing that repression was actually a homeostatic mechanism in biosynthesis. Gorini and Maas clearly recognized the relevance of this phenomenon for an understanding of the induction process and also its relevance for the generalized repression model of Vogel.

It is interesting that although the arginine pathway provided some of the earliest evidence for the negative control model of gene regulation, the genes of the arginine pathway do not constitute an operon in the classical Jacob/Monod sense. From the work of Luigi Gorini and Werner Maas, it was subsequently found that the arginine genes are not clustered on the chromosome in a single transcriptional unit, but are scattered throughout the chromosome. Despite this, they are subject to classical negative regulation. It is now known that the arginine repressor protein, coded for by the *arg*R gene, interacts with a family of slightly dissimilar operators in association with each arginine gene (Glansdorff, 1987), a type of regulatory system that has been called a *regulon*.

Repression as a General Phenomenon

Vogel noted in his paper that the arginine pathway was not an isolated incident of end-product inhibition of enzyme synthesis. Enzymes involved in the synthesis of methionine, tryptophan, valine, and pyrimidines had also been shown to be subject to control by their end products. Actually, Monod's laboratory had already reported the inhibition of the synthesis of tryptophan synthetase (Monod and Cohen-Bazire, 1953a) and of methionine synthase (Cohn, Cohen, and Monod, 1953). The specificity especially of the methionine inhibition was emphasized by the fact that methionine had no effect on the synthesis of β-galactosidase. However, the decreases noted by Monod were much smaller than those Vogel described. Monod's experiments at this time were motivated by his earlier demonstration that β-galactosides inhibited the *constitutive* synthesis of β-galactosidase (Monod and Cohen-Bazire, 1953b). Although Monod recognized that these specific inhibitions were comparable to the specific inhibition by glucose in the β-galactosidase system, his experiments stood in isolation until Vogel's paper brought them into the general concept of repression. However, even after Vogel's paper, Monod continued to think in terms of an internally synthesized inducer responsible for constitutive synthesis that was being antagonized by the specific amino acids.

Independently of Vogel, Yates and Pardee (1957) reported inhibition of the synthesis of enzymes of the pyrimidine pathway by uracil. Three enzymes of the orotic acid pathway increased markedly in amount when pyrimidine-requiring mutants were deprived of pyrimidine; the effect was specific for these three enzymes. They also showed that pyrimidine precursors were unable to act as inducers. Thus, the Monod internal inducer model did not seem to apply. Having received a copy of Vogel's paper for the Johns Hopkins symposium before publication, Yates and Pardee (1957) realized that their results could be explained by Vogel's repression model. They also noted that the term "constitutive enzyme" might not have a clear meaning, since the three enzymes of the pyrimidine pathway were clearly constitutive, yet their levels varied markedly depending on growth conditions. Another point of the Yates and Pardee work was that the effect of metabolites on enzyme synthesis (repression) and their effects on enzyme activity (feedback inhibition, see later) were distinct processes.

At about this same time, extensive radioisotope competition experiments of Roberts, Cowie, Abelson, Bolton, and Britten (1955) had shown that internal syntheses of a variety of essential cell constituents: amino acids, purines, and pyrimidines, were turned off when the exogenous building blocks were present in the medium. The data of the Roberts group also provided an early indication of the phenomenon of *feedback inhibition*. An early review of work on regulation of enzymes can be found in Pardee (1959).

10.9 THE PAJAMO EXPERIMENT

We now come to the famous Pardee, Jacob, and Monod experiment (Pardee, Jacob, and Monod, 1958, 1959; Schaffner, 1974), called variously PaJaMa, PaJaMo, and PyJaMa. This experiment involved measuring the synthesis of β-galactosidase during mating experiments in which the z gene was being transferred from Hfr to F$^-$ cells. The experiment was done in such a way that enzyme synthesis could not occur in either Hfr or F$^-$ separately, but only in zygotes. The PaJaMo experiment thus built on the knowledge of the conjugation process that had been worked out by Wollman and Jacob (see Section 5.7), but added the interesting modification that the enzyme itself was assayed.

The PaJaMo experiment depended on several preceding developments (Jacob, 1979; Pardee, 1979, 1985a): (1) the availability of constitutive mutants and the need to know how constitutivity was expressed; (2) the availability of Hfr strains and an understanding of the kinetics of the mating process (Wollman, Jacob, and Hayes, 1956; Hayes, 1957); (3) the development of the interrupted mating technique, which permitted separation of mating pairs after gene transfer had occurred; (4) the knowledge from zygotic induction (see Section 7.5) that transfer of genes from an Hfr to an F^- resulted, under certain conditions, in immediate expression of the transferred genes; and (5) the availability of a sensitive assay for β-galactosidase (ONPG hydrolysis). It was also at about this time that the collaboration between Francois Jacob and Jacques Monod began (Jacob, 1979). Jacob brought the insights of a geneticist to the problems of induced enzyme synthesis, thus effectively complementing Monod's more physiological background.

The Basic PaJaMo Experiment

Although the PaJaMo experiment provided the insights that led to the negative control mechanism of enzyme induction and to the concept of messenger RNA, the motivation for the experiment was quite different: to detect evidence for an internal inducer produced by constitutive mutants. As discussed earlier, Monod's "generalized induction model" stated that a constitutive mutant produced under all conditions an inducer that in wild-type cells had to be provided exogenously. Thus, measurement of β-galactosidase synthesis after the gene for constitutivity was inserted into an F^- would permit analysis of the synthesis of the internal inducer.

The various parts of the PaJaMo experiment are given in Table 10.3. Mating of Hfr with F^- could be used to transfer the z^+ gene from inducible donor

Table 10.3 Various Parts of the PaJaMo Experiment, the Results, and the Interpretation

Conditions	Mating	
	$z^+i^+ \rightarrow z^-i^-$	$z^-i^- \rightarrow z^+i^+$
Early transfer		
inducer present	enzyme formed	enzyme formed
inducer absent	enzyme formed	*no* enzyme
interpretation	z^+ is expressed in recipient; i^- has no effect on expression	z^+ is dominant over i^-; no internal inducer
Late transfer		
inducer present	continued enzyme synthesis	
inducer absent	enzyme synthesis ceases!	
interpretation	repressor is synthesized from i^+ and inhibits further expression of z^+	

(Summarized from data of Pardee, Jacob, and Monod, 1959.)

cells to constitutive z^- recipient cells (thus, $z^+i^+ \rightarrow z^-i^-$). The use of i^+ and i^- to signify inducibility and constitutivity, respectively, was probably because the experiments themselves showed i^+ to be dominant. In the earlier work on constitutive mutants (Cohen-Bazire and Jolit, 1953; Monod, 1956), this explicit terminology was not used.

To eliminate enzyme synthesis by the donor, streptomycin was used with a SM^R recipient and SM^S donor, since streptomycin did not inhibit gene transfer (see Section 5.5), but promptly blocked enzyme induction. Thus, neither strain could synthesize enzyme alone, the Hfr because streptomycin was present, the F^- because it lacked the structural gene for β-galactosidase (z^-). In this experimental arrangement, the internal inducer presumed to be already present in the cytoplasm of the recipient would be expected to act upon the transferred z^+ gene, and zygotes should thus produce the enzyme in the *absence* of externally added inducer. Wollman and Jacob had shown that the i gene and the z gene were closely linked (95% cotransduced) and because of this close linkage, only rare recombinants would develop. Thus, any enzyme synthesis obtained could not be due to such recombinants, an important condition if the experiment was to work.

The experiment (top part, left column, Table 10.3) showed that enzyme appeared promptly upon gene transfer, even in the absence of added inducer. Thus, conditions within the recipient cell were "constitutive." It was concluded, therefore, that the introduced z gene found inducer already present in the F^- cell, presumably arising from the i^- (constitutivity) gene (Pardee, 1979).

As a control, the experiment was done with the genes *reversed*: $z^-i^- \rightarrow z^+i^+$ (right side, Table 10.3). No enzyme synthesis occurred in the absence of inducer, although synthesis would have been expected if i^- coded for an inducer. Thus, the i^+ gene present in the recipient was *dominant* to the i^- gene derived from the donor. An additional problem was that although in the $z^+i^+ \rightarrow z^-i^-$ cross, enzyme synthesis began within a few minutes after the z^+ gene entered the recipient, in the absence of inducer, enzyme synthesis eventually ceased, suggesting that some factor controlled by one of the transferred genes was becoming gradually *expressed* (bottom part, Table 10.3, and Fig. 10.5). Taken together, these two sets of results showed that despite their close linkage, the genes z and i determined separate functions that were capable of cooperating in the cytoplasm, independently of their physical (chromosomal) linkage (Fig. 10.6). That is, following Benzer (see Section 6.10), z and i were in separate *cistrons*.

The *rapid* expression of the z^+ gene in the i^- cytoplasm was to become one of the motivations for the development of the messenger RNA model (see later).

In the initial (French) paper describing the PaJaMo experiment, the authors concluded:

> This suggests, contrary to expectation, that the dominant allele is not the constitutive one, but the inducible. If so, the dominance of the i^+ allele is only evident in the zygotes $z^+i^+ \rightarrow z^-i^-$ after a certain period of time...[Enzyme] synthesis stops *because the zygotes become phenotypically inducible*...
> Up until now it had seemed reasonable to postulate that constitutive mutants synthesized an endogenous inducer which was absent in inducible cells. The results described here suggest an exactly opposite hypothesis. The facts can be

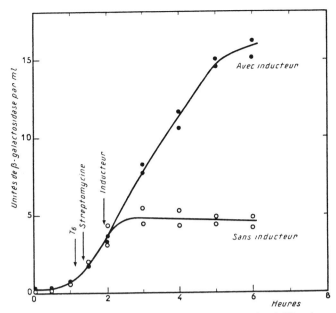

Formation de β-galactosidase chez des zygotes ♂ z^+i^+/♀ z^-i^-.

Figure 10.5 The basic PaJaMo experiment. Hfr strain (SM^{S}T6^S) was mated with F$^-$ (SM^{R}T6^R) in a glycerol synthetic medium: $z^+i^+ \rightarrow z^-i^-$. At the times indicated, T6 (multiplicity of 20) and streptomycin (1 mg/ml) were added. The inducer TMG (2×10^{-3} M) was added to the culture as indicated. At intervals, samples were removed and assayed for recombinants on lactose-streptomycin agar and for β-galactosidase with ONPG. (Reprinted, with permission, from Pardee, Jacob, and Monod, 1958.)

explained by the supposition that the *i* gene determines (via an enzyme inter-mediate) the synthesis, not of an inducer, but of a "repressor" which *blocks* the synthesis of β-galactosidase, and the exogenous inducer displaces this repressor and restores enzyme synthesis. With the *i*$^-$ allele, present in an inactive form in the constitutives, the repressor is not formed, and β-galactosidase is synthesized, the exogenous inducer therefore being without effect. This hypothesis, although at first surprising, is in agreement with many other facts. It has been well known for many years that the synthesis of certain constitutive enzymes is inhibited by "repressors," exogenous or endogenous[12]...It is true that other interpretations, not involving a repressor, are not excluded by the present observations. But all these hypotheses depend upon the postulate of an endogenous substance, inducer or repressor. Whether a system is inducible or not, in the future we can in general speak of "repressible" systems (translated from Pardee, Jacob, and Monod, 1958).

In the full version of this paper, published in English, a new term, *cyto-plasmic messenger*, was introduced, to refer to the product of the *i* gene (Pardee, Jacob, and Monod, 1959). Note that this use of the term "messenger" is quite different from that used in the term "messenger RNA," coined later. In the present instance, the term "cytoplasmic messenger" refers to what would later be called "repressor." The reluctance of Monod and colleagues to use the term repressor probably stems from the reluctance to abandon completely the *internal inducer* hypothesis:

While proving that the interaction of the *i* and *z* factors involves a specific cytoplasmic messenger, the data presented here do not, by themselves, give any

[12]This statement refers to the work of Vogel and Yates and Pardee discussed above.

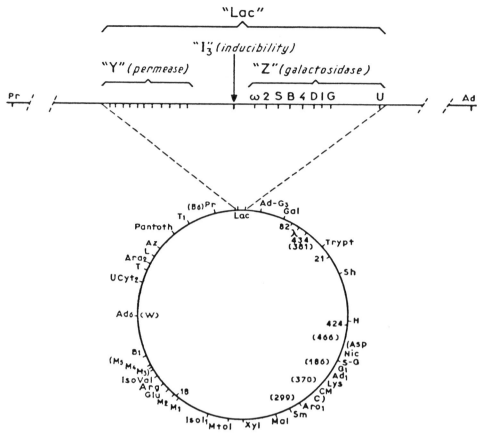

Figure 10.6 The genetic map of the *lac* region, as published by Pardee, Jacob, and Monod. (Reprinted, with permission, from Pardee, Jacob, and Monod, 1959.) It was later found that the *i* locus mapped to the right of *z*.

indication as to the mode of action of this compound. Two alternative models of this action can be considered.

According to one, which we shall call the "inducer" model, the activity of the galactosidase-forming system requires the presence of an inducer, both in the constitutive and in the inducible organism. Such an inducer (a galactoside) is synthesized by *both* types of organisms. The i^+ gene controls the synthesis of an enzyme that destroys or inactivates the inducer: hence the requirement for external inducer in the wild type. The i^- mutation inactivates the gene (or its product, the enzyme) allowing accumulation of endogenous inducer. This model accounts for the dominance of inducibility over constitutivity, and for the kinetics of conversion of the zygotes.

According to the other, or "repressor," model the activity of the galactosidase-forming system is inhibited in the wild type by a specific "repressor" (probably also involving a galactosidic residue) synthesized under the control of the i^+ gene. The inducer is required only in the wild-type as an *antagonist* of the repressor. In constitutive cells (i^-) the repressor is not formed, or is inactive, hence the requirement for an inducer disappears. This model accounts equally well for the dominance of i^+ and for the kinetic relationships (Pardee, Jacob, and Monod, 1959).

This, then, is the clearly formulated statement of the negative control model of enzyme regulation. It should be noted that at this time ideas about the chemistry of the repressor were quite vague. The suggestion that it contained a galactosidic residue indicates that some version of the internal inducer model was still being retained. The authors recognized that the repressor hypothesis was somewhat "ad hoc" but was in keeping with the numerous cases of enzyme repression now known to occur in biosynthetic pathways (see above). Additionally, the repressor model was simpler, since it did not require an independent inducer-synthesizing system.

Basically, the PaJaMo experiment constituted a complementation or *cis/trans* test in which the *i* gene was placed in opposition to the *z* gene. Because F-*lac* plasmids were not yet available, the test had to be done using Hfr and F^- strains, using the kinetics of enzyme synthesis to observe the expression of the complementing genes. Because of the close linkage between *z* and *i*, strains were not yet available in which *z* could be put in *cis* with either i^+ or i^- in a cytoplasm *lacking* repressor. Such experiments would eventually be done and would fully confirm the repression model of induction, but at the time of the PaJaMo experiment, *kinetics* had to be used to interpret the results, a definite limitation. Such kinetics would, of course, be familiar to Monod from his extensive physiological and growth studies, and would not seem to offer a particular limitation.

Another aspect not accounted for by the inducer model was the inhibition of enzyme synthesis by glucose, exhibited in the phenomenon of *diauxie*. Shortly before the above work, Neidhardt and Magasanik (1957) had shown that this "glucose effect" was comparable to a nonspecific repression. This was considered a strong argument in favor of the repressor model.

The PaJaMo experiment elicited considerable excitement in the growing molecular biology community, and the ideas implicit in it were quickly confirmed. The approach was applied in numerous other systems (reviewed by Riley and Pardee, 1962). In one case, β-galactosidase synthesis was followed after F-mediated gene transfer. In other cases, high-frequency transduction with defective *lambda* was used to introduce the galactokinase or β-galactosidase genes, enzyme synthesis then being followed. In these cases, DNA was the only substance entering the recipient cell, thus proving that the DNA of the structural gene contains the specificity necessary to dictate the synthesis of a specific protein.

The results of the PaJaMo experiment also provided the final evidence that there was no cytoplasmic mixing when Hfr and F^- strains mated. Indeed, if a conventional mating had occurred in *Escherichia coli*, such as occurs in yeast and other fungi, the results of the PaJaMo experiment would have been quite different.

10.10 THE DEVELOPMENT OF THE OPERON CONCEPT

The operon concept developed gradually over the years between 1958 and 1960 as more genetic and physiological data became available. Although the concept had its background in the physiological studies of Monod, it was the genetics of Francois Jacob that provided the essential insights (Jacob, 1979). Jacob had been developing ideas about the regulation of bacteriophage *lambda*, and it was his realization that *lambda* and β-galactosidase regulation had important parallels that led to the idea of the operon (Jacob and Wollman, 1961; Jacob, 1979, 1988).

By the time of the PaJaMo experiment, numerous *lambda* mutants had been isolated and their interactions in the host had been studied. Although the correct model for *lambda* integration was not to be described until a few years later (in 1962 by Campbell, see Section 7.6), *lambda* mutants incapable of lysogenization provided important insights to the whole repression story. There was an important analogy between the phenomenon of zygotic induction (see Section 7.5) and the PaJaMo experiment. Both involved measurement of gene expression after transfer of genes from one cytoplasmic background (in the Hfr) to another (in the F$^-$). Jacob realized that "in both cases, a group of normally silent genes could be triggered and become expressed at will; in both cases, this silence was due to a single, distinct gene: C_I in phage λ, *i* in the *lac* system; in both cases, genetic analysis showed that the wild type allele of this gene was expressed by a cytoplasmic product, a repressor blocking in some way the expression of the other genes" (Jacob, 1979).

Starting from this base, Jacob developed the idea that in both systems, regulation was based on a repression system that operated like a switch, by a mechanism that involved only two states, *on* or *off*. Furthermore, genes did not exist simply as independent entities, but genetic units of a higher order existed, "units of activity" that contained several genes subject to unitary expression. In the case of *lambda*, this would be the whole set of genes involved in viral production. In the case of the *lac* system, it would be β-galactosidase as well as the permease system. The *lambda* and the *lac* systems were thus developed simultaneously, with parallels being sought in one for phenomena found in the other. The coordinate regulation idea from *lambda* led to the discovery that synthesis of β-galactosidase and permease (and thiogalactoside transacetylase when this was discovered) occurred at the same rates after addition of inducer, and that in *i*$^-$ mutants the coordinate constitutive synthesis of all *lac* proteins occurred. In the *lambda* version of the model, the C_I gene was considered to produce the cytoplasmic repressor that in *lac* was coded by the *i* gene.

On the basis of these ideas and observations, Jacob and Monod (1959) developed the concept of two kinds of genes, *structural*, which coded for the synthesis of proteins, and *regulatory*, which did not. Regulatory genes were postulated to govern the expression of proteins through the intermediary of repressors. The striking aspect of repressor genes was their pronounced pleiotropic effects, often affecting the synthesis of whole blocks of structural genes.

In considering how a regulatory gene might control simultaneously the synthesis of more than one protein, after rejecting several other models, Jacob and Monod (1959) postulated that there was a unique genetic structure, sensitive to the action of the repressor, that was associated with the group of structural genes under the control of a single regulator. This unique site was called the "*operator*," and mutations in this site would lead to a loss of sensitivity to the repressor.

An important aspect of this regulatory model was the combination of the repressor (whose proteinaceous nature at this time was not suspected) with an acceptor site in the genetic region of the switch. Such an acceptor site should, itself, be alterable by mutation. A mutant in the acceptor site should no longer be sensitive to the repressor produced by the *i* gene. Although an acceptor mutant was not known in the *lac* system, such a mutant had already been found in *lambda*, the so-called "virulent" (*v*) mutant, which grew on bacteria lysogenic for *lambda*. Genetic analysis of the *v* mutant had shown that only those *lambda* genes that were in a position *cis* with respect to *v* were expressed in a *lambda* lysogen.

Jacob therefore turned to the isolation of the predicted mutant in *lac* that would be analogous to *v*. In cells containing two copies of the *lac* genes (that is, partial diploids or heterogenotes), such a mutant should be *dominant* only in the *cis* position, since the repressor would be no longer able to bind. Mutants in the regulator gene (i^-), on the other hand, were known to act as *recessive* in *cis* (that is, the wild-type allele was dominant in *trans*), consistent with the model that the repressor produced by the i^+ allele was present in the cytoplasm. By this time, the F-lac system had been developed (see Section 5.12) and could be used to construct partial diploids that could be used in the isolation of such an o^c mutant. (Jacob and Monod first called such mutants A^-, but when the term "operon" was coined, the designation of these mutants was changed to o^c, for "operator constitutive.") Constitutive lactose fermenters were therefore selected in bacteria carrying an F-*lac* factor homozygous for *i* and heterozygous for *z* ($i^+z^-/F^-i^+z^+$). These constitutive mutants were indeed o^c mutants, having the genotype $i^+o^cz^-/F^-i^+o^cz^+$ and synthesizing β-galactosidase (and also permease) constitutively, even when present in a heterogenote with an o^+ allele, so long as o^c was in the *cis* position (Jacob, Perrin, Sanchez, and Monod, 1960).

The construction of strains with particular genetic backgrounds to test a model followed exactly the approach of Thomas Hunt Morgan, Hermann Muller, and colleagues earlier in the century with *Drosophila*. It was the discovery of the o^c mutants and their genetic analysis that actually led to the development of the operon concept.

Another interesting mutant in the *lac* system isolated at this time was the so-called "super-repressed" mutant, designated i^s (Jacob and Wollman, 1961). This mutant had lost the ability to synthesize *both* β-galactosidase and permease, but it did not carry a deletion, since the *z* and *y* genes could be recovered when it was crossed with z^- or y^- mutants. The i^s mutant was dominant in partial diploids, and its properties could best be interpreted as due to an alteration in the repressor so that it no longer could be antagonized by external inducers. (It was subsequently shown that this mutant lacked the allosteric site where β-galactosides bound.)

In the *lambda* system, a parallel mutant to i^s was isolated by Jacob and Campbell (1959). This mutant, which was uninducible (ind^-) by ultraviolet radiation or any other means, mapped in the C_I gene that was known to be the gene for the cytoplasmic *lambda* repressor. The most likely interpretation of this ind^- mutant was that it was a result of loss of sensitivity to the repressor of whatever the (unknown) substance was that was produced by UV irradiation. This mutant was thus considered to be the *lambda* parallel of the i^s mutant in the *lac* system.

In 1960, the first proposal for the *operon* concept was published (Jacob, Perrin, Sanchez, and Monod, 1960). The sought-for operator mutants had been found and mapped in the region between the *i* gene and the *z* gene, and the position of *i* had been clarified. The term "coordinate expression," analogous to "coordinate repression" of Ames and Garry (1959), was used to describe the regulatory function of the operator gene. The repressor, whose chemical nature was still unknown, was postulated to act on the operator. By now the distinction between the regulator gene, which brings about the production of a repressor, and the operator, upon which the repressor acts, had been made. Repression was postulated to be exerted either directly at the level of the gene or indirectly at the level of a "cytoplasmic replica" of the operon (Jacob, Perrin, Sanchez, and Monod 1960). The hypothesis of a cytoplasmic replica of the repressor may have been based on the theoretical discussions of Crick (1958), who had advanced the idea of an RNA intermediate involved in protein synthesis. These ideas would be clarified within the year by the development of the messenger RNA concept (see later).

One reason it was possible to recognize operator mutants was that the resolving power of recombinational analysis in bacteria was high enough to allow the mapping of mutations occurring at different sites within the same locus. The concepts that Benzer was developing from studies on the *r*II locus of phage T4 (see Section 6.10) were quickly applied to *lac* and *lambda*. Mapping thus showed that mutations in *o* and *i* were closely linked to *z*. Even conjugation had sufficient resolving power to map mutants in these loci, but transductional analysis with phage P1 soon added a further level of refinement.

10.11 NATURE OF THE REPRESSOR

Although genetics provided the foundation for the operon concept, it could not reveal the chemical nature of the repressor. When Jacob and Monod (1961) published their important review paper on gene regulation, they concluded that the repressor would be RNA rather than protein. There were, indeed, two explicit experiments that seemed to show that the repressor was not protein, that of Pardee and Prestidge (1959) on *lac*, and that of Jacob and Campbell (1959) on *lambda*. Both were based on the demonstration that inhibitors of protein synthesis added during the induction process did not prevent the development of repression. In retrospect, it appears that insufficient concentrations of inhibitor were used in these experiments (Horiuchi and Ohshima, 1966).

The hypothesis that repressor was protein had apparently been rejected a priori, probably because at the time it could not be visualized how a protein could interact in a *specific* way with DNA. However, soon evidence that the repressor *was* a protein accumulated. The most telling experiment involved a mutant that contained a temperature-sensitive *lac* repressor (Horiuchi, Horiuchi, and Novick, 1961). Since proteins are much more heat-sensitive than nucleic acids, it seemed reasonable that this mutant formed a temperature-sensitive protein repressor. If this mutant was heated briefly to a temperature of 43.5°C, it lost its repressibility and made β-galactosidase even in the absence of inducer, whereas at 14°C, β-galactosidase was fully inducible. Although several *ad hoc* alternate explanations could be advanced, the most reasonable explanation was that the repressor was protein rather than RNA.

Additional evidence suggesting that the repressor was protein included the discovery of suppressor mutations in both the *lac* and *lambda* systems that affected the expression of the repressor gene. Since suppressor mutations were considered to act at the level of translation, they could only be intervening at the level of the synthesis of a protein. A strain of *Escherichia coli* carrying such a suppressor was able to restore the functional activity of mutants of *lambda* that were virulent because of insensitivity to the *lambda* repression system (Jacob, Sussman, and Monod, 1962). Similarly, suppressible *amber* mutants of the *i* gene (i^-) were also found (Bourgeois, Cohn, and Orgel, 1965).

Within several years, the concept of the allosteric protein (see later) would be developed (Monod, Changeux, and Jacob, 1963), a concept that explained how repressor or inducer might act. Detailed studies from Novick's group provided measurements of the affinity of the repressor for inducer, providing important insights into how the repressor might be isolated biochemically (Sadler and Novick, 1965). The final proof that the repressor was protein was an heroic experiment by Gilbert and Müller-Hill (1966) (see also Gilbert and Müller-Hill, 1970) in which the *lac* repressor was completely purified, using the one property most central to its action, its ability to bind (measured by equilibrium dialysis) radioactive isopropyl-thio-galactoside (IPTG), a potent inducer. To enhance the ability to detect the presumed repressor, a mutant was isolated (called i^t, for *tight-binding*) that coded for a repressor protein which bound more tightly to the inducer. Because calculations had shown that a cell had only a few molecules of repressor, it was essential to blindly concentrate the presumed repressor protein by a factor of at least 100 before any binding activity could be detected experimentally. The Gilbert/Müller-Hill experiments were greatly aided by the availability of a large collection of mutants in the *lac* region that could be used as controls. Using a defective phage containing the *lac* region, Gilbert and Müller-Hill then showed that the *lac* repressor bound specifically to the *lac* region, that its binding was weakened in o^c mutants, and that inducer was able to bring about the release of repressor bound to operator (Gilbert and Müller-Hill, 1967).

Within a short time, the *lambda* repressor had also been purified, and it was shown that it bound specifically to that region of the *lambda* DNA where the operator region had been mapped (Ptashne, 1967).

10.12 RNA AND PROTEIN SYNTHESIS

The story of the discovery of messenger RNA and its central role in protein synthesis has been told numerous times, perhaps most extensively by Judson (1979). I do not intend to review this whole story here, but merely to highlight the central role that bacterial and phage genetics played in the final solution. The idea that the gene might act through a cytoplasmic replica was one that had been often discussed, although in the early days it was considered in relation to cytoplasmic inheritance (the plasmagene model of Spiegelman, see Section 10.2; see also Monod, 1947). During the period under consideration here, it was assumed that bacteria were like higher organisms, only smaller. That bacteria did not have a well-defined nucleus was not realized until *after* the messenger RNA concept had been developed (see Cairns, 1963 and Section 3.9).

The best early evidence for the role of RNA in protein synthesis was based on cytochemical studies on eucaryotes. The theory of an RNA template for protein synthesis motivated numerous research projects throughout the 1950s. Radioactive tracers made possible the study of reactions that occurred only at very tiny rates. The idea that RNA might play a role in protein synthesis preceded the Watson/Crick structure of DNA, but once this structure became available, a molecular rationale for RNA involvement became clear. Obviously, if DNA replicated by complementary base pairing, then it was easy to understand that its base sequence could also be copied into RNA. An enthusiastic supporter of a role for RNA in protein synthesis was Sol Spiegelman, whose work on inducible enzymes had carried him from yeast to bacteria (Spiegelman, 1957). In the bacterial world, the most active laboratories studying cell-free protein and RNA synthesis were Spiegelman's at the University of Illinois (Spiegelman, 1956) and Ernest Gale's at the University of Cambridge (Gale, 1956; Gale and McQuillen, 1957).

Cell-free work in bacteria was rife with artifacts, however, and the results led to little solid information. More conventional biochemistry actually provided the most important insights. Among the most significant developments at this time was the discovery of the process of amino acid activation. Concepts of energetics and intermediary metabolism required that an amino acid must first be *activated* before it could enter into a peptide-bond-forming reaction. Although cell-free protein synthesis was a messy business, amino acid activation was readily studied in cell extracts. Enzyme systems that would convert amino acids to amino acyl-AMP derivatives were readily detected in bacteria, yeast, and animal tissues. The activation of an amino acid was analogous to that of the fatty acid activating system already known. It was satisfying that a specific activating enzyme was present for each of the amino acids. Although the initial product of the enzyme was identified as an amino acyl-AMP derivative, it was subsequently shown that the amino acyl group was transferred to a small-molecular-weight RNA, the so-called "soluble RNA" (reviewed by Hoagland, 1959, 1960). To express the perceived role of soluble RNA in protein synthesis, the term "transfer RNA" was coined (Hoagland, 1960). The role of transfer RNA in protein synthesis became defined through cell-free protein-synthesizing systems derived primarily from animal tissues. The early history of this research has been presented by Zamecnik (1979) and Hoagland (1989).

The fact that protein synthesis occurred in the cytoplasm in eucaryotes focused attention on the RNA components of the cytoplasm. Ribosomes were detected in both mammalian systems and bacteria (Schachman, Pardee, and Stanier, 1952). In bacteria most of the RNA of the cell was in ribosomal particles. The involvement of ribosomes in protein synthesis was shown in several ways (reviewed by Hoagland, 1960): (1) Cells actively synthesizing protein had higher RNA content, and much of this RNA was in the ribosomal fraction; (2) incorporation of radioactive amino acids in whole cells took place initially into the ribosomal fraction; (3) amino acid incorporation studies in cell-free systems showed that after amino acid activation it was also the ribosome particle to which the radioactive amino acid first became bound.

The review on protein synthesis by Crick (1958) was extremely influential in focusing attention on the intermediate steps in protein synthesis. Among other important features of Crick's paper was a discussion of information transfer from DNA to protein by way of RNA. Presciently, Crick had conceived of a nucleic acid "adaptor" that would intervene between the genetic code in the DNA and the amino acid and put the amino acid in the correct place in the polypeptide chain of the protein. Although Crick's biochemical details were wrong, the adaptor hypothesis was correct, and the details were soon filled in. It was found that once the amino acid was attached to its transfer RNA, the specificity then resided in the transfer RNA.

By 1960, the accepted model of protein synthesis involved *two* kinds of RNA, soluble (transfer) and ribosomal. Because of the demonstrated role of the ribosome in protein synthesis, and the presence of RNA in the ribosome, it seemed reasonable that the template was actually the ribosomal RNA itself, and the DNA was thought to be involved in the manufacture of this RNA template (Hoagland, 1960). It should be noted that at this time, the only enzyme known that made RNA was *polynucleotide phosphorylase*, which did not require either a primer or a template for its activity (reviewed in Khorana, 1960). It was the messenger RNA model, derived from genetics and physiology, that forced a search for an enzyme that would copy DNA sequences into RNA (see below).

The virulent T-even bacteriophages (see Section 6.12) provided a unique opportunity to study the biochemical events involved in information transfer from DNA, and several key links in the development of the messenger RNA concept came from such phage studies. Because infection with a T-even phage leads to cessation of host-cell macromolecular synthesis, all nucleic acid made after infection is viral (Cohen, 1968). Hershey had shown the commanding role of the phage DNA in phage replication (see Section 6.13).

A very important experiment was performed by Volkin and Astrachan (1956, 1957), who showed that when ^{32}P is incorporated during T2 infection, a small fraction (1–3% of the P of the total RNA) is incorporated into an RNA fraction, even though there is no net RNA synthesis. When the base composition of the radioactive RNA formed after phage infection was determined, it was found that the four bases were present in the same ratios as the bases in the phage DNA, which were significantly different from the base ratios of the host DNA (Table 10.4). Although there were various complications to a straightforward interpretation of the Volkin/Astrachan experiment, these complications were not perceived, and this experiment became a major step on the way to messenger RNA.

Although the model that placed the template in the ribosome had considerable experimental support, a number of difficulties became evident. These were laid out by Brenner (1961) shortly after the messenger RNA model had

Table 10.4   Base Composition of Phage T2 and Host DNA and the RNA That Becomes Rapidly Labeled after Infection

Base	DNA host[a]	DNA phage[b]	Rapidly labeled RNA[c]
Adenine	25	32	33
Thymine	25	32	29 (uracil)
Guanine	25	18	20
Cytosine	25	17	18

In the phage DNA, cytosine is replaced by 5-hydroxy methyl cytosine, but this does not affect the base-pairing properties. RNA rapidly labeled after T2 infection has a base composition similar to that of phage rather than host DNA. According to Belozersky and Spirin (1960), the base composition of host RNA is about 26A, 19U, 29G, and 25C.

[a]Belozersky and Spirin (1960).
[b]Wyatt (1953).
[c]Volkin and Astrachan (1957).

been devised. Although some of these difficulties were biochemical, most significant were the facts of enzyme induction, especially as seen in the β-galactosidase system. The greatest difficulty arose from the facts of the PaJaMo experiment, which showed that induction and deinduction took place at very rapid rates. To fit these results into the ribosome model current at that time, it had to be assumed that there was a constant number of ribosomes present in the cell for each enzyme and that the small controlling molecules (inducers and repressors) turned the ribosomes on and off. This model had first been presented by Vogel (1957) even before the role of the ribosome in protein synthesis was known. As the biochemical facts of protein synthesis became established, Vogel's model was applied to the ribosome.

As Brenner (1961) pointed out, although Vogel's model could account for many aspects of β-galactosidase synthesis, it could not account for the results of the PaJaMo experiment. It should be recalled that in the PaJaMo experiment the structural gene for β-galactosidase was transferred into a cell that did not contain it, and enzyme synthesis was initiated at *maximum rate* very soon after the entry of the gene. With the ribosome-as-template model, it would have to be assumed that *new* ribosomes would have to be synthesized after the gene was transferred, and since this would require a period of time, the rate of enzyme synthesis should increase gradually. However, after a short lag there was *no* observable change in the differential rate of β-galactosidase synthesis.

If regulation occurred on the ribosome, it would also have to be assumed that each gene made the same number of ribosomes. With the number of known enzymes in *Escherichia coli*, there could be no more than 5 or 10 ribosomes for each enzyme, and with the rapid rate of synthesis of β-galactosidase after induction, it would have to be assumed that each ribosome was capable of functioning at an extraordinary rate. Thus, the stability of the template presented great difficulties in explaining enzyme induction, although it could readily explain constitutive enzyme synthesis. Therefore, the ribosomal RNA could not be the message.

> De Vries...long ago suggested that the primary product [of gene action] is much like the gene itself, and produced by a process akin to its duplication, and he thought that this product then migrates into the cytoplasm to do its work (Muller, 1947).

The circumstances that led to the messenger RNA insight have been told by several sources (Jacob, 1979; Judson, 1979; Crick, 1988). Early research papers that provided important ideas were those of Riley, Pardee, Jacob, and Monod (1960) and Pardee and Prestidge (1961). The key insight was the realization by Sydney Brenner that the rapidly turning over RNA seen by Volkin and Astrachan after T2 infection (which mirrored the base composition of the infecting phage) was a different *species* of RNA than ribosomal RNA. What Volkin and Astrachan had seen was *phage* messenger RNA. Thus, the ribosome was simply a nonspecific translation machine, something like a computer whose behavior depended on what software it contained. The kinetics seen in the PaJaMo experiment could then be explained by assuming that the messenger RNA was used only a few times before being destroyed.

The argument for the existence of a rapidly turning over messenger RNA (mRNA) was then spelled out in a seminal paper of Jacob and Monod, a paper that was also to present the details of the operon concept (Jacob and Monod, 1961a). Direct evidence for the existence of messenger RNA was soon obtained by Sydney Brenner and Francois Jacob in collaboration with Matthew Meselson in Max Delbrück's laboratory at CalTech, using phage T4 infection (Brenner, Jacob, and Meselson, 1961), and by Spiegelman in the related phage T2 (Spiegelman, 1961) (see also Gros, Gilbert, Hiatt, Attardi, Spahr, and Watson, 1961). To present the details of these important biochemical experiments would take us too far afield.

The idea of mRNA came from genetic experiments, but once the concept had been advanced, its reality was shown by biochemical means. This is an excellent example of the importance of a model in science.

One of the important contributions of Spiegelman's laboratory was to show that messenger RNA was able to hybridize specifically with the gene (Spiegelman, 1964). This hybridization procedure was to have far-reaching significance in the era of recombinant DNA research, but at the time it was developed, its main significance was to provide direct *physical* evidence of mRNA.

Spiegelman's hybridization procedure derived originally from studies on the experimental formation of double-stranded DNA hybrids that were done by Julius Marmur and Paul Doty (see Section 9.15) (Doty, Marmur, Eigner, and Schildkraut, 1960; Marmur and Lane, 1960). These observations suggested to Hall and Spiegelman (1961) that double-stranded hybrid structures could also be formed from mixtures of single-stranded DNA and RNA, if the base sequences were complementary or nearly so. The detection of hybrids made use of the density-gradient centrifugation technique. RNA has a slightly higher density than DNA, and the two species can be separated by cesium chloride density-gradient centrifugation. A major technical advance was made when it was found that cellulose nitrate filters bind denatured DNA and RNA/DNA hybrids, while allowing free RNA to pass through (Nygard and Hall, 1964; Gillespie and Spiegelman, 1965). Filter hybridization provides an extremely simple and sensitive way of pulling complementary nucleic acid molecules out of cell extracts and nucleic acid mixtures.

An important biochemical advance that related directly to messenger RNA was the discovery of RNA polymerase, an enzyme that copies DNA into RNA. Although an enzyme that copies DNA had been known since 1956 (Kornberg, 1957), the only enzyme known that made RNA was polynucleotide phosphorylase (for review, see Ochoa and Heppel, 1957), which did not require RNA either as a primer or as a template. With the advance of the messenger RNA concept, it became critical to find the enzyme that copied DNA into RNA. Suddenly, in 1960, three laboratories reported the *real* RNA polymerase, an enzyme that required *DNA* for activity (Hurwitz, Bresler, and Diringer, 1960; Weiss, 1960; Hurwitz, Furth, Anders, Ortiz, and August, 1961; Chamberlain and Berg, 1962). Although it would be many years before the details of RNA polymerase action would be determined (which strand is copied, direction of copying, role of promoters, etc.), the discovery of RNA polymerase in 1960 opened up an extensive new field for biochemical research.

Soon experiments on messenger RNA were being carried out throughout the molecular biological community. The 1963 Cold Spring Harbor symposium on structure and synthesis of macromolecules had several major sections dealing with messenger RNA and related subjects. Although genetics would continue to play an important role, most of the critical solutions would come from biochemistry.

Another area of research that flourished as a result of the mRNA concept was the study of cell-free protein synthesis. This led almost immediately to the startling discovery by Marshall Nirenberg that synthetic polynucleotides could function as artificial messengers. Nirenberg employed a cell-free system and measured the incorporation of individual amino acids as directed by synthetic messenger RNA molecules (Nirenberg and Matthaei, 1961). Soon this system would lead to a complete deciphering of the genetic code. Nirenberg's approach, a direct application of the messenger RNA model, ushered in another revolution in molecular biology and was perhaps the greatest contribution of "pure" biochemistry to genetics. Nirenberg's code would be quickly verified from mutation studies. (Nirenberg's experiment was one that seems in forethought unlikely to have worked. It required a faith that a code existed and knowledge of the concept of messenger RNA.)

10.13 THE CHANGING CONCEPT OF THE OPERON

The operon model provided an exciting framework for experiments on gene expression, but in retrospect it suffered from being too specific and detailed, yet at the same time too all-encompassing. The operon concept has undergone numerous revisions since its first inception.

In 1961, the model that Jacob and Monod presented for gene expression involved two parts, one dealing with the overall flow of information (Fig. 10.7), the second with the manner by which expression was regulated. The contrast between gene replication and gene function was clearly delineated and three major stages were described: (1) *Replication*, in which double-stranded DNA was copied. (2) *Transcription I*, in which the information of one strand of DNA (which strand not specified) was copied into messenger RNA: "The synthesis of messenger RNA is supposed to be a sequential and

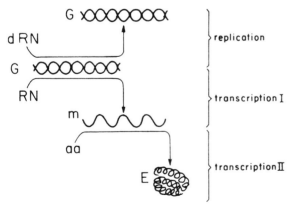

Figure 10.7 The 1961 Monod and Jacob model for the flow of information during replication and transcription. Transcription II was subsequently called *translation*. G: gene; dRN: deoxyribonucleotides; RN: ribonucleotides; m: messenger; aa: amino acids; E: enzyme. (Reprinted, with permission, from Jacob and Monod, 1961b.)

oriented process which can be initiated only at certain regions, or *operators*, on the DNA strands...The operon may therefore be defined as the unit of primary transcription." (3) *Transcription II*, in which the information in RNA is decoded in the ribosomes: "The second transcription takes place on ribosomes, and the messenger is destroyed in the process" (Jacob and Monod, 1961b).

The term "transcription" was apparently first used by Jacob and Monod in their 1961 *Journal of Molecular Biology* article. The term "transcription II," used above for what is now called translation, was clearly infelicitous and not especially accurate. As Spiegelman recognized, the transfer of information from RNA to protein was not a transcription but a translation (Spiegelman and Hayashi, 1963).

The second part of the Jacob/Monod model concerned the regulation of gene expression. This was clearly the principal concern of the 1961 *Journal of Molecular Biology* paper, but by the time of the 1961 Cold Spring Harbor symposium, evidence for messenger RNA was stronger, and their review at that meeting goes into the *mechanics* of protein synthesis in more detail.

The operon model was defined, essentially, on the basis of genetic evidence. One problem with the operon model was that it was difficult to distinguish operator mutants from mutants in the proximal gene of the operon that affected transcription (or even mutants that affected translation, such as reading frame mutants). Very soon after the operon model was presented, the concept of *polarity* of transcription was developed. In the *lac* operon, it was recognized that certain z mutants had reduced levels of *permease*, whereas mutants in permease never showed reduced β-galactosidase levels (Jacob and Monod, 1961a). "It is clear, therefore, that some polarity exists in the $o \rightarrow z \rightarrow y$ operon in such a way that some rare blocks in the z region decrease the rate of transcription in the distal part of the operon, whereas further blocks in the y region cannot alter the rate of initial transcription in the z region" (Jacob and Monod, 1961a).

Among other things, the discovery of polarity led to the realization that in the *lac* system, permease and transacetylase were not identical. Some mutants in the permease (y) gene abolished synthesis of transacetylase because of polarity. This was an important concept, since it showed the danger of equating two functions to a single gene based on gain or loss by mutation. Another class of operator mutants in which all genes of the operon were not expressed were termed by Jacob and Monod (1961a) *operator-negative* mutants (o^0), but these were subsequently found to be polarity mutants in which transcription of the whole operon was abolished (reviewed by Hayes, 1964).

Transduction analysis of genes involved in amino acid biosynthesis by Demerec and associates at Cold Spring Harbor had shown that often the genes involved in various steps in a single biosynthetic pathway were closely linked. It was initially thought that the order of the genes was the same as the order in which the enzymes *acted* in the biochemical pathway (see Section 8.5). However, this correlation soon proved to exist only for a few pathways. Enzyme repression was being observed in numerous systems. In 1959, an important discovery by Ames and Garry (1959) on the histidine pathway enzymes was announced. These authors found that four enzymes in the histidine pathway were coordinately repressed, that is, all were repressed to the same extent by a given concentrate of histidine. Hartman had shown that the whole series of histidine enzymes was closely linked (Ames and Hartman, 1963; see also Section 8.5) and on the basis of this knowledge, Ames and Garry (1959) postulated the existence of a repressor that blocked enzyme synthesis at the gene level. It is interesting that this model was published before the operon model of Jacob and Monod. Ames and Garry (1959) do not cite the PaJaMo experiments of Pardee, Jacob, and Monod (1958, 1959), although they should have been aware of them. Likewise, Jacob and Monod do not cite Ames and Garry when advancing the operon concept. These ''elegant experiments'' were cited by Jacob and Wollman (1961), although with the note that operator mutants had not been found. In actuality, classical operator mutants in the *his* system do not exist, since the operon is not controlled by the type of repression system being postulated by Jacob and Monod (Beckwith, 1987b), but by an attenuation system (Winkler, 1987).

The term *promoter* was first coined in 1964 to explain the behavior of a group of mutants (Jacob, Ullmann, and Monod, 1964). The evidence for the promoter was based on the isolation of certain mutants that contained a deleted region overlapping part of the *z* gene and the *o* region, but although the *o* was mutant, the entire operon was not expressed. This implied to Jacob, Ullmann, and Monod (1964) that there was another indispensable element, the *promoter*, which was needed for expression. As reviewed by Miller (1970), these so-called promoter mutants turned out to be something else, but true promoter mutants were isolated within several years in Beckwith's laboratory. These mutations, which were *cis*-dominant pleiotropic negatives, mapped in the appropriate region between *o* and *z* and supported the argument for a region essential for transcription. Promoters have, of course, become a major area of study and are now known to be the site of binding of RNA polymerase.

It is interesting that the meaning of the term *operon* has changed over the years since it was first created. As seen above, the operon was defined by genetic experiments, and central to its definition was the detection of mutants in the operator locus. In the original definition of the operon (Jacob, Perrin, Sanchez, and Monod, 1960) and its subsequent enhancements (Jacob and Monod, 1961a), the operon was a ''coordinated unit of expression...constituted of an operator and a group of structural genes linked to the operator'' (translated from Jacob, Perrin, Sanchez, and Monod, 1960). The site of action of the repressor was thus the operator locus, defined as a *cis*-acting region that was *dominant* in the mutated form. The repressor, on the other hand, was a *trans*-acting element (later shown to be protein). Although it originally appeared that the operon model would explain everything about regulation of gene expression, through the years it has become obvious that the *lac* operon is virtually a special case, and that numerous other mechanisms exist (Beckwith, 1987a). In many systems, regulation is by attenuation rather than at the start of transcription (Yanofsky, 1981, 1988). In some cases, positive rather than negative regulator genes exist, such as the arabinose system (Beckwith, 1987a). Therefore, the term ''operon'' has acquired a completely different meaning. Today, an operon is sometimes defined simply as a related gene cluster transcribed into a single messenger RNA molecule, without any statement regarding operator genes. However, in the arginine pathway the genes are not even linked, although subject to the action of a single repressor protein that acts on an operator locus adjacent to each site. The term *regulon* is often used for such *global* regulatory systems (Ingraham, Maaløe, and Neidhardt, 1983).

In *Escherichia coli*, at least 75 different operons have been recognized. In all these cases, regulation is at the level of DNA transcription, and the genetic *linkage* of the genes is crucial to this transcriptional regulation. It is transcriptional regulation, rather than negative control through an operator/repressor system, that defines an operon.

10.14 COLINEARITY OF GENE AND PROTEIN

In the 1950s, procedures for determining the amino acid sequences of proteins were perfected and were being applied to a variety of proteins. This whole approach was aided by developments of new column chromatographic methods for purifying proteins. With highly purified proteins, it then became possible to prepare protein crystals that could be subjected to X-ray diffraction analysis. Most of the proteins studied, at least in the early days, were from animal tissues, and the linkage of this work to genetics only came later when bacterial genetics was developed. The history of this research area, which is linked with the names of Max Perutz and John Kendrew, has been described in detail by Judson (1979). Anfinsen (1959) presented an early review.

The discovery of a link between protein structure and genetics was first made by V.M. Ingram (1957) through studies on sickle-cell anemia. Ingram showed that sickle-cell hemoglobin differed from normal hemoglobin in a single tryptic peptide in which a glutamic acid residue had been replaced by a valine. This exceedingly small change was sufficient to alter the properties of the protein. Genetic studies had shown that a single mutational step led to the difference between normal and sickle-cell hemoglobin. At this time, fine-structure genetics studies in phage by Benzer (see Section 6.10) and by Demerec and co-workers using transduction (see Section 8.5) had been published. Thus, the precise relationship between DNA base sequence and amino acid sequence was being hypothesized.

In the late 1950s, a promising approach connecting protein changes to gene changes came from studies on altered enzymes in bacterial mutants (Yanofsky, Helinski, and Maling, 1961). Although such mutant enzymes lacked function, they could be detected by immunological procedures. To avoid any preconception about the nature of the substance immunologically related to the enzyme, the material that reacted with antibody prepared against the purified enzyme was called *cross-reacting material* (CRM). In a sense, the study of immunologically cross-reacting proteins in mutants was an extension of the work of Melvin Cohn on the protein called Pz, which cross-reacted with β-galactosidase (see Section 10.7).

Numerous CRMs were detected, including those for the enzymes tryptophan synthetase, β-galactosidase, and alkaline phosphatase in *Escherichia coli*, glutamic dehydrogenase in *Neurospora*, and lysozyme of phage T4. In each case, the mutants producing CRMs mapped in small, unique genetic regions that corresponded to the structural gene for the enzyme. In addition, reverse mutants could be isolated, some of which mapped to the original mutational site, others to close but different sites.

By comparing amino acid and base sequences in CRM mutants and wild type it would have been possible, eventually, to arrive empirically at a genetic code, but this approach was soon supplanted by the direct approach of Nirenberg and Matthaei (1961) (see Section 10.12). However, an important question that could only be answered by a combination of genetic and biochemical procedures was whether there was a one-to-one correlation between nucleotide sequence and amino acid sequence. The tools were available: peptide mapping such as Ingram had used, genetically modified proteins (CRMs), and fine-structure genetic analysis. The problem was to select the right protein for the analyses.

The enzyme that was to prove the most suitable for determining colinearity of gene and protein was *tryptophan synthetase*, which had been the favorite enzyme of David Bonner and his students for many years. One of Bonner's students, Charles Yanofsky, perfected the tryptophan synthetase system in *Escherichia coli* and embarked on an extensive study that combined sophisticated genetic analysis with protein structure determination.

The tryptophan synthetase in *Escherichia coli* carries out the final reaction leading from indoleglycerol phosphate and serine to tryptophan. The enzyme consists of two proteins, designated components A and B, which combine to catalyze the overall reaction, although each component can be assayed separately. The A protein, with a molecular weight of only 26,500, was studied in some detail (Yanofsky, Helinksi, and Maling, 1962). A collection of mutants was obtained and subjected to both fine-structure mapping (with phage P1) and peptide mapping (with two-dimensional electrophoresis of trypsin-generated peptides). As the genetic code was being unraveled by Nirenberg and co-workers by biochemical means, Yanofsky was correlating mutational alterations with amino acid replacements (Yanofsky, 1963, 1967a,b; Yanofsky, Ito, and Horn, 1966). These were the days before it was possible to sequence DNA, and the only approach to the colinearity question was through a correlation of the genetic map with the amino acid sequence. Since the fine-structure work of Benzer had shown that point mutations could be attributed to single base substitutions, Yanofsky's work led to the important conclusion that there is a colinearity between the codons in DNA and the amino acids in a polypeptide chain. Furthermore, specific mutational changes at single sites in the tryptophan A gene could be related through the code to specific amino acid substitutions. In the admirable book by Judson (1979), certain lapses are found. One of these is clearly the neglect of the work of Charles Yanofsky, whose demonstration of the colinearity between gene and protein was of great importance.

Colinearity was also shown, in less detail, for one of the head proteins of phage T4 by Sydney Brenner and associates (Sarabhai, Stretton, Brenner, and Bolle, 1964). Although this study was the first to show colinearity, it was the Yanoksky work on tryptophan synthetase that was to provide the specifics of amino acid substitutions with gene mutations.

10.15 FEEDBACK INHIBITION AND ALLOSTERIC INTERACTIONS

Studies in bacterial genetics resulted in recognition of another important cellular regulatory mechanism that came to be called *feedback inhibition*. Feedback inhibition is a process that operates within a particular biosynthetic pathway. The end product of this pathway inhibits the activity of an enzyme *early* in the pathway. Thus, the inhibitor is *chemically unrelated* to either the substrate or the product of the enzyme being inhibited. The term "feedback inhibition" is derived from electronic control theory, where a signal at the end of a circuit is fed back to the beginning of the circuit. (Umbarger [1961], one of the discoverers of this process, preferred the term *endproduct inhibition*, but the term *feedback inhibition* seems to have become accepted.)

During the years when auxotrophic mutants were being isolated in large numbers as rich sources of biosynthetic intermediates (Davis, 1950), it was universally recognized that accumulation of intermediates did not occur in the presence of an excess of the end product. For some years, the reasons for this interesting phenomenon were not pursued, perhaps because those working in this area were more concerned with the pathways and enzymes themselves than with their regulation.

In early work on the dynamics of bacterial growth processes, Novick and Szilard (1954) obtained evidence suggestive of an effect of tryptophan on the tryptophan pathway. The work of Roberts, Cowie, Abelson, Bolton, and Britten (1955), mentioned in Section 10.8, also suggested end-product regulation of biosynthetic pathways. In 1956, a paper from Pardee's laboratory described the feedback inhibition of aspartic transcarbamylase, the first enzyme in the pyrimidine pathway, by uracil, an end product of the pathway (Yates and Pardee, 1956a,b). Independently, Umbarger (1956) reported on feedback inhibition in the isoleucine pathway.

Although repression and induction operate to control the *synthesis* of enzymes, feedback inhibition affects the *function* of preexisting enzymes. In addition to its inherent interest, an understanding of how feedback inhibition worked led to the concept of *allosteric proteins* (Changeux, 1979; Pardee, 1985b). The rationale for feedback inhibition, which has been well established by a large number of studies in numerous biosynthetic pathways (Cohen, 1965), is that the end product inhibits the *first* enzyme in the pathway that is unique to its own biosynthesis (Gerhart and Pardee, 1962, 1964; Pardee, 1971). The site on the enzyme at which the feedback inhibitor acts is *different* from the site where the substrate binds.

The understanding of the mechanism of feedback inhibition was to play an important role in the discovery of the mechanism by which repressors worked. The most remarkable feature of end-product-inhibited enzymes is that the inhibitor is not a steric analog of the substrate. This led Monod and Jacob (1961) to propose the concept of "allosteric inhibition" to account for this effect. Enzymes subject to allosteric inhibition presumably arose during evolution because they had selective advantages as efficient regulatory devices (Monod and Jacob, 1961). Although many of the biochemical details that Monod hypothesized for allosteric proteins were incorrect, the basic concept was central. In an important review paper published in 1963 (Monod, Changeux, and Jacob, 1963), the essential features of the allostery model were exposed. Allosteric proteins were postulated to possess at least two stereospecifically different sites, the *active site*, which binds the substrate, and the *allosteric site*, which binds specifically to the *allosteric effector*. Since the allosteric effector binds at a site distinct from the active site, it need not bear any particular chemical or metabolic relation with the substrate.

The phenomenon of feedback inhibition can thus be viewed as only one class of allosteric effects. Allostery is a particularly interesting phenomenon because it links the units of a genetic program with their products via the process of organic evolution. Allostery would never have been conceived of by a chemist, but only by an investigator steeped in Darwinian evolution (Davis, 1988). The concept of allostery was a major biological generalization and one of the most important ideas to develop out of the study of bacterial regulatory mechanisms. In a sense, allostery could explain any mysterious biological phenomenon. This, in fact, is one of the weak points of the model. On the other hand, the concept of a separation between the substrate and the regulatory metabolite explained induction and repression so well that the idea *had* to be correct. Soon, Monod had turned his whole laboratory over to the study of allosteric interactions (Changeux, 1979).

10.16 CATABOLITE REPRESSION

We have discussed diauxie in Section 10.4 and have shown that this was the phenomenon that brought Monod into the study of induced enzymes. Actually, Monod was not the first to observe the inhibition of a specific protein by glucose. Stark, Sherman, and Stark (1928) had described such an effect for enzymes in *Clostridium botulinum*, and Karström (1930) had observed the inhibition of invertase in lactobacillus. However, Monod's diauxie was a quantitative and precisely measurable phenomenon and suitable for experimental study. For many years, Monod had sought a unitary model that would explain both the positive role of the inducer and the negative role of glucose on β-galactosidase synthesis. Gradually, it became clear that glucose inhibition, often called the "glucose effect," was not directly related to the mechanism of induction. Among the important pieces of information that distinguished the two phenomena was the discovery that certain inducible enzymes were fully formed in the presence of glucose, and the demonstration that the synthesis of β-galactosidase in certain constitutive mutants was inhibited by glucose. Additionally, the inhibitory effects were not entirely specific for glucose, since in some systems other compounds, such as gluconic acid, mannitol, and galactose, had effects equal to glucose. Eventually, it was realized that any compound that could serve efficiently as a source of intermediary metabolites, and of energy, could reduce the rate of formation of glucose-sensitive enzymes. Even substrates that were not normally considered to cause induced enzyme inhibition, such as glycerol, could, under conditions where they were rapidly metabolized, inhibit β-galactosidase formation.

The clarification of the glucose effect came from the work of Frederick Neidhardt and Boris Magasanik working on induced enzyme synthesis in *Aerobacter aerogenes* (Neidhardt and Magasanik, 1956). Neidhardt and Magasanik concluded that repression of enzyme synthesis occurred when intermediary catabolites of a rapidly utilizable substrate such as glucose accumulated in the cell. Because such catabolites were not always derived from glucose, Magasanik (1961) coined the term "catabolite repression." Eventually, it would be shown that the role of glucose was to promote the breakdown or inhibit the synthesis of cyclic 3',5' AMP (cyclic AMP). Cyclic AMP (cAMP) had been identified first as a compound involved in hormone control in animals (it has often been called a "second messenger" in such systems), and Ullmann and Monod (1968) showed that addition of cyclic AMP to the medium could overcome catabolite repression of β-galactosidase synthesis. Cyclic AMP was also found to affect the synthesis of enzymes other than β-galactosidase (Magasanik, 1970). It was later shown that cyclic AMP interacted with a positive control protein, the catabolite activator protein (CAP), to promote transcription of the β-galactosidase operon (Magasnik and Neidhardt, 1987). Thus, CAP and *lac* repressor proteins have *opposite* effects on transcription of the *lac* operon. When glucose concentrations are high, the cyclic AMP concentration inside the cell is low, and CAP does not promote transcription. However, the situation is actually much more complicated than this, involving aspects of a global regulatory system that has still not been completely defined (Magasanik and Neidhardt, 1987).

It is interesting that although it was the glucose inhibition of β-galactosidase synthesis that brought Monod into the study of induced enzymes, the basis of the glucose effect is actually a positive rather than a negative control protein.

10.17 CONCLUSION

In the late 1950s and early 1960s, work on genetics, biochemistry, and physiology of bacteria and phage led to the development of a major series of hypotheses that were to become the paradigm for the new molecular biology. As a result of these theoretical constructs, the science of molecular biology as we know it came into existence. Although much of molecular biology could have developed without bacterial genetics and the study of enzyme induction, the whole field would have been greatly delayed. Although many of the ideas advanced in this period were oversimplified, some even wrong, they were in general so "right," so central, and, most importantly, so testable experimentally, that they provided the research framework for a whole generation of scientists.

In 1960, absolutely no details of the genetic code were known, but in 1966, a whole symposium on this topic was held at Cold Spring Harbor. Although the code itself was cracked by biochemical rather than by genetic techniques, it required the concept of messenger RNA to focus on the proper way to perform the biochemistry. Although gene expression in bacteria was eventually to prove much more complex than first envisaged, regulation of the genetic material in bacteria at the level of transcription is the chief mechanism controlling gene expression in bacteria.

A key insight was that regulation of gene expression for an induced enzyme involved a *negative* control mechanism. Among other things, this model focused on the mechanism of protein synthesis, which by this time was known to involve ribosomes and amino acid activating enzymes. However, the rapid rate of response to the addition and removal of inducer presented problems because of the known stability of ribosomes. Although the negative control model did not actually *require* the postulate of messenger RNA, the kinetics of induction could only be explained if the information from DNA was being transferred to an unstable (rapidly turning over) species. Fortuitously, evidence for such a rapidly turning over RNA already existed from studies on phage infection. Thus, phage and induced enzymes came together to provide the evidence for the development of the model of protein synthesis.

Although it is now clear that the operon model in its original form applies almost exclusively to *lac*, the concept of an operon as a cluster of closely linked genes all transcribed into a single polycistronic messenger RNA remains intact. Although this mechanism applies primarily to bacteria, the search (albeit unsuccessful) for similar systems in eucaryotes has led to an understanding of how genes are regulated in eucaryotes.

Beckwith (1987a) has reviewed briefly the history of the operon concept and has indicated the difficulty that workers had applying this model to some other systems. As we noted earlier, the *his* operon is regulated primarily by attenuation. There are *no* repressor genes, and o^c mutants are actually alterations in the attenuator region. Even more at variance with the original operon model were the studies of Ellis Englesberg in the arabinose system (Beckwith, 1987a). Here, there is no negative control but rather a system in which the product of the regulatory gene is a *positive* factor, the *ara*C protein. Together with catabolite activator protein (CAP) and cyclic AMP, *ara*C protein binds to a site adjacent to the arabinose promoter and stimulates the binding of RNA polymerase. Beckwith (1987a) has described the difficulty that Englesberg experienced in getting his positive-control model accepted in the 1960s, in an era when all attention was riveted on the negative-control model.

It is reasonable that the models of an earlier era seem to us oversimplified and specialized; what is not reasonable has been the reluctance to accept alternative models when the evidence is at hand. *Escherichia coli* is a complicated organism, with numerous mechanisms for regulating gene expression. Thus, although the original operon model has been tremendously fruitful, it is no longer possible to generalize from it to the world. Each system will have to be studied of and for itself, starting with an open mind as to how it might work. At least at the level of gene regulation, there are no unifying themes.

What *Escherichia coli* has provided is the experimental material to connect biochemistry to genetics. Although the revelations of Jacob and Monod seemed revolutionary when they first were announced around 1960, their ideas were only the initial step toward a much more spectacular discovery: recombinant DNA. Bacterial genetics was the essential forerunner of the biotechnology revolution. In the next and last chapter, we describe briefly how the findings that have been presented in this book led to this revolution.

REFERENCES

Ames, B.N. and Garry, B. 1959. Coordinate repression of the synthesis of four histidine biosynthetic enzymes by histidine. *Proceedings of the National Academy of Sciences* 45: 1453–1461.

Ames, B.N. and Hartman, P.E. 1963. The histidine operon. *Cold Spring Harbor Symposia on Quantitative Biology* 28: 349–356.

Anfinsen, C.B. 1959. *The Molecular Basis of Evolution.* John Wiley, New York, 228 pp.

Bachmann, B.J. 1987. Linkage map of *Escherichia coli* K-12, edition 7. pp. 807–876 in Neidhardt, F.C. (ed.), Escherichia coli *and* Salmonella typhimurium. American Society for Microbiology, Washington, D.C.

Beckwith, J. 1987a. The operon: an historical account. pp. 1439–1443 in Neidhardt, F.C. (ed.), Escherichia coli *and* Salmonella typhimurium. American Society for Microbiology, Washington, D.C.

Beckwith, J. 1987b. The lactose operon. pp. 1444–1452 In Neidhardt, F.C. (ed.), Escherichia coli *and* Salmonella typhimurium. American Society for Microbiology, Washington, D.C.

Belozersky, A.N. and Spirin, A.S. 1960. Chemistry of the nucleic acids of microorganisms. pp. 147–185 in Chargaff, E. and Davidson, J.N. (ed)., *The Nucleic Acids*, volume III. Academic Press, New York.

Benzer, S. 1953. Induced synthesis of enzymes in bacteria analyzed at the cellular level. *Biochimica et Biophysica Acta* 11: 383–395.

Benzer, S. 1966. Adventures in the rII region. pp. 157–165 in Cairns, J., Stent, G.S., and Watson, J.D. (ed.), *Phage and the Origins of Molecular Biology*. Cold Spring Harbor Laboratory of Quantitative Biology, Cold Spring Harbor, New York.

Bourgeois, S., Cohn, M., and Orgel, L.E. 1965. Suppression of and complementation among mutants of the regulatory gene of the lactose operon of *Escherichia coli*. *Journal of Molecular Biology* 14: 300.

Breinl, F. and Haurowitz, F. 1930. Chemische Untersuchung des Präzipitates aus Hämoglobin und Anti-Hämoglobin-Serum und Bemerkungenen über die Natur der Antikörper. *Hoppe-Seyler's Zeitschrift für physiologische Chemie* 192: 45–57.

Brenner, S. 1961. RNA, ribosomes, and protein synthesis. *Cold Spring Harbor Symposia on Quantitative Biology* 26: 101–110.

Brenner, S., Jacob, F., and Meselson, M. 1961. An unstable intermediate carrying information from genes to ribosomes for protein synthesis. *Nature* 190: 576–581.

Buchner, E. 1897. Alkoholische Gährung ohne Hefezellen. *Berichte der Deutschen Chemischen Gesellschaft* 30: 117–124.

Cairns, J. 1963. The chromosome of *Escherichia coli*. *Cold Spring Harbor Symposia on Quantitative Biology* 28: 43–46.

Chamberlin, M. and Berg, P. 1962. Deoxyribonucleic acid-directed synthesis of ribonucleic acid by an enzyme from *Escherichia coli*. *Proceedings of the National Academy of Sciences* 48: 81–94.

Changeux, J.-P. 1979. A Ph.D. with Jacques Monod: prehistory of allosteric proteins. pp. 191–202 in Lwoff, A. and Ullmann, A. (ed.), *Origins of Molecular Biology. A Tribute to Jacques Monod*. Academic Press, New York.

Cohen, G.N. 1965. Regulation of enzyme activity in microorganisms. *Annual Review of Microbiology* 19: 105–126.

Cohen, G.N. 1979. Permeability as an excuse to write what I feel. pp. 89–93 in Lwoff, A. and Ullmann, A. (ed.), *Origins of Molecular Biology. A Tribute to Jacques Monod*. Academic Press, New York.

Cohen, G.N. and Monod, J. 1957. Bacterial permeases. *Bacteriological Reviews* 21: 169–194.

Cohen, S.S. 1968. *Virus-Induced Enzymes*. Columbia University Press, New York. 315 pp.

Cohen-Bazire, G. and Jolit, M. 1953. Isolement par sélection de mutants d'*Escherichia coli* synthétisant spontanément l'amylomaltase et la β-galactosidase. *Annales de l'Institut Pasteur* 84: 937–945.

Cohn, M. 1957. Contributions of studies on the β-galactosidase of *Escherichia coli* to our understanding of enzyme synthesis. *Bacteriological Reviews* 21: 140–168.

Cohn, M. 1978. In Memoriam. pp. 1–9 in Miller, J.H. and Reznikoff, W.S. (ed.), *The Operon*. Cold Spring Harbor Laboratory, Cold Spring Harbor, New York.

Cohn, M. 1989. The way it was: a commentary by Melvin Cohn. *Biochimica et Biophysica Acta* 1000: 109–112.

Cohn, M. and Monod, J. 1953. Specific inhibition and induction of enzyme biosynthesis. *Society for General Microbiology Symposium* 3: 132–149.

Cohn, M. and Torriani, A.-M. 1952. Immunochemical studies with β-galactosidase and structurally related proteins of *Escherichia coli*. *Journal of Immunology* 69: 471–491.

Cohn, M. and Torriani, A.-M. 1953. The relationships in biosynthesis of the β-galactosidase- and PZ-proteins in *Escherichia coli*. *Biochimica et Biophysica Acta* 10: 280–289.

Cohn, M., Cohen, G.N., and Monod, J. 1953. L'effet inhibiteur spécifique de la méthionine dans la formation de la méthionine-synthase chez *Escherichia coli*. *Comptes rendus des Academie des Sciences* 236: 746–748.

Cohn, M., Monod, J., Pollock, M.R., Spiegelman, S., and Stanier, R.Y. 1953. Terminology of enzyme formation. *Nature* 172: 1096.

Crick, F.H.C. 1958. On protein synthesis. *Symposium of the Society for Experimental Biology* 12: 138–163.

Crick, F.H.C. 1988. *What Mad Pursuit. A Personal View of Scientific Discovery*. Basic Books, New York.

Davis, B.D. 1950. Studies on nutritionally deficient bacterial mutants isolated by means of penicillin. *Experientia* 6: 41–50.

Davis, B.D. 1956. Relation between enzymes and permeability (membrane transport) in bacteria. pp. 509–522 in Gaebler, O.H. (ed.), *Enzymes: units of biological structure and function*. Academic Press, New York.

Davis, B.D. 1958. On the importance of being ionized. *Archives of Biochemistry and Biophysics* 78: 497–509.

Davis, B.D. 1986. On the importance of being ionized. *Biologist* 33: 291–295.

Davis, B.D. 1988. Allostery, information and reductionism. *Trends in Biochemical Sciences* 13: 371–378.

Dienert, F. 1900. Sur la fermentation du galactose et sur l'accoutumance des levures a ce sucre. *Annales de l'Institut Pasteur* 14: 139–189.

Doty, P., Marmur, J., Eigner, J., and Schildkraut, C. 1960. Strand separation and specific recombination in deoxyribonucleic acids: physical chemical studies. *Proceedings of the National Academy of Sciences* 46: 461–476.

Dunn, L.C., ed. 1951. *Genetics in the 20th Century*. Macmillan, New York. 640 pp.

Emerson, S. 1945. Genetics as a tool for studying gene structure. *Annals of the Missouri Botanical Garden* 32: 243–249.

Gaebler, O.H., ed. 1956. *Enzymes: Units of Biological Structure and Function*. Academic Press, New York. 624 pp.

Gale, E.F. 1947. *The Chemical Activities of Bacteria*. University Tutorial Press, London. 199 pp.

Gale, E.F. 1956. Nucleic acids and enzyme synthesis. pp. 49–66 in Gaebler, O.H. (ed.), *Enzymes: Units of Biological Structure and Function*. Academic Press, New York.

Gale, E.F. and McQuillen, K. 1957. Nitrogen metabolism. *Annual Review of Microbiology* 11: 283–316.

Gerhart, J.C. and Pardee, A.B. 1962. The enzymology of control by feedback inhibition. *Journal of Biological Chemistry* 237: 891–896.

Gerhart, J.C. and Pardee, A.B. 1964. Aspartate transcarbamylase, an enzyme designed for feedback inhibition. *Federation Proceedings* 23: 727–735.

Gilbert, W. and Müller-Hill, B. 1966. Isolation of the *lac* repressor. *Proceedings of the National Academy of Sciences* 56: 1891–1899.

Gilbert, W. and Müller-Hill, B. 1967. The lac operator is DNA. *Proceedings of the National Academy of Sciences* 58: 2415–2421.

Gilbert, W. and Müller-Hill, B. 1970. The lactose repressor. pp. 93–109 in Beckwith, J.R. and Zipser, D. (ed.), *The Lactose Operon*. Cold Spring Harbor Laboratory, Cold Spring Harbor, New York.

Gillespie, D. and Spiegelman, S. 1965. A quantitative assay for DNA-RNA hybrids with DNA immobilized on a membrane. *Journal of Molecular Biology* 12: 829–842.

Glansdorff, N. 1987. Biosynthesis of arginine and polyamines. pp. 321–344 in Neidhardt, F.C. (ed.), Escherichia coli *and* Salmonella typhimurium. American Society for Microbiology, Washington, D.C.

Gorini, L. and Maas, W.K. 1957. The potential for the formation of a biosynthetic enzyme in *Escherichia coli*. *Biochimica Biophysica Acta* 25: 208–209.

Gros, F., Gilbert, W., Hiatt, H.H., Attardi, G., Spahr, P.F., and Watson, J.D. 1961. Molecular and biological characterization of messenger RNA. *Cold Spring Harbor Symposia on Quantitative Biology* 26: 111–132.

Hall, B.D. and Spiegelman, S. 1961. Sequence complementarity of T2-DNA and T2-specific RNA. *Proceedings of the National Academy of Sciences* 47: 137–146.

Hayes, W. 1957. The kinetics of the mating process in *Escherichia coli*. *Journal of General Microbiology* 16: 97–119.

Hayes, W. 1964. *The Genetics of Bacteria and their Viruses*. John Wiley, New York. 740 pp.

Hershey, A.D. and Bronfenbrenner, J. 1936. Dissociation and lactase activity in slow lactose-fermenting bacteria of intestinal origin. *Journal of Bacteriology* 31: 453–464.

Hoagland, M.B. 1959. The present status of the adaptor hypothesis. pp. 40–46 in *Structure and Function of Genetic Elements*, publication BNL 558 (C-29). Brookhaven National Laboratory, Upton, New York.

Hoagland, M.B. 1960. The relationship of nucleic acid and protein synthesis as revealed by studies in cell-free systems. pp. 349–408 in Chargaff, E. and Davidson, J.N. (ed.), *The Nucleic Acids*, volume 3. Academic Press, New York.

Hoagland, M.B. 1989. Commentary on 'Intermediate reactions in protein synthesis.' *Biochimica et Biophysica Acta* 1000: 103–105.

Hogness, D.S., Cohn, M., and Monod, J. 1955. Studies on the induced synthesis of β-galactosidase in *Escherichia coli*: The kinetics and mechanism of sulfur incorporation. *Biochimica et Biophysica Acta* 16: 99–116.

Horiuchi, T. and Ohshima, Y. 1966. Inhibition of repressor formation in the lactose system of *Escherichia coli* by inhibitors of protein synthesis. *Journal of Molecular Biology* 20: 517–526.

Horiuchi, T., Horiuchi, S., and Novick, A. 1961. A temperature-sensitive regulatory system. *Journal of Molecular Biology* 3: 703–704.

Hurwitz, J., Bresler, A., and Diringer, R. 1960. The enzymic incorporation of ribonucleotides into polyribonucleotides and the effect of DNA. *Biochemical and Biophysical Research Communications* 3: 15–19.

Hurwitz, J., Furth, J.J., Anders, M., Ortiz, P.J., and August, J.T. 1961. The enzymatic incorporation of ribonucleotides into RNA and the role of DNA. *Cold Spring Harbor Symposia on Quantitative Biology* 26: 91–100.

Ingraham, J.L., Maaløe, O., and Neidhardt, F.C. 1983. *Growth of the Bacterial Cell.* Sinauer Associates, Sunderland, Massachusetts, 435 pp.

Ingram, V.M. 1957. Gene mutations in human haemoglobin: the chemical difference between normal and sickle cell haemoglobin. *Nature* 180: 326–328.

Jacob, F. 1966. Genetics of the bacterial cell. *Science* 152: 1470–1478.

Jacob, F. 1979. The switch. pp. 95–107 in Lwoff, A. and Ullmann, A. (ed.), *Origins of Molecular Biology. A Tribute to Jacques Monod.* Academic Press, New York.

Jacob, F. 1988. *The Statue Within.* Basic Books, New York. 326 pp.

Jacob, F. and Campbell, A. 1959. Sur le système de répression assurant l'immunité chez les bactéries lysogènes. *Comptes rendus des Academie des Sciences* 248: 3219–3221.

Jacob, F. and Monod, J. 1959. Gènes de structure et gènes de règulation dans la biosynthèse des protéines. *Comptes rendus des Academie des Sciences* 249: 1282–1284.

Jacob, F. and Monod, J. 1961a. Genetic regulatory mechanisms in the synthesis of proteins. *Journal of Molecular Biology* 3: 318–356.

Jacob, F. and Monod, J. 1961b. On the regulation of gene activity. *Cold Spring Harbor Symposia on Quantitative Biology* 26: 193–211.

Jacob, F. and Wollman, E.L. 1961. *Sexuality and the Genetics of Bacteria.* Academic Press, New York. 374 pp.

Jacob, F., Perrin, D., Sanchez, C., and Monod, J. 1960. L'opéron: group de gènes á expression coordonnée par un opérateur. *Comptes rendus des Academie des Sciences* 250: 1727–1729.

Jacob, F., Sussman, R., and Monod, J. 1962. Sur la nature du répresseur assurant l'immunité des bactéries lysogènes. *Comptes rendus des Academie des Sciences* 254: 4214–4216.

Jacob, F., Ullmann, A., and Monod, J. 1964. Le promoteur, élément génétique nécessaire à l'expression d'un opéron. *Comptes rendus des Academie des Sciences* 258: 3125–3128.

Judson, H.F. 1979. *The Eighth Day of Creation. The Makers of the Revolution in Biology.* Simon and Shuster, New York. 686 pp.

Karström, H. 1930. *Über die Enzymbildung in Bakterien und über einige physiologische Eigenschaften der untersuchten Bakterienarten.* Suomalainen Tiedeakatemia, Helsinki. 147 pp. Also published as Ann. Acad. Scient. Fennicae, A. XXXIII, Number 2, 1930.

Karström, H. 1937. Enzymatische Adaptation bei Mikroorganismen. *Ergebnisse Enzymforschung* 7: 350–376.

Kennedy, E.P. 1970. The lactose permease system of *Escherichia coli*. pp. 49–92 in Beckwith, J.R. and Zipser, D. (ed.), *The Lactose Operon.* Cold Spring Harbor Laboratory, Cold Spring Harbor, New York.

Khorana, H.G. 1960. Chemical and enzymic synthesis of polynucleotides. pp. 105–146 in Chargaff, E. and Davidson, J.N. (ed.), *The Nucleic Acids*, volume 3, Academic Press, New York.

Kluyver, A.J. and van Niel, C.B. 1956. *The Microbe's Contribution to Biology*. Harvard University Press, Cambridge. 182 pp.

Kornberg, A. 1957. Pathways of enzymatic synthesis of nucleotides and polynucleotides. pp. 579–608 in McElroy, W.D. and Glass, B. (ed.), *Chemical Basis of Heredity*. Johns Hopkins Press, Baltimore.

Lederberg, E.M. 1948. The mutability of several lac⁻ mutants of *Escherichia coli*. *Genetics* 33: 617.

Lederberg, E.M. 1950. Genetic control of mutability in the bacterium *Escherichia coli*. Ph.D. thesis, University of Wisconsin-Madison. 78 pp.

Lederberg, E.M. 1952. Allelic relationships and reverse mutation in *Escherichia coli*. *Genetics* 37: 469–483.

Lederberg, J. 1947. Gene recombination and linked segregations in *Escherichia coli*. *Genetics* 32: 505–525.

Lederberg, J. 1948. Gene control of β-galactosidase in *Escherichia coli*. *Genetics* 33: 617–618.

Lederberg, J. 1950. The beta-D-galactosidase of *Escherichia coli*, strain K-12. *Journal of Bacteriology* 60: 381–392.

Lederberg, J. 1951. Genetic studies with bacteria. pp. 263–289 in Dunn, L.C. (ed.), *Genetics in the 20th Century*. Macmillan, New York.

Lederberg, J. 1955. Comments on gene-enzyme relationship. pp. 161–174 in Gaebler, O.H. (ed.), *Enzymes: Units of Biological Structure and Function*. Academic Press, New York.

Lindegren, C.C. and Lindegren, G. 1946. The cytogene theory. *Cold Spring Harbor Symposia on Quantitative Biology* 11: 115–129.

Lwoff, A. 1946. Some problems connected with spontaneous biochemical mutations in bacteria. *Cold Spring Harbor Symposia on Quantitative Biology* 11: 139–155.

Lwoff, A. 1977. Jacques Lucien Monod, 9 February 1910—31 May 1976. *Biographical Memoirs of the Fellows of the Royal Society* 23: 385–412.

Lwoff, A. and Ullmann, A. 1978. *Selected Papers in Molecular Biology by Jacques Monod*. Academic Press, New York. 753 pp.

Lwoff, A. and Ullmann, A. 1979. *Origins of Molecular Biology. A Tribute to Jacques Monod*. Academic Press, New York. 246 pp.

Magasanik, B. 1961. Catabolite repression. *Cold Spring Harbor Symposia on Quantitative Biology* 26: 249–256.

Magasanik, B. 1970. Glucose effects: inducer exclusion and repression. pp. 189–219 in Beckwith, J.R. and Zipser, D. (ed.), *The Lactose Operon*. Cold Spring Harbor Laboratory, Cold Spring Harbor, New York.

Magasanik, B. and Neidhardt, F.C. 1987. Regulation of carbon and nitrogen utilization. pp. 1318–1325 in Neidhardt, F.C. (ed.), Escherichia coli *and* Salmonella typhimurium. American Society for Microbiology, Washington, D.C.

Marmur, J. and Lane, D. 1960. Strand separation and specific recombination in deoxyribonucleic acids: biological studies. *Proceedings of the National Academy of Sciences* 46: 453–461.

Miller, J.H. 1970. Transcription starts and stops in the *lac* operon. pp. 173–188 in Beckwith, J.R. and Zipser, D. (ed.), *The Lactose Operon*. Cold Spring Harbor Laboratory, Cold Spring Harbor, New York.

Monod, J. 1941a. Croissance des populations bactériennes en fonction de la concentration de l'aliment hydrocarboné. *Comptes rendus des Academie des Sciences* 212: 771–773.

Monod, J. 1941b. Sur un phénomène nouveau de croissance complexe dans les cultures bactériennes. *Comptes rendus des Academie des Sciences* 212: 934–936.

Monod, J. 1942. *Recherches sur la croissance des cultures bactériennes*. Hermann and Co., Paris. 211 pp.

Monod, J. 1945. Sur la nature du phénomène de diauxie. *Annales de l'Institut Pasteur* 71: 37–40.

Monod, J. 1947. The phenomenon of enzymatic adaptation and its bearings on problems of genetics and cellular differentiation. *Growth Symposium* 11: 223–289.

Monod, J. 1949. The growth of bacterial cultures. *Annual Review of Microbiology* 3: 371–394.

Monod, J. 1950. Adaptation, mutation and segregation in the formation of bacterial enzymes. *Biochemical Society Symposia* 4: 51–58.

Monod, J. 1956. Remarks on the mechanism of enzyme induction. pp. 7–28 in Gaebler, O.H. (ed.), *Enzymes: Units of Biological Structure and Function*. Academic Press, New York.

Monod, J. 1966. From enzymatic adaptation to allosteric transitions. *Science* 154: 475–483.

Monod, J. and Audureau, A. 1946. Mutation et adaptation enzymatique chez *Escherichia coli*-mutabile. *Annales de l'Institut Pasteur* 72: 868–878.

Monod, J. and Cohen-Bazire, G. 1953a. L'effet d'inhibition spécifique dans la biosynthèse de la tryptophane-desmase chez *Aerobacter aerogenes*. *Comptes rendus des Academie des Sciences* 236: 530–532.

Monod, J. and Cohen-Bazire, G. 1953b. L'effet d'inhibition spécifique des β-galactosides dans la biosynthèse 'constitutive' de la β-galactosidase chez *Escherichia coli*. *Comptes rendus des Academie des Sciences* 236: 417–419.

Monod, J. and Cohn, M. 1952. La biosynthèse induite des enzymes (adaptation enzymatique). *Advances in Enzymology* 13: 67–119.

Monod, J. and Jacob, F. 1961. General conclusions: teleonomic mechanisms in cellular metabolism, growth, and differentiation. *Cold Spring Harbor Symposia on Quantitative Biology* 26: 389–401.

Monod, J. and Wollman, E. 1947. L'inhibition de la croissance et de l'adaptation enzymatique chez les bactéries infectées par le bactériophage. *Annales de l'Institut Pasteur* 73: 937–956.

Monod, J., Changeux, J.-P., and Jacob, F. 1963. Allosteric proteins and cellular control systems. *Journal of Molecular Biology* 6: 306–329.

Monod, J., Cohen-Bazire, G., and Cohn, M. 1951. Sur la biosynthese de la β-galactosidase (lactase) chez *Escherichia coli*. La specificite de l'induction. *Biochimica et Biophysica Acta* 7: 585–599.

Monod, J., Pappenheimer, A.M., Jr., and Cohen-Bazire, G. 1952. La cinétique de la biosynthèse de la β-galactosidase chez *Escherichia coli* considérée comme fonction de la croissance. *Biochimica et Biophysica Acta* 9: 648–660.

Muller, H.J. 1947. The gene. *Proceedings of the Royal Society* B134: 1–37.

Neidhardt, F.C. and Magasanik, B. 1956. Inhibitory effect of glucose on enzyme formation. *Nature* 178: 801–802.

Neidhardt, F.C. and Magasanik, B. 1957. Reversal of the glucose inhibition of histidase biosynthesis in *Aerobacter aerogenes*. *Journal of Bacteriology* 73: 253–259.

Nirenberg, M.W. and Matthaei, J.H. 1961. The dependence of cell-free protein synthesis in *Escherichia coli* upon naturally occurring or synthetic polyribonucleotides. *Proceedings of the National Academy of Sciences* 47: 1588–1602.

Novick, A. and Szilard, L. 1954. Experiments with the chemostat on the rates of amino acid synthesis in bacteria. pp. 21–32 in Boell, E.J. (ed.), *Dynamics of Growth Processes* Princeton University Press, Princeton, New Jersey.

Nygard, N.P. and Hall, B.D. 1964. Formation and properties of RNA-DNA complexes. *Journal of Molecular Biology* 9: 125–142.

Ochoa, S. and Heppel, L.A. 1957. Polynucleotide synthesis. pp. 615–638 in McElroy, W.D. and Glass, B. (ed.), *Chemical Basis of Heredity*. Johns Hopkins Press, Baltimore.

Pappenheimer, A.M. Jr. 1979. Whatever happened to Pz? pp. 55–60 in Lwoff, A. and Ullmann, A. (ed.), *Origins of Molecular Biology. A Tribute to Jacques Monod*. Academic Press, New York.

Pardee, A.B. 1957. An inducible mechanism for accumulation of melibiose in *Escherichia coli*. *Journal of Bacteriology* 73: 376–385.

Pardee, A.B. 1959. The control of enzyme activity. pp. 681–716 in Boyer, P.D., Lardy, H., and Myrbäck, K. (ed.), *The Enzymes*, 2nd edition. Academic Press, New York.

Pardee, A.B. 1971. Control of metabolic reactions by feedback inhibition. *Harvey Lectures* 65: 59–71.

Pardee, A.B. 1979. The pajama experiment. pp. 109–116 in Lwoff, A. and Ullmann, A. (ed.), *Origins of Molecular Biology. A Tribute to Jacques Monod*. Academic Press, New York.

Pardee, A.B. 1985a. Molecular basis of gene expression: origins from the Pajama experiment. *BioEssays* 2: 1–4.

Pardee, A.B. 1985b. Molecular basis of biological regulation: origins from feedback inhibition and allostery. *BioEssays* 2: 37–40.

Pardee, A.B. and Prestidge, L.S. 1959. On the nature of the repressor of β-galactosidase synthesis in *Escherichia coli*. *Biochimica et Biophysica Acta* 36: 545–547.

Pardee, A.B. and Prestidge, L.S. 1961. The initial kinetics of enzyme induction. *Biochimica et Biophysica Acta* 49: 77–88.

Pardee, A.B., Jacob, F., and Monod, J. 1958. Sur l'expression et le role des allèles "inductible" et "constitutif" dans la syntheèse de la β-galactosidase chez les zygotes d'*Escherichia coli*. *Comptes rendus des Academie des Sciences* 246: 3125–3127.

Pardee, A.B., Jacob, F., and Monod, J. 1959. The genetic control and cytoplasmic expression of "inducibility" in the synthesis of β-galactosidase by *E. coli*. *Journal of Molecular Biology* 1: 165–178.

Pasteur, L. 1860. Mémoire sur la fermentation alcoölique. *Annales de Chimie et de Physique* 58: 323–426.

Pauling, L. 1940. A theory of the structure and process of formation of antibodies. *Journal of the American Chemical Society* 62: 2643–2657.

Pollock, M.R. 1953. Stages in enzyme adaptation. *Society for General Microbiology Symposium* 3: 150–183.

Ptashne, M. 1967. Isolation of the λ phage repressor. *Proceedings of the National Academy of Sciences* 57: 306–313.

Rickenberg, H.V., Cohen, G.N., Buttin, G., and Monod, J. 1956. La galactoside-perméase d'*Escherichia coli*. *Annales de l'Institut Pasteur* 91: 829–857.

Riley, M. and Pardee, A.B. 1962. Gene expression: its specificity and regulation. *Annual Review of Microbiology* 16: 1–34.

Riley, M., Pardee, A.B., Jacob, F., and Monod, J. 1960. On the expression of a structural gene. *Journal of Molecular Biology* 2: 216–225.

Roberts, R.B., Cowie, D.B., Abelson, P.H., Bolton, E.T., and Britten, R.J. 1955. *Studies of Biosynthesis in* Escherichia coli. Carnegie Institution of Washington Publication 607, Washington, D.C. 521 pp.

Rotman, B. and Spiegelman, S. 1954. On the origin of the carbon in the induced synthesis of β-galactosidase in *Escherichia coli*. *Journal of Bacteriology* 68: 419–429.

Sadler, J.R. and Novick, A. 1965. The properties of the repressor and the kinetics of its action. *Journal of Molecular Biology* 12: 305–327.

Sapp, J. 1987. *Beyond the Gene. Cytoplasmic Inheritance and the Struggle for Authority in Genetics*. Oxford University Press, New York. 266 pp.

Sarabhai, A.S., Stretton, A.O.W., Brenner, S., and Bolle, A. 1964. Colinearity of the gene with the polypeptide chain. *Nature* 201: 13–17.

Schachman, H.K., Pardee, A.B., and Stanier, R.Y. 1952. Studies on macromolecular organizations of microbial cells. *Archives of Biochemistry and Biophysics* 38: 245–260.

Schaffner, K. 1974. Logic of discovery and justification in regulatory genetics. *Studies in the History and Philosophy of Science* 4: 349–385.

Seidman, M. and Link, K.P. 1950. *o*-Nitrophenyl β-D-galactopyranoside and its tetraacetate. *Journal of the American Chemical Society* 72: 4324.

Spiegelman, S. 1945. The physiology and genetic significance of enzymatic adaptation. *Annals of the Missouri Botanical Garden* 32: 139–258.

Spiegelman, S. 1946. Nuclear and cytoplasmic factors controlling enzymatic constitution. *Cold Spring Harbor Symposia on Quantitative Biology* 11: 256–277.

Spiegelman, S. 1956. On the nature of the enzyme-forming system. pp. 67–89 in Gaebler, O.H. (ed.), *Enzymes: Units of Biological Structure and Function*. Academic Press, New York.

Spiegelman, S. 1957. Nucleic acids and the synthesis of proteins. pp. 232–267 in McElroy, W.D. and Glass, B. (ed.), *Chemical Basis of Heredity*. Johns Hopkins Press, Baltimore.

Spiegelman, S. 1961. The relation of informational RNA to DNA. *Cold Spring Harbor Symposia on Quantitative Biology* 26: 75–90.

Spiegelman, S. 1964. Hybrid nucleic acids. *Scientific American* 210: 48–56.

Spiegelman, S. and Hayashi, M. 1963. The present status of the transfer of genetic information and its control. *Cold Spring Harbor Symposia on Quantitative Biology* 28: 161–181.

Spiegelman, S. and Kamen, M.D. 1946. Genes and nucleoproteins in the synthesis of enzymes. *Science* 104: 581–584.

Spiegelman, S., Lindegren, C.C., and Lindegren, G. 1945. Maintenance and increase of a genetic character by a substrate-cytoplasmic interaction in the absence of the specific gene. *Proceedings of the National Academy of Sciences* 31: 95–102.

Stanier, G. (Cohen-Bazire). 1979. Remembrance of things past. pp. 49–53 in Lwoff, A. and Ullmann, A. (ed.), *Origins of Molecular Biology. A Tribute to Jacques Monod.* Academic Press, New York.

Stanier, R.Y. 1951. Enzymatic adaptation in bacteria. *Annual Review of Microbiology* 5: 35–56.

Stark, C.N., Sherman, J.M., and Stark, P. 1928. Glucose inhibition of extracellular toxin-producing enzymes of *Clostridium botulinum*. *Journal of Infectious Diseases* 43: 566–568.

Stephenson, M. 1930. *Bacterial Metabolism*. Longmans, Green, and Co., London. 320 pp.

Stephenson, M. and Gale, E.F. 1937. CLXV. The adaptability of glucozymase and galactozymase in *Bacterium coli*. *Biochemical Journal* 31: 1311–1315.

Stephenson, M. and Yudkin, J. 1936. LXXVI. Galactozymase considered as an adaptive enzyme. *Biochemical Journal* 30: 506–514.

Szilard, L. 1960. The control of the formation of specific proteins in bacteria and in animal cells. *Proceedings of the National Academy of Sciences* 46: 277–292.

Ullmann, A. and Monod, J. 1968. Cyclic AMP as an antagonist of catabolite repression in *Escherichia coli*. *Federation of European Biological Societies Letters* 2: 57.

Umbarger, H.E. 1956. Evidence for a negative-feedback mechanism in the biosynthesis of isoleucine. *Science* 123: 848.

Umbarger, H.E. 1961. Feedback control by endproduct inhibition. *Cold Spring Harbor Symposia on Quantitative Biology* 26: 301–312.

Vogel, H.J. 1957. Repression and induction as control mechanisms of enzyme biogenesis: the "adaptive" formation of acetylornithinase. pp. 276–289 in McElroy, W.D. and Glass, B. (ed.), *Chemical Basis of Heredity*. Johns Hopkins Press, Baltimore.

Vogel, H.J. 1961. Aspects of repression in the regulation of enzyme synthesis: pathway-wide control and enzyme-specific response. *Cold Spring Harbor Symposia on Quantitative Biology* 26: 163–172.

Vogel, H.J. and Davis, B.D. 1952. Adaptive phenomena in a biosynthetic pathway. *Federation Proceedings* 11: 485.

Volkin, E. and Astrachan, L. 1956. Phosphorus incorporation in *Escherichia coli* ribonucleic acid after infection with bacteriophage T2. *Virology* 2: 149–161.

Volkin, E. and Astrachan, L. 1957. RNA metabolism in T2-infected *Escherichia coli*. pp. 686–695 in McElroy, W.D. and Glass, B. (ed.), *Chemical Basis of Heredity*. Johns Hopkins Press, Baltimore.

Weiss, S.B. 1960. Enzymatic incorporation of ribonucleoside triphosphates into the interpolynucleotide linkages of ribonucleic acid. *Proceedings of the National Academy of Sciences* 46: 1020–1030.

Winkler, M.E. 1987. Biosynthesis of histidine. pp. 395–411 in Neidhardt, F.C. (ed.), *Escherichia coli and Salmonella typhimurium*. American Society for Microbiology, Washington, D.C.

Wollman, E.L., Jacob, F., and Hayes, W. 1956. Conjugation and genetic recombination in *Escherichia coli* K-12. *Cold Spring Harbor Symposia on Quantitative Biology* 21: 141–162.

Wyatt, G.R. 1953. The quantitative composition of deoxypentose nucleic acid as related to the newly proposed structure. *Cold Spring Harbor Symposia on Quantitative Biology* 18: 133–134.

Yanofsky, C. 1963. Amino acid replacements associated with mutation and recombination in the A gene and their relationship to in vitro coding data. *Cold Spring Harbor Symposia on Quantitative Biology* 28: 581–588.

Yanofsky, C. 1967a. Gene structure and protein structure. *Harvey Lectures* 61: 145–168.

Yanofsky, C. 1967b. Gene structure and protein structure. *Scientific American*, 216: 80–94.

Yanofsky, C. 1981. Attentuation in the control of expression of bacterial operons. *Nature* 289: 751–758.

Yanofsky, C. 1988. Transcription attenuation. *Journal of Biological Chemistry* 263: 609–612.

Yanofsky, C., Helinski, D.R., and Maling, B.D. 1961. The effects of mutation on the composition and properties of the A protein of *Escherichia coli* tryptophan synthetase. *Cold Spring Harbor Symposia on Quantitative Biology* 26: 11–24.

Yanofsky, C., Ito, J., and Horn, V. 1966. Amino acid replacements and the genetic code. *Cold Spring Harbor Symposia on Quantitative Biology* 31: 151–162.

Yates, R.A. and Pardee, A.B. 1956a. Pyrimidine biosynthesis in *Escherichia coli*. *Journal of Biological Chemistry* 221: 743–756.

Yates, R.A. and Pardee, A.B. 1956b. Control of pyrimidine biosynthesis in *Escherichia coli* by a fee-back mechanism. *Journal of Biological Chemistry* 221: 757–770.

Yates, R.A. and Pardee, A.B. 1957. Control by uracil of formation of enzymes required for orotate synthesis. *Journal of Biological Chemistry* 227: 677–692.

Yudkin, J. 1932. CCXXI. Hydrogenlyases. II. Some factors concerned in the production of the enzymes. *Biochemical Journal* 26: 1859–1871.

Yudkin, J. 1937. Enzyme variation in micro-organisms. *Biological Reviews* 12: 93–106.

Zabin, I., Kepes, A., and Monod, J. 1959. On the enzymic acetylation of isopropyl-β-D-thiogalactoside and its association with galactoside-permease. *Biochemical and Biophysical Research Communications* 1: 289–292.

Zamecnik, P.C. 1979. Historical aspects of protein synthesis. *Annals of the New York Academy of Sciences* 325: 269–301.

11

FROM BACTERIAL GENETICS TO RECOMBINANT DNA

We are now in the midst of a revolution in biology that has been aptly called the *biotechnology revolution*; at the center of this revolution are the techniques of *recombinant DNA*. Without bacterial genetics, recombinant DNA would not have developed. Thus, it seems appropriate to end this book with a short discussion of those discoveries of bacterial genetics that have provided the foundations of recombinant DNA.

Recombinant DNA research comprises a whole group of techniques that permit the manipulation of the genetic material. It involves the construction of hybrid DNA molecules with the purpose of inserting these hybrids into living organisms, where the hybrids can replicate. Thus transplanted, the genes of the hybrid may impart new hereditary properties on the recipients, or the genes may merely be replicated in the recipients. There are two major goals of recombinant DNA research: (1) to produce clones of specific DNA sequences that can be used in biochemical and genetic research; (2) to obtain expression of the cloned genes, with the goal of producing large amounts of the gene product, generally for commercial purposes. Although certain aspects of recombinant DNA research could be done without the use of bacteria, some of the most critical procedures can only be done using bacterial systems.

By permitting the facile study of the genetic material of a wide variety of organisms, recombinant DNA research has revolutionized biological research. Among other things, it has led to completely unexpected findings about the organization and expression of eucaryotic organisms. Applications of recombinant DNA research exist in such disparate fields as virology, oncology, immunology, endocrinology, and developmental biology, as well as in genetics.

In the rest of this chapter, the key discoveries of bacterial genetics that have permitted the development of recombinant DNA research will be briefly discussed. Some implications of bacterial genetics for recombinant DNA research have already been alluded to in earlier chapters.

At the basis of recombinant DNA research are the unique properties of *restriction enzymes*. Such enzymes have been found so far only in bacteria.

Restriction enzymes were discovered as a result of research on a phenomenon in bacteriophage termed *host-controlled modification* or *phenotypic modification* (see Section 6.15). This phenomenon was discovered almost simultaneously in the early 1950s in four separate laboratories, but it was first studied in detail by Luria and Human (1952). Soon after its discovery, host-controlled modification had been detected in numerous phage/host systems (Luria, 1953). The most extensive studies on restriction were done with the *lambda* system. Work of Arber (1965) and others showed that a host which restricted *lambda* had a *restriction enzyme* that degraded *lambda* DNA, but this restriction enzyme did not act if the DNA had been modified by methylation. Thus, restriction and modification exist as a paired system, the function of which is to protect host DNA but destroy foreign DNA.

The unique property of restriction enzymes, as far as recombinant DNA research is concerned, is that they recognize specific, relatively short, base sequences of DNA and then hydrolyze this DNA. There are two kinds of restriction enzymes, termed type I and type II (for a review, see Szalay, Mackey, and Langridge, 1979). Type I restriction enzymes recognize specific nucleotide sequences but cut at random sites away from recognition sites. Thus, type I restriction enzymes produce DNA fragments of various sizes and with various end sequences, and hence are of no value for recombinant DNA research.

Type II restriction enzymes make single-stranded cuts in staggered locations on the two strands, thus generating complementary single-stranded ends. The first type II restriction enzyme was described by Smith and Wilcox (1970) from *Hemophilus influenzae*, and over the next few years a large number of such enzymes were isolated from various bacterial species, each enzyme having its own unique specificity (Smith and Nathans, 1973; Roberts, 1983).

The recognition that cleavage of DNA by a type II restriction enzyme led to the production of cohesive ends (Mertz and Davis, 1972) soon led to the idea that restriction enzymes could be used to construct biologically functional hybrid molecules (Cohen, Chang, Boyer, and Helling, 1973). If two disparate DNA molecules were cleaved with the same restriction enzyme and then mixed, hybrids could be formed after hydrogen bonding of the single-stranded ends. The joined fragments could then be covalently connected through the action of DNA ligase. Cohen, Chang, Boyer, and Helling (1973) constructed a biologically functional hybrid plasmid that contained antibiotic-resistance genes and inserted this plasmid into *Escherichia coli* by transformation. The procedure used had virtually unlimited research potential and could be used to insert DNA of many organisms, even eucaryotes, into *Escherichia coli*. Since the plasmid was an independently replicating element, it would also replicate the foreign DNA that had been inserted into it. The Cohen/Boyer procedure, subsequently to be called gene cloning, became the foundation of modern biotechnology (Cohen, 1975). It also led to the Asilomar Conference on the potential hazards of recombinant DNA (Berg, Baltimore, Brenner, Roblin, and Singer, 1975).

Although the above brief history is widely known, there are numerous unrecognized contributions that bacterial genetics has made to the development of recombinant DNA technology. Many of these contributions have become so engrained in modern research that they are ignored. These various contributions are discussed briefly in the rest of this chapter.

In *bacterial transduction*, a virus is the agency by which a gene or genes are transferred from one organism to another (see Chapter 8). The practical potential of transduction for manipulation of the genetic material was gradually realized. Because the transducing fragment represents only a fragment of the genome, transduction is a recombinant process, although occurring in vivo instead of in vitro. Transduction also requires homology between the transducing fragment and the chromosome, something not required in the recombinant DNA technique, but the idea of transduction points the way to fragmentation and reconstitution of the genome.

Soon after it was discovered that the *F factor* played a major role in fertility of *Escherichia coli*, it was found that under certain conditions the F factor could be made to pick up extraneous genes, which could then be transferred to a new organism (see Section 5.12). Initially, the F factor had been viewed merely as a carrier of genetic elements involved only in its own replication, but Jacob and Adelberg (1959) isolated a modified F^+ that had incorporated the chromosomal *lac* genes. This variant had arisen from an Hfr in which the F factor had become incorporated close to the *lac* locus. When F spontaneously detached, it excised *lac* with it and hence transferred the chromosomally derived *lac* gene, and *lac* alone, at high frequency. Such modified F particles came to be called F *prime*. Jacob and Adelberg (1959) concluded that any segment of the bacterial chromosome and the F factor could associate together and form a unit of replication and transmission. They recognized that this process of genetic transfer was comparable to that of transduction (see also Adelberg and Burns, 1960).

The idea that F *prime* arose by recombination with the bacterial chromosome led to a facile method for isolating a large number of different F *primes* by interrupting mating early but selecting for a marker known to be transferred late. In this way, F *prime* plasmids could be isolated that include genes from virtually the entire *Escherichia coli* chromosome. This approach was essentially in vivo cloning since a defined portion of the genome was isolated and attached to a vector—the F particle. Thus, this approach foreshadowed the development of in vitro cloning methods in the 1970s and 1980s.

Studies on the F factor and bacterial mating led to the discovery of *resistance transfer factors* and other *plasmids* (see Section 5.17). It soon became clear that plasmids were circular DNA molecules that behaved as independent genetic elements, controlling their own replication (see Section 5.18). Using an F factor containing *lac*, Marmur, Rownd, Falkow, Baron, Schildkraut, and Doty (1961) were able to detect the F factor DNA after it had been transferred to *Serratia*, because the base composition of *Serratia* DNA was different from that of *Escherichia coli* DNA. This study led to the isolation and purification of F factor DNA. The first plasmid shown to be circular was the *Col* E1 colicinogenic factor isolated from *Proteus mirabilis* (Roth and Helinski, 1967). This work on the purification and characterization of plasmid DNA was essential for the development of suitable vectors for recombinant DNA.

Although in generalized transduction any gene of the donor could be transferred to the host, in certain *temperate phage* only a restricted range of host genes could be transferred. Studies on bacteriophage *lambda* (see Sections 7.5 and 7.6) led to detailed understanding of the mechanism by which a foreign genetic element can become integrated into the host genome. Furthermore, discovery of specialized transduction in *lambda* (see Sections 8.6

and 8.7) showed that under certain conditions, phage genes could be replaced with host genes. Although such transducing particles were defective, they could be rendered infectious by use of helper phage. This whole research area led to the development of derivatives of *lambda* for recombinant DNA research (Blattner et al., 1977).

Although restriction enzymes and DNA ligase provided the essential tools for putting together hybrid DNA molecules, without *transformation*, the replication of the hybrids would have been difficult or impossible. As noted in Section 9.9, transformation in *Escherichia coli* is an extremely inefficient process, but by the use of antibiotic-resistance markers, it is possible to select even rare transformants. Once appropriate transformants have been selected, they can be checked for presence of nonselected markers.

As discussed in Section 5.3, the *prototrophic selection procedure* developed by Lederberg made possible the study of rare genetic events. Because the efficiency of gene transfer in recombinant DNA research is generally low, the concepts developed by Lederberg were essential for the perfection of recombinant DNA technology. The resolution of genetic analysis using microbes is vastly superior to that of higher organisms because enormous populations can be analyzed (see Section 6.10 and Pontecorvo, 1958).

The *mutation research* of Luria and Delbrück and the *indirect selection procedure* developed by Lederberg and Lederberg also played important roles in the perfection of recombinant DNA techniques. The understanding of the clonal nature of mutation by Luria and Delbrück (see Section 4.8) made it possible to handle bacteria as genetic systems. The indirect selection procedure (see Section 4.10) and the *penicillin enrichment procedure* (see Section 4.12) pointed the way for the development of numerous techniques for finding mutants of whatever type needed. The modern biotechnologist still depends on these ancient concepts for the isolation of appropriate recombinant strains.

In the development of practical processes using recombinant organisms, it is essential to obtain high-level gene expression and to have expression under experimental control. Without the development of ideas about *operon* and *repressor* (see Sections 10.10 and 10.11), effective expression could not be approached. Recombinant DNA technology not only depends on understanding gene regulation, it also helps in developing effective regulatory systems, since the ability to move genes into other genetic regions where control of gene expression is under the control of the experimenter makes it possible to turn on expression at the appropriate time.

Research on *messenger RNA* led to the idea of information transfer from one nucleic acid molecule to another. This was an important forerunner of one of the major developments of recombinant DNA era, the use of *nucleic acid probes* for determining the presence of particular base sequences. Probes are used to detect specific sequences by artificial hybridization on gels or membrane filters. Nucleic acid hybridization derived originally from studies on the experimental formation of double-stranded DNA hybrids by Julius Marmur and Paul Doty (see Section 9.15) (Doty, Marmur, Eigner, and Schildkraut, 1960; Marmur and Lane, 1960). It was subsequently found that double-stranded hybrid structures could also be formed from mixtures of single-stranded DNA and RNA, if the base sequences were complementary or nearly so. The hybridization technique was adapted to membrane filters, where it can be carried out efficiently and with small amounts of material. Filter hybridization provides an extremely simple and sensitive way of pulling complementary nucleic acid molecules out of cell extracts and nucleic acid mixtures.

The history of the recombinant DNA era has not yet been written. Although not as revolutionary as the discernment of the Watson/Crick structure for DNA, recombinant DNA has provided a whole set of extremely versatile tools for investigating numerous problems of modern biology. Thus, recombinant DNA technology can be placed in the same category as X-ray diffraction, electron microscopy, radioisotope technology, and genetic analysis, tools that have permitted major advances in our understanding of the nature of life itself. As the present chapter shows, recombinant DNA technology has been built on the fundamental discoveries of bacterial and phage genetics, discoveries that are so central to modern biological research that their importance is often not even realized.

REFERENCES

Adelberg, E.A. and Burns, S.N. 1960. Genetic variation in the sex factor of *Escherichia coli. Journal of Bacteriology* 79: 321–330.

Arber, W. 1965. Host-controlled modification of bacteriophage. *Annual Review of Microbiology* 19: 365–378.

Berg, P., Baltimore, D., Brenner, S., Roblin, R.O., and Singer, M.F. 1975. Summary statement of the Asilomar Conference on recombinant DNA molecules. *Proceedings of the National Academy of Sciences* 72: 1981–1984.

Blattner, F.R., Williams, B.G., Blechl, A.E., Denniston-Thompson, K., Faber, H.E., Furlong, L.-A., Grunwald, D.J., Kiefer, D.O., Moore, D.D., Schumm, J.W., Sheldon, E.L., and Smithies, O. 1977. Charon phages: safer derivatives of bacteriophage lambda for DNA cloning. *Science* 196: 161–169.

Cohen, S.N. 1975. The manipulation of genes. *Scientific American* 233: 25–33.

Cohen, S.N., Chang, A.C.Y., Boyer, H.W., and Helling, R.B. 1973. Construction of biologically functional bacterial plasmids in vitro. *Proceedings of the National Academy of Sciences* 70: 3240–3244.

Doty, P., Marmur, J., Eigner, J., and Schildkraut, C. 1960. Strand separation and specific recombination in deoxyribonucleic acids: physical chemical studies. *Proceedings of the National Academy of Sciences* 46: 461–476.

Jacob, F. and Adelberg, E.A. 1959. Transfert de caractères génétiques par incorporation au facteur sexuel d'*Escherichia coli. Comptes rendus des Séances de l' Académie des Sciences* 249: 189–191.

Luria, S.E. 1953. Host-induced modifications of viruses. *Cold Spring Harbor Symposia on Quantitative Biology* 18: 237–244.

Luria, S.E. and Human, M.L. 1952. A non-hereditary host-induced variation of bacterial viruses. *Journal of Bacteriology* 64: 557–569.

Marmur, J. and Lane, D. 1960. Strand separation and specific recombination in deoxyribonucleic acids: biological studies. *Proceedings of the National Academy of Sciences* 46: 453–461.

Marmur, J., Rownd, R., Falkow, S., Baron, L.S., Schildkraut, C., and Doty, P. 1961. The nature of intergeneric episomal infection. *Proceedings of the National Academy of Sciences* 47: 972–979.

Mertz, J.E. and Davis, R.W. 1972. Cleavage of DNA by RI restriction endonuclease generates cohesive ends. *Proceedings of the National Academy of Sciences* 69: 3370–3374.

Pontecorvo, G. 1958. *Trends in Genetic Analysis.* Columbia University Press, New York. 145 pp.

Roberts, R.J. 1983. Restriction and modification enzymes and their recognition sequences. *Nucleic Acids Research* 11: r135–r167.

Roth, T.F. and Helinski, D.R. 1967. Evidence for circular DNA forms of a bacterial plasmid. *Proceedings of the National Academy of Sciences* 58: 650–657.

Smith, H.O. and Nathans, D. 1973. A suggested nomenclature for bacterial host modification and restriction systems and their enzymes. *Journal of Molecular Biology* 81: 419–423.

Smith, H.O. and Wilcox, K.W. 1970. A restriction enzyme from *Hemophilus influenzae.* I. Purification and general properties. *Journal of Molecular Biology* 51: 379–391.

Szalay, A.A., Mackey, C.J., and Langridge, W.H.R. 1979. Restriction endonucleases and their applications. *Enzyme and Microbial Technology* 1: 154–164.

Author Citation Index

Subject Index

CHRONOLOGY

Year	Name	Event	Chapter
1865	G. Mendel	Discovery of laws of heredity	2.1
1870	W. Flemming	Discovery of mitosis	3.9
1875	C. Darwin	Gemmules: first ideas of mechanism of inheritance	2.2
1880	L. Pasteur	Attenuation of virulence of a bacterial pathogen	3.5
1881	R. Koch	Agar plate technique for isolating pure cultures of bacteria	3.3
1889	H. de Vries	Hypothesis of intracellular pangenesis	2.3
1900	H. de Vries	Mutation theory and rediscovery of Mendel's laws	2.3
1900	M. Beijerinck	First discussion of mutation theory as applied to bacteria	4.1
1900	F. Dienert	First adaptive enzyme	10.1
1901	E. Wildiers	First growth factor in a microorganism	3.8
1904	A.F. Blakeslee	Discovery of sexual differentiation in a microorganism	5.1
1907	R. Massini	Discovery of *Escherichia coli-mutabile*, first *lac*-negative mutant	4.4
1908	A.E. Garrod	Inborn errors of metabolism, first recognition of role of genetics in biochemistry	2.6
1909	W. Johannsen	Differentiation of genotype and phenotype	2.5
1910	T.H. Morgan	Location of genes on chromosomes	2.7
1915	F.W. Twort	Discovery of a bacterial virus	6.1
1916	L.J. Cole and W.H. Wright	The "pure line" concept in bacteriology	4.2
1917	F. d'Herelle	Independent discovery of a bacterial virus, coinage of the term "bacteriophage"	6.1
1921	J.A. Arkwright	Description of smooth/rough variation in bacteria	4.5
1922	M. Lisbonne and L. Carrère	Development of an indicator strain for bacterial lysogeny	7.2
1922	H.J. Muller	Bacteriophage may be a free-living gene	1.1
1923	M. Heidelberger and O.T. Avery	The first nonprotein antigen, pneumococcus polysaccharide	9.1
1925	J. Bordet	Coinage of the term "lysogeny"	7.2
1927	H.J. Muller	Artificial induction of mutation with radiation	4.11
1928	F. Griffith	Discovery of pneumococcus transformation	9.3
1929	F.M. Burnet and M. McKie	Studies on the nature of lysogeny	7.2
1929	M.H. Dawson and R.H.P. Sia	First in vitro transformation	9.4
1930	H. Karström	Coinage of the terms "adaptive and constitutive enzymes"	10.2
1931	L.E. den Dooren de Jong	Discovery of the *Bacillus megatherium* lysogenic system	7.3
1932	J.L. Alloway	First transformation with a cell-free extract	9.4
1934	I.M. Lewis	Proof that acquisition of *lac* positivity in *E. coli mutabile* is due to a mutation	4.4
1934	M. Schlesinger	Purification of phage and demonstration of the presence of DNA	6.12
1935	W.M. Stanley	Crystallization of tobacco mosaic virus	6.12
1935	M. Delbrück	Target theory and the nature of the gene	6.3
1936	F.M. Burnet and D. Lush	First study of phage mutations	6.2
1936	E. Wollman and E. Wollman	Lysogeny is due to incorporation of phage into host cell	7.3
1939	E.L. Ellis and M. Delbrück	First one-step growth curve of phage	6.3
1941	G.W. Beadle and E.L. Tatum	The one-gene/one-enzyme hypothesis	5.1
1941	J. Monod	Discovery of the phenomenon of diauxie	10.4
1943	S.E. Luria and M. Delbrück	Beginnings of bacterial genetics. The fluctuation test: first quantitative study of mutation in bacteria	4.8
1944	O.T. Avery, C. MacLeod, and M. McCarty	The transforming principle is DNA	9.5
1944	G.W. Beadle and V.L. Coonradt	First complementation test	5.1
1944	C.H. Gray and E.L. Tatum	First biochemical mutants in *E. coli* K-12	5.2
1945	M. Demerec and U. Fano	Description of the T system of phages	6.6
1945	E.L. Tatum and G.W. Beadle	Biochemical genetics in *Neurospora*	5.1
1945	E.L. Tatum	First double mutants in *E. coli*	5.2
1946	M. Delbrück	Mixed phage infection leads to genetic recombination	6.9
1946	A.D. Hershey	Genetic recombination studies with phage	6.9
1946	J. Lederberg and E.L. Tatum	First crosses in *E. coli* K-12	5.3
1946	J. Monod	Acquisition of *lac* positivity can be due to either adaptive enzyme formation or mutation	10.2
1947	J. Lederberg	First genetic map in *E. coli* K-12	5.3
1948	B.D. Davis	Penicillin technique for isolation of biochemical mutants of *E. coli*	4.12
1948	J. Lederberg and N. Zinder	Penicillin technique for isolation of biochemical mutants of *E. coli*	4.12
1948	A.D. Hershey and R. Rotman	First phage genetic map	6.9
1949	R.D. Hotchkiss	First chemical analysis of bases in transforming DNA	9.8
1949	S.S. Cohen	First studies on phage biochemistry	6.12
1950	L.L. Cavalli	Isolation of the first Hfr strain of *E. coli*	5.4
1950	E. Chargaff	Chemical analyses of different DNAs showing definite proportions of different bases	9.9
1950	J. Lederberg	Description of ONPG and first studies of β-galactosidase	10.6
1950	A. Lwoff and A. Gutmann	Discovery of the nature of lysogeny; coinage of the word "prophage"	7.4
1950	A. Lwoff, L. Siminovitch, and N. Kjeldgaard	Lysogenic cultures can be induced by radiation	7.4
1950	J. Lederberg	First constitutive mutants of *lac*	10.6
1951	V. Freeman	Discovery of lysogenic conversion in *Corynebacterium diphtheriae*	8.9
1951	E.M. Lederberg	Discovery of bacteriophage *lambda*	7.5
1952	J. Lederberg and E. Lederberg	Replica plating for indirect selection	4.10
1952	A.H. Doermann	Discovery of the eclipse phase in phage reproduction	6.12
1952	A.D. Hershey and M. Chase	Only the DNA of phage enters the cell in significant amounts	6.13
1952	S.E. Luria and M.L. Human	Discovery of the restriction phenomenon in bacteriophage	6.15